SECOND EDITION

Ecosystems
and
HUMAN
HEALTH
Toxicology and Environmental Hazards

SECOND EDITION

Ecosystems
and
HUMAN
HEALTH
Toxicology and Environmental Hazards

RICHARD B. PHILP

LEWIS PUBLISHERS
Boca Raton London New York Washington, D.C.

Library of Congress Cataloging-in-Publication Data

Philp, Richard B.
   Ecosystems and human health : toxicology and environmental
hazards / Richard B. Philp – 2nd ed.
      p.   cm.
Includes bibliographical references and index.
   ISBN 1-56670-568-1
   1.   Environmental toxicology.   2.   Environmental health.   I.
Title.
   RA1226 .P48 2001
   615.9'02--dc21

2001001149
CIP

Visit the CRC Press Web site at www.crcpress.com

© 2001 by CRC Press LLC

No claim to original U.S. Government works
International Standard Book Number 1-56670-568-1
Library of Congress Card Number 2001001149
Printed in the United States of America 1 2 3 4 5 6 7 8 9 0
Printed on acid-free paper

Dedication

Elizabeth, Brendan, Douglas, Danielle, William, Nathan,
Danny, Anders, Margaret, Matthew, Jemma, Lauren,
and kids everywhere. Perhaps this book will help them to look
after this place better than we did. Also for my wife Joan,
who is my calm harbor in a stormy world.

Foreword

There is a commonly held myth in our society that anything that is "natural" is good, wholesome, and healthful, whereas anything that is "synthetic" is bad, toxic, and harmful. The mere mention of the word *chemical* is enough to strike terror into the heart of any food faddist. This attitude is, at best, naïve and, at worst, dangerous. Toxic substances abound in nature, ranging from inorganic heavy metals such as arsenic and mercury, through organic substances such as hydrocyanic acid, to complex enzymes and other proteins of the neurotoxins and coagulant-anticoagulants present in venoms and toxins. One of the more serious environmental hazards may be natural radon gas, and cancer from solar radiation is a real concern.

Increasingly, it is becoming necessary for students of environmental sciences to know something of toxicology and for students of toxicology to know something of the environment. This text is designed to bridge these fields by acquainting the student with the major environmental hazards — both man-made and natural — and with the risks to human health that they pose. It is designed such that topics are generally introduced in the early chapters and covered in greater detail in subsequent ones. This is neither an environmentalist's handbook nor does it deal exclusively with toxicology; rather, it attempts to strike a balance between the extremes of opinion and to indicate where information is inconclusive. Examples of major accidental exposures of humans to chemical toxicants are used liberally and case studies taken from reported incidents are provided. Historical background of the development of a class of chemicals or a particular environmental problem is often provided in the belief that an educated student should know more than merely the technical aspects of the field. It is hoped that this text will assist students in acquiring the information and judgmental skills necessary to differentiate between real and perceived risks, as well as acquaint them with the toxicology of important chemicals in the environment. Because most people spend 8 hours daily, 5 days weekly in the workplace, it constitutes an important component of our environment and it will be considered as such.

Richard B. Philp, D.V.M., Ph.D.

About the Author

Richard B. Philp, D.V.M., Ph.D., is an emeritus professor and former Chair of the Department of Pharmacology and Toxicology at the University of Western Ontario. After graduating from the Ontario Veterinary College, he practiced veterinary medicine in Illinois and in Ontario and also served as a public health officer in a small Ontario town. He obtained his Ph.D. in pharmacology from the University of Western Ontario and did postdoctoral studies at the Royal College of Surgeons of England in London. He has served on advisory committees to Canadian federal and provincial governments regarding the use of antibiotics in agriculture. He was Honorary Visiting Professor in the School of Pathology, University of New South Wales, and has authored or co-authored over 90 scientific papers, two books, and several book chapters. His current research involves the study of pollution along the Florida Gulf Coast and its effects on a species of marine sponge.

Contents

chapter one

Principles of pharmacology and toxicology

The right dose differentiates a poison and a remedy.

— Paracelsus, 1493–1541

Introduction

The past century has seen a tremendous expansion in the number of synthetic chemicals employed by humankind as materials, drugs, preservatives for foods and other products, pesticides, cleaning agents, and even weapons of war. An estimated 64,000 chemicals are currently in use commercially, with 5 billion tons being produced annually in the world. Some 4000 chemicals are used as medicinals and at least 1200 more as household products. An estimated 700 new chemicals are synthesized each year. Add to this the numerous natural substances, both inorganic and organic, that possess toxic potential, and it is little wonder that the public expresses concern and even, sometimes, panic about the harmful effects these agents may exert on their health and on the environment. Many of these agents, perhaps 50,000 of them, have never been subjected to a thorough toxicity testing.

Approximately 500 chemicals have been evaluated for carcinogenic potential. Some 44 have been designated as possible human carcinogens on the basis of evidence, either limited or conclusive, obtained from human studies. Of these, 37 tested positive for carcinogenicity in animal tests and were later shown to be carcinogens for humans. There are, however, numerous other agents that have been shown to be carcinogenic in rodents but which have yet to be identified as human carcinogens. This creates significant problems regarding the legislative and regulatory decisions that need to be made about their use. Some of the areas of uncertainty that surround the extrapolation of data from the animal setting to the human setting are discussed in the following chapter. The process of extrapolation requires input from many different disciplines that may include engineering, physics, biology, chemistry, pathology, pharmacology, physiology, public health, immunology, epidemiology, biostatistics, and occupational health. The field of toxicology thus depends on all of these, but perhaps draws most heavily on pharmacology, biochemistry, and pathology. It is the identification of the degree of risk to which individuals or groups are exposed in a given set of circumstances that directs all of this activity.

Other forms of toxicity, hepatotoxicity, nephrotoxicity, and neural toxicity, for example, may be more important in acute exposures such as might occur in the industrial setting. Reproductive and fetal toxicity has been frequently demonstrated experimentally, but their significance for the general population exposed to low levels of toxicants in the environment remains unclear.

The (U.S.) Agency for Toxic Substances and Disease Registry, and the (U.S.) Environmental Protection Agency jointly maintain a priority list of 275 toxic substances. The "top 20" include arsenic, lead, metallic mercury, vinyl chloride, benzene, polychlorinated biphenyls (PCBs), cadmium, benzo[a]pyrene, benzo(b)fluoranthene, polycyclic aromatic hydrocarbons (PAHs), chloroform, aroclor 1254, P'P'-DDT, aroclor 1260, trichloroethylene, chromium(+6), and dibenz[a,h]anthracene. The complete list can be viewed on the internet at http://atsdrl.atsdr.cdc.gov:8080/97list.html.

Considerable difficulty attends efforts to extrapolate the results of toxicity tests in experimental animals to humans exposed to very low levels in their environment, especially with regard to the risk of cancer. Current legislation requires testing in two species with sufficient numbers for reliable statistical analysis. Rats and mice are generally used, as hamsters are resistant to many carcinogens and primates are too expensive and, in the case of some species, too environmentally threatened. For statistical purposes, cancer includes all tumors, whether benign or malignant. A 2-year carcinogen study — one for analysis (pathology, etc.) and one for documentation and statistics — employing two species cost, in 1991, at least $1,000,000 plus the costs of 1 year for preparation. Because it is not practical to test every chemical, several factors should be considered in selecting test chemicals. These include the frequency and severity of observed effects, the extent to which the chemical is used, its persistence in the environment (examples of persistent chemicals include chlorinated hydrocarbons), and whether transformations to more toxic agents occur.

Heavy metals, the by-products of most mining and ore extraction processes, are examples of ubiquitous toxicants with almost infinite half-lives. Mercury (Hg), for example, is present in all canned tuna at about 5 ppm, mostly from natural sources. Aquatic bacteria can transform mercury to methylmercury. This has a different toxicity profile. Cadmium (Cd) enters the environment at about 7000 tons/year and is concentrated by livestock because they recycle it in feces used for fertilizer. It is then passed on to forage grasses. Radioactive isotopes of cesium and iodine entered the food chain after Chernobyl.

The estimation of the degree of risk associated with the presence of a potentially toxic substance in the environment is the basis for all decisions relating to the legislative controls over that chemical, including its industrial use and eventual disposal. Pharmacological/toxicological principles are essential in understanding the processes involved in toxicity testing.

Pharmacology can be defined as the science of drugs. It includes a study of their sources (*materia medica*); their actions in the living animal organism (pharmacodynamics); the manner in which they are absorbed, moved around in the body, and excreted (pharmacokinetics); their use in medicine (therapeutics); and their harmful effects (toxicology). In this context, a *drug* is any substance used as a medicine but pharmacology generally includes the study of substances of abuse, and in the broadest sense deals with the interactions of *xenobiotics* (literally, substances foreign to living organisms) whether they be natural or man-made (*anthropogenic*), therapeutic, or toxic. In this sense, toxicology can be considered to be a branch of pharmacology. Xenobiotics can also be exploited as research tools to reveal mechanisms underlying physiological processes.

Toxicology is the study of the harmful effects of xenobiotics on living organisms, the mechanisms underlying those effects, and the conditions under which they are likely to occur. *Environmental toxicology* is the study of the effects of incidental or accidental exposure of organisms, including human beings (the focus of this text), to toxins in the environment (i.e., air, water, and food). While the greatest concern today centers on pollutants of human origin, it should not be forgotten that toxic substances, including carcinogens, abound in nature. The subject of environmental toxicology embraces the study of the causes, conditions, environmental impact, and means of controlling pollutants in the environment. It can also be extended to include the environment of the workplace (industrial hygiene). The related term *ecotoxicology* deals with the harmful effects of chemicals, usually of anthropogenic origin, on ecosystems.

Economic toxicology is the study of chemicals that are developed expressly for the purpose of improving economic gain by selectively eliminating a species (insecticides and herbicides), by improving health and productivity (drugs), by preserving foodstuffs (food additives), or for the manufacture of a marketable product (industrial solvents, cleaning agents, etc.).

Forensic toxicology refers to the medico-legal aspects of the harmful effects of drugs and poisons administered or taken deliberately or accidentally. Detection of xenobiotics in tissues and fluids and in, or on, objects is an important aspect of this field as is the preparation of evidence for submission in court.

Pharmacokinetics

There has been a trend in recent years to attempt to separate toxicology from pharmacology by the use of such terms as *toxicokinetics* and *toxico-dynamics*. The distinction is largely semantic as the principles are the same in both cases. Throughout this text, *pharmacological* can also be taken to represent *toxicological*.

The response of organisms to drugs and chemicals is governed by natural laws. One of these is the *Law of Mass Action* which dictates that, in the absence of a transport system, chemicals in solution will move from an area of high concentration to one of low concentration. If a semi-permeable membrane is interposed between these areas, the chemical will move across it, assuming the chemical can penetrate the membrane. In reality, molecules wander randomly across the barrier, but the frequency of transfers will be greater from the area of high concentration to that of the low one until equilibrium is established. Cell walls and other biological membranes function as semi-permeable membranes, and the Law of Mass Action influences the uptake of most drugs and toxicants by living organisms. The concentration of a toxicant in the environment (water, air, and soil) is thus an important determinant affecting its uptake. Transport mechanisms are dealt with in the "Absorption" and "Distribution" subsections of this chapter.

The *partition coefficient* is the ratio of a chemical's relative solubility in two different phases. The ratio of solubility in oil (often n-octanol) to that in water is frequently used to predict the distribution of a xenobiotic between the aqueous and lipid phases in the body.

Absorption

Whether or not a xenobiotic is toxic, and how that toxicity is manifested, depends primarily on how the body deals with it. Substances that are not absorbed from the gastrointestinal tract have no systemic toxicity. This fact allows barium to be used as an X-ray contrast medium, despite its toxicity by other routes of administration. The selective toxicity of most insecticides depends solely on a greater ability to penetrate the chitin of the insect's exoskeleton than to penetrate human skin. A substance that is not readily excreted by the body (usually through the kidneys or in the feces) will accumulate to toxic levels.

The primary routes of absorption for toxicants are the skin, the lungs, and the gastrointestinal tract. The latter two are important for the population at large, but the skin may be a very significant site in certain industrial settings. The site of absorption, more commonly called the *portal of entry* in toxicology, can have a significant influence on the toxicity of a substance.

Larger molecules require a degree of lipid solubility to cross biological barriers because cell membranes consist of a fluid phospholipid matrix with embedded proteins that can penetrate part way or all the way through the membrane. Factors that influence the lipophilicity of a chemical will therefore affect its absorption. Many chemicals are weak acids or bases that may exist in an ionized (polar) or a non-ionized (nonpolar) state with equilibrium established between them. For example:

$$R\text{-}H \longrightarrow R\text{-}^- + H^+$$
$$\text{Nonpolar} \quad \text{Polar}$$

The polar form is water soluble, whereas the nonpolar form is lipid-soluble. The pH will influence the equilibrium and hence the amount of the lipid-soluble form available for absorption. The *dissociation constant (pKa)* of a substance is defined as the pH at which 50% of it will exist in each state. Weakly acidic drugs are shifted to the nonpolar state in an acid medium and to the polar state in an alkaline medium. The reverse is true for weakly alkaline drugs. Because the pH of the stomach and upper small bowel is acidic (pH 2–4), acidic chemicals will be absorbed here. Alkaline substances tend to be absorbed in the lower small bowel and the upper colon which are more alkaline, whereas the descending colon, becomes acidic again.

Lipid solubility is not essential for the passage of all molecules across membranes. There is the bulk transfer of water across the cell membrane that can carry very small (less than 200 Daltons), water-soluble molecules with it. Metallic ions such as calcium, sodium, and potassium, as well as chlorine, can pass through special channels, some of which are regulated by the trans-membrane potential (voltage regulated) and others by specific receptors (receptor activated). Specialized exchangers also exist; for example, the sodium pump.

Active transport is an energy-consuming process by which a substance can be moved against a concentration gradient. Active transport is important in the kidney and the liver. In addition to energy consumption, it is also characterized by saturability, selectivity for specific chemical configurations, and the ability to move substances against an electrochemical gradient. *Facilitated diffusion* is similar except that no energy is consumed and it cannot occur against an electrochemical gradient.

Pinocytosis is a process whereby a segment of the plasma membrane of a cell invaginates to form a sack in which extracellular fluid and colloidal particles can be taken into the cell by pinching off the "mouth" of the sack. This is an important mechanism by which the mucosal cells of the intestinal tract take up nutrients and some drugs and chemicals.

Distribution

Once absorbed, the agent can be distributed throughout various compartments in the body. Serum albumin possesses many nonspecific binding sites for xenobiotics, especially weakly acidic ones, and it therefore becomes a transport system for many substances. The balance between dissociated (polar) and undissociated (nonpolar) states affects the distribution of a chemical as well, because pH changes from the extracellular fluid (pH 7) to the plasma (pH 7.4). The partition coefficient of a substance also influences its distribution, determining, for example, the extent to which it will be sequestered in fat. Highly lipid-soluble substances will be sequestered in body fat where they may remain for long periods. Everyone has DDT and its metabolites dissolved in his/her fat. The amount varies with age and location.

The use of DDT in North America was drastically reduced in the 1970s and a complete ban was legislated in Canada in 1990. Substances such as DDT that are sequestered in fat can be released during periods of fat loss (starvation, extreme dieting), as a result of illness, and even during lactation when lipids are transferred to milk. The released toxicant may reach concentrations at target sites sufficient to cause a toxic response. Figure 1 illustrates these relationships among storage fat, blood, and target organ.

The rate of distribution of a substance is a function of the rate of blood flow through the tissues (tissue perfusion). Highly vascular organs will accumulate it first; organs that are poorly perfused will accumulate it last. The substance is thus distributed initially on the basis of tissue perfusion; then as equilibrium states are reached, it will redistribute on the basis of its solubility. Following the intravenous injection of a chemical with a high partition coefficient, equilibrium will be established instantly with the kidney and liver because of their high vascularity, almost as quickly with the brain, with muscle in about 30 min, and with fat in about 3 hr. The membranes surrounding the brain and separating it from its blood vessels constitute the *blood-brain barrier* that generally will pass only quite lipid-soluble agents, such as all anesthetics.

Thus, tissue perfusion and the partition coefficient may play important roles in determining the onset and termination of either a therapeutic or a toxic response. Sodium thiopental, an ultrashort-acting barbiturate, is used for anesthetic induction. The rate of biotransformation is so slow as to have little effect on recovery. The drug readily penetrates the blood-brain barrier because of its high lipid solubility and the brain, which is richly perfused, rapidly takes it up and anesthesia ensues. This effect is terminated because the drug is redistributed to other tissues, including depot fat, which is poorly perfused. New equilibria are established among blood, brain, and other tissues so that, while initial recovery is rapid, a state of sedation may persist for several hours. Figure 2 shows the effects of perfusion and partition coefficient on the biological half-life ($t_{1/2}$) of thiopental in different tissues.

Biotransformation

Biotransformations of xenobiotics are classified as either Phase I reactions or Phase II reactions. *Phase I reactions*, also known as nonsynthetic biotransformations, convert a lipophilic (fat-soluble) substance to a more polar and, hence, more water-soluble substance. This metabolite is excreted more readily by the kidneys than the parent compound, but it usually retains

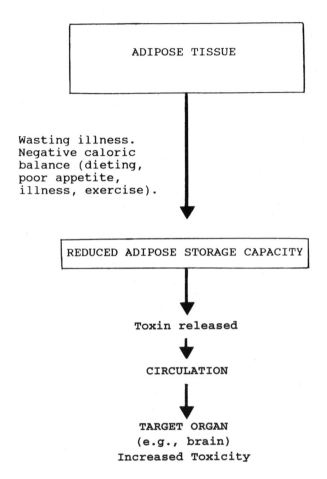

Figure 1 Disposition of lipid-soluble chemicals in adipose tissue and the effects of weight loss.

significant bioactivity. It may be more active, or less active, than the parent substance. If the parent chemical is nontoxic but the metabolite is toxic, this is a toxication reaction. A drug that requires biotransformation to become active is referred to as a *pro-drug*. Figure 3 shows some examples of Phase I reactions and their consequences.

Phase I chemical reactions include oxidation, reduction, and hydrolysis and generally unmask or introduce a functional (reactive) group such as -NH$_2$, -OH, -SH, or -COOH. The oxidation reactions include N- and O-dealkylations, side-chain and aromatic hydroxylations, N-oxidation and hydroxylation, sulfoxide formation, and desulfuration. Hydrolysis of esters and amides also occurs. Reduction reactions may involve azo (RN = NR) or nitro (RNO$_2$) groups.

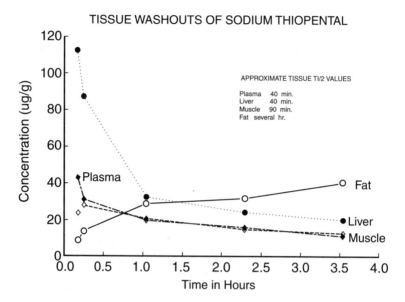

Figure 2 Tissue $t_{1/2}$ values of sodium thiopental, a highly lipid-soluble drug, in various tissues.

Figure 3 Some examples of Phase I reactions. The product may be more or less active than the parent chemical, or it may be inactive.

Many oxidation reactions are under the control of a group of mixed-function oxidases for which cytochrome P-450 (CYP450) serves as a catalyst. These are located primarily in the smooth endoplasmic reticulum (SER) of hepatic cells, but they exist in many tissues as well as many species, including single-celled organisms. The CYP450 monooxygenases have tremendous substrate versatility, being able to oxidize lipophilic xenobiotics plus fatty acids, fat-soluble vitamins, and various hormones. This is, in part, because there are at least 20 variants of the enzyme (isoenzymes) and because each is capable of accepting many substrates. CYPs 1, 2, and 3 are isozymes especially involved in xenobiotic transformations. It should be noted that pro-carcinogens are converted to carcinogens by Phase I reactions. Examples of this include benzo[a]pyrene, the fungal toxin aflatoxin B_1, and the synthetic estrogen diethylstilbestrol. This process often involves the formation of an epoxide compound, as it does in the three examples given in Figure 3. An epoxide has the chemical configuration shown in Figure 4, making it highly nucleophilic and chemically reactive. Many epoxides are carcinogens. Figure 4 shows this chemical transformation for stilbestrol and benzo[a]pyrene, which is an example of a polyaromatic hydrocarbon (PAH). Many of these are carcinogenic and are environmental pollutants. Other enzymes, called epoxide hydrolases, can detoxify the epoxides.

Phase II reactions are conjugation (synthetic) reactions that render the agent not only more water-soluble, but biologically inactive, with a very few exceptions. A common conjugation reaction is with glucuronic acid. Conjugation also occurs with sulfuric acid, acetic acid, glycine, and glutathione. Many Phase I metabolites are still too lipophilic (fat soluble) to be excreted by the kidneys and are subjected to Phase II conjugation. All chemicals need not be subjected first to Phase I transformations. Many, if they possess the necessary functional groups (e.g., -OH, -NH$_2$), are conjugated directly.

DIETHYLSTILBESTROL   CARCINOGENIC EPOXIDE

EPOXIDE   BENZO [A] PYRENE (BP)   BP-7 , 8-DIOL-9 , 10-EPOXIDE
(CARCINOGEN)

Figure 4 Examples of epoxide formation to potentially carcinogenic metabolites.

An important concept in understanding toxication and detoxication of xenobiotics is *enzyme induction*. Hepatic enzymes of the smooth endoplasmic reticulum can be stimulated to a higher level of activity by many highly lipophilic agents. Because these enzymes are nonspecific, this has consequences for many other agents transformed by the same enzymes. Induction is accomplished by the increased synthesis of more enzyme, so the SER actually increases in density. The result may be increased detoxication of a chemical or the increased synthesis of a toxic metabolite. Cigarette smoke contains many inducers and may increase the breakdown of many drugs (theophylline, phenacetin, etc.) but, conversely, it may act through this mechanism as a promoter or as a co-carcinogen.

Elimination

Every secretory or excretory site in the body is a potential route of elimination for xenobiotics. Thus, they may be excreted in saliva, sweat, milk, tears, bile, mucus, feces, and urine. Of these, the most significant site is urine, followed by feces and bile. The kidney is the principal organ for the elimination of natural waste metabolites. Most of these are toxic if they exceed normal levels. The kidney is also the main organ for maintaining fluid and electrolyte balance. It is therefore not surprising that the kidney also is the main site of elimination of xenobiotics, including drugs. Although it constitutes only 0.4% of total body weight, it takes 24% of the cardiac output. It is a highly efficient filter of blood.

The basic physiological unit of the kidney is the nephron (see Figure 5), which is composed of the glomerulus (a tightly wound bundle of blood vessels) and the tubule, which is closed at the glomerular end to provide a semi-permeable membrane. The tubule is composed of several segments with different functions. Substances smaller than 66,000 Daltons (Da) are passed through the glomerulus. They may be reabsorbed further down the tubule and even re-secreted. This occurs with uric acid, which is completely passed through the filter, 98% reabsorbed, and further secreted. The pH of urine will determine the degree of dissociation of acids and bases and, hence, influence their movement across the reabsorption sites. Passive diffusion across the distal tubule depends on the degree of ionization in the plasma and extracellular fluid as only the lipid-soluble form will be diffused. Thus, the concentration gradient is also an important rate-limiting factor. Very water-soluble agents are passed through the glomerulus if they are small enough, and this is the reason why most biotransformations result in increased water solubility. Other substances are actively secreted (an energy-consuming process) at tubular sites (see Figure 5).

It should be noted that the lungs are a very important site of elimination for volatile substances, including solvents, alcohols, and volatile and gaseous anesthetics. These can, in fact, be smelled on the breath, which can be an important first-aid procedure to determine the cause of unconsciousness or stupor. Ketoacidosis in diabetics can also be detected by the acetone-like

Figure 5 The nephron is the basic renal unit.

odor on the breath. Young diabetics have been suspected of glue sniffing when brought to an emergency department in a stupor or coma because of this fact.

Many drugs and chemicals are excreted into the bile. These tend to be polar agents, both cationic and anionic, the latter including glucuronide conjugates. Nonselective active transport systems, similar to those in the kidneys, are involved in the excretory processes. Once they enter the small intestine, these chemical metabolites can be excreted in the feces or reabsorbed back into the bloodstream. Enzymatic hydrolysis of glucuronide conjugates favors a return to the more lipid-soluble state and hence reabsorption.

The excretion of xenobiotics in mother's milk may not be an important route of elimination, but it can have significance for toxicity in the infant. The chloracne rash associated with the now-obsolete bromide sedatives appears to be related to the secretion of this halogen in sweat. It is distributed in the body as chloride ion.

Extensive batteries of enzymes in the body may render the chemical nontoxic (detoxication), more water soluble, and hence more easily excreted, or they may activate it to a toxic form (toxication). The liver is the primary site of xenobiotic biotransformation in the mammalian body but it is by no means the only one. Indeed, significant biotransformation can occur at the portal of entry. The chemical pathways are often the same. The response of the body to chemical insult also depends on the mitotic activity of the target tissue. Rapidly dividing tissues allow little time for repair to occur before cell division, so that the chance of a mutation is increased. Moreover, tissues that regenerate poorly are vulnerable to permanent damage by toxicants.

Pharmacodynamics

Ligand binding and receptors

Because only the molecules that are free in solution contribute to the concentration gradient, their binding to tissue components or their chemical alteration by tissue enzymes will contribute to the maintenance of the gradient. The nature and strength of the chemical bond determines how easily the xenobiotic will dissociate when the concentration gradient is reversed. Drugs interact with specific sites (receptors) on proteins such as plasma membrane proteins, cytosolic enzymes, membranes on cell organelles, and in some cases, nucleic acids (e.g., certain antineoplastic drugs). Membrane receptors and enzymes have molecular configurations that will react only with certain molecules in a kind of "lock-and-key" manner. Ease of reversibility is an important characteristic for most drugs, so that as concentration of the free substance falls, the drug comes off the receptor and its effect is terminated. This is often expressed by the equation:

$$\text{Drug (D)} + \text{Receptor (R)} \longleftrightarrow \text{DR complex} \longleftrightarrow \text{Response}$$

The magnitude of the response is determined by the number (percentage) of receptors occupied at any given time. Neither the drug nor the receptor is altered by the reaction, which is defined as *pharmacodynamic*.

In many cases, drugs and toxicants interact with receptors that normally accept physiological ligands such as neurotransmitters, hormones, ions, and nutritional elements. The proteins of cell surface receptors can penetrate to the interior of the cell in the case of ion channels and exchangers, or they can connect with other proteins in the membrane to transduce signals. Many neurotransmitters operate through a family of receptors that share the property of connecting to a protein having seven membrane-spanning peptide chains. These G proteins (G for guanosine triphosphate or GTP) are transducers that interact with enzymes such as adenylcyclase or phospholipase C to initiate intracellular second messengers. G proteins may be inhibitory

(G_i), stimulatory (G_s), or operate through other, unidentified mechanisms (G_o). The neurotransmitters noradrenaline, acetylcholine, dopamine, serotonin, histamine, gamma-aminobutyric acid (GABA), glycine, and glutamic acid have been shown to act through G-protein receptors. Many centrally acting drugs work through these receptors.

Steroid receptors also exist. These are soluble cytosolic receptors that bind to the steroid after it diffuses into the cell and carry it to the nucleus. Opioid receptors in the CNS (central nervous system) accept the endogenous peptide endorphins and enkephalins. These receptors are the site of action of the narcotic analgesics.

Any receptor is a potential target for a toxicant interaction. A special case is the aryl hydrocarbon, or Ah, receptor. This cytosolic receptor binds to aromatic hydrocarbons such as dioxins and it is believed that it is involved in their toxicity. No natural ligand for this receptor has yet been identified in mammals. This subject is discussed in detail in Chapter 5 on halogenated hydrocarbons.

The chemical bond with the target receptor can involve covalent bonds, as well as non-covalent bonds including ionic, hydrogen, and van der Waal's forces. If the xenobiotic interacts irreversibly with a component of a cell, the effect may be long-lasting. Indeed, irreversibility of effect is an important characteristic of many toxicants (organophosphorus insecticides are examples of irreversible inhibitors of the enzyme acetylcholinesterase). If a chemical reacts irreversibly with DNA, a mutation may result in carcinogenesis or teratogenesis. This effect is sometimes described as "hit-and-run" because it is unrelated to any measurable concentration of the agent in the serum (see below).

Irreversibility of binding does not always mean irreversibility of effect. The drug acetylsalicylic acid (aspirin) is an irreversible inhibitor of the enzyme cyclooxygenase, which accounts for many of its pharmacological actions. Provided that exposure to aspirin is terminated, the effect declines as new enzyme is synthesized.

Biological variation and data manipulation

Within any given population of organisms, there will be some that will respond to a drug or toxicant at the lowest concentration, others that only respond at the very highest concentration, but most subjects will be grouped around the mean response. This is true of all organisms, including human beings and single-celled ones. It is even true of populations of like cells (liver cells, kidney cells, and blood cells) within the body, and may partly explain why some cells may become malignant while others do not. It is the existence of biological variation that necessitates the use of large populations of test subjects and the development of mathematical treatments of data to permit the comparison of different populations of test subjects. If the responses of the species in question are grouped symmetrically about the mean response, a "normal" or Gaussian distribution curve is obtained (see Figure 6). In this

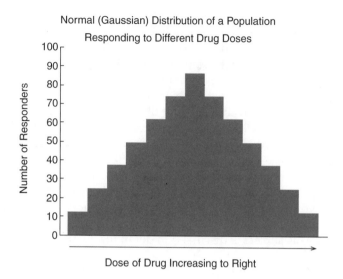

Figure 6 A theoretical normal, or Gaussian, distribution.

case, 68.3% of the population will fall within plus or minus one standard deviation (±1 SD) of the arithmetic mean, 95.5% between ±2 SD, and 99.7% between ±3 SD. A data point lying outside these limits is assumed not to belong to the test population. Sometimes the population is skewed, with more subjects falling on one side of the mean than on the other. Factors accounting for variability could include differences in the rate and degree of uptake, distribution, biotransformation, excretion, and even the nature and number of binding sites and receptors for the agent. These factors may be under genetic control or they may be due to environmental differences in such things as temperature, nutrition, disease, the presence of other xeno-biotics including medications, etc. They also tend to vary with age and sex.

Dose response

A population distribution in response to a drug or chemical applies even to like cells within the same organism or cell culture because some cells will be more aged than others or may be defective in some way. Therefore, it is impossible, from a single dose of a drug or toxicant, to draw any conclusions about its potency because it is not known whether it is recruiting only the most sensitive cells or nearly all of them. For this reason, it is necessary to construct a dose response curve, whether testing a new drug for its effective dose or a chemical for its toxicity. Typically, the response rises rapidly once the threshold is exceeded, and then flattens out as fewer and fewer cells remain to be recruited (see Figure 7). This type of response is a graded or *dose-dependent response*. There is another type of response that can be described as yes/no or all-or-nothing, and this is a *quantal response*; lethality is an example. For graded responses, it is important to establish standard

Figure 7 A typical dose–response curve plotted arithmetically.

points of comparison because comparing a dose at the low end of the response for one chemical with one at the top end for another is not statistically reliable. The point usually chosen is that dose which produces 50% of the maximum effect, the effective dose (or concentration) 50% (ED_{50} or EC_{50}). "Dose" is used when the test substance is administered individually to the test subjects and "concentration" when it is added to the surrounding medium, such as the water in an aquarium or the fluid bathing an isolated tissue in an organ bath.

A quantal response can be converted to a graded one using several test groups, each receiving a different dose of the agent being tested. The percent of animals showing the expected response can then be plotted against dose. Thus, one can calculate the dose that, on average, will kill 50% of the test animals (lethal dose 50%, LD_{50}). If the response is a toxic one (liver necrosis, kidney failure), the value is the toxic dose 50% (TD_{50}). Variations on this approach include the LD_{10}, TD_{10}, etc. The LD_1 is also called the *Minimum Lethal Dose*. Values such as the LD_1 and the LD_{10} are replacing the LD_{50} in many jurisdictions.

In attempting to compare responses to two different chemicals, it is useful to perform a mathematical manipulation on the data so that differences or similarities in the shapes of the dose response curves are more obvious. This involves plotting the logarithm of the dose against the response, and this converts the exponential curve shown in Figure 7 to the sigmoidal one (S-shaped) shown in Figure 8.

It is now much easier to interpolate to the EC_{50} (because the selected doses might not have included it) and to compare these points. Using the log of the dose tends to overcome the fact that large increases in dose result

Figure 8 A semi-logarithmic plot (response vs. log-dose) of the dose–response curve shown in Figure 7.

in small increases in response on the right side of the curve, whereas small increases in dose result in large increases in response on the left side of the curve. Thus, the midpoint, the EC_{50} (or TD_{50}), provides the greatest statistical reliability. Parallel slopes of curves suggest similar mechanisms of action, and comparisons based on molar concentrations provide information on relative potencies. Toxicity comparisons can be done by calculating the Therapeutic Index (TI) if the substance is a therapeutic agent. This is the LD_{50}/ED_{50}. The higher the number, the safer the agent. Other estimates of safety, more appropriate to toxicity studies of nontherapeutic agents, involve comparisons of the TD_1 and EC_{99}.

It is important to note that all toxicity tests contain a temporal factor in that the determination of toxic effects is conducted at a specific time after exposure. Acute toxicity studies generally involve determinations made 72 hr after a single high dose, whereas long-term toxicity requires multiple exposures with measurements made at least 28 days later. These studies are defined by government regulations in jurisdictions where there is a legal requirement for testing new chemicals.

Another value that is frequently used is NOEL (or NOAEL), the No Observable (Adverse) Effect Level. The NOEL includes effects, such as minor weight loss, that are not considered to be adverse. These values are applicable only to that species in which the test was conducted. Extrapolation to other species will require dosage adjustment.

Probit analysis

It is often desirable to compare the toxicity of one xenobiotic to that of another. This information may help to determine whether a substance used commercially or industrially can be replaced with a safer one, or whether a metabolite of a parent compound is more or less toxic than the compound itself. For this purpose, probit analysis is often used. When a toxic reaction is expressed as the number of experimental animals in a group displaying that reaction (e.g., kidney failure), the percent of a group responding to a given dose (or exposure) can be expressed as units of deviation from the mean. These are called normal equivalent deviations (NEDs). The NED for the group in which there were 50% responders would be zero because it lies right on the mean. A NED of +1 corresponds to 84.1% responders. NEDs are positive or negative relative to the mean; thus, a value of 5 is added to each to make them all positive. The result is called a probit (for probability unit). Table 1 shows the equivalent probits and NEDs for given percent responses.

Table 1 Conversion of Percent Responders to Probit Units

% Responding	NED	Probit
0.1	−3	2
2.3	−2	3
15.9	−1	4
50.0	0	5
84.1	+1	6
97.7	+2	7
99.9	+3	8

When quantal data are plotted as probit units against the log of the dose, a straight line results, regardless of whether the original data were distributed normally or skewed. The method, in fact, assumes that the data were distributed normally. It is now easier to compare the quantal data for two different xenobiotics exhibiting the same toxic manifestation (or their lethality). These concepts apply equally to toxicological studies in mammals and in nonmammalian species. The following example illustrates these concepts using hypothetical toxicity data (see Table 2) for two toxicants tested in fathead minnows (0.25–0.5 g). Each test group consisted of 100 fish. Values listed are mg/L concentration in water. Tables are available for conversion to probits.

When using aquatic or marine organisms for toxicity studies, it is important to remember that they are continuously exposed to a given concentration of the test substance, but they may not take it up instantly or even rapidly. A consistent time of exposure must therefore be incorporated into the experimental design. Figures 9 through 11 illustrate arithmetic, semi-logarithmic, and probit plots for these data.

Table 2   Lethality Data in Fathead Minnows for Two Toxic Chemicals

Lethality	Fluorine (mg/L)	Naphthalene (mg/L)
10%	25.0	0.5
20%	50.0	1.0
60%	100.0	2.0
84.1%	200.0	4.0

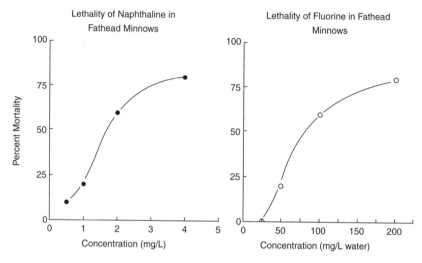

Figure 9   Arithmetic plot of the data shown in Table 2.

Figure 10   Semi-log plot of data shown in Table 2 and Figure 9.

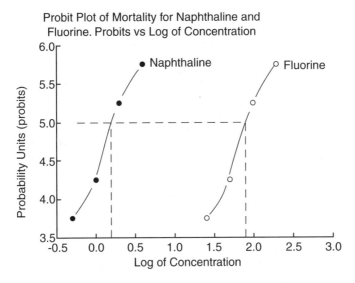

Figure 11 Probit plot of the data shown in Table 2 and Figures 9 and 10.

Cumulative effects

It may be that the manifestation of toxicity does not occur until the individual has been exposed continuously or repetitively (as with repeated daily injections) for a prolonged period, perhaps days or weeks. This generally occurs with agents that are metabolized or eliminated very slowly, so that the rate of intake slightly exceeds the rate of detoxification and the drug slowly accumulates until a toxic level is reached. It is analogous to filling a bathtub with a faulty drain. The tub will fill, but only slowly because the water is running out almost as fast as it comes in. This involves the concept of *biological half-life* or $t_{1/2}$. The plasma $t_{1/2}$ of a drug or chemical is the time required for the plasma concentration to fall 50%. It is important to note that in most cases, this value is a constant for a given xenobiotic; that is, it remains the same regardless of the initial level of the chemical. This is because the rates of biotransformation and excretion are usually concentration-driven, increasing or decreasing according to the plasma level. Biotransformations are enzymatic processes that generally obey first-order kinetics; that is, the conversion rate is dependent upon the initial concentration of substrate. $t_{1/2}$ values can also be established for other tissue compartments in the body. If the exposure interval greatly exceeds the $t_{1/2}$ of a substance, it can be virtually eliminated between exposures. It requires about 5× the plasma $t_{1/2}$ to achieve virtual elimination. If the exposure interval is equal to, or less than, the $t_{1/2}$ and the dose is constant, the plasma steady-state concentration will be achieved in five plasma $t_{1/2}$s. The xenobiotic may be sequestered in organs and tissues. If no detoxification or elimination occurs, of course, the chemical will accumulate significantly with each exposure. Cumulative effects can occur as the result of repeated exposures although no detectable levels of the toxicant accumulate, as when genetic damage is induced by carcinogens. Figure 12 illustrates the relationship of frequency of exposure and $t_{1/2}$ with respect to clearance from tissues. These factors influence whether a toxic reaction is defined as *acute* (within 48 hr), *subacute* (in 7 to 9 days), *subchronic* (about 90 days), or *chronic* (>90 days).

Factors influencing responses to xenobiotics

It should be evident that anything that influences the absorption, distribution, metabolism, or excretion of a xenobiotic will affect its toxicity. Many such factors exist.

Age

In general, biotransformation and excretion are less efficient at the extremes of life. Although drug-metabolizing enzymes are detectable at mid-gestation, they do not become fully developed until 6 to 12 months of age. Thus, the toxicity of many substances is higher in neonates. An example is the antibiotic chloramphenicol, which can accumulate to toxic levels because the

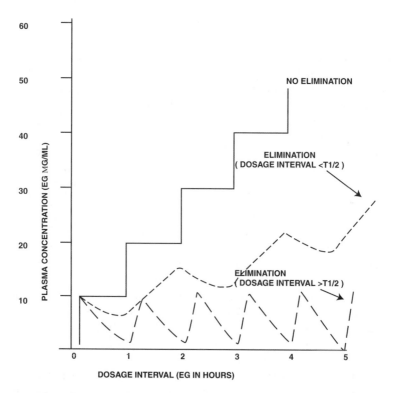

Figure 12 The influence of frequency of exposure or dose on tissue concentration of a chemical.

glucuronide-conjugase enzyme is lacking. Renal function is also underdeveloped so that excretion of drugs is impaired. The $t_{1/2}$ for inulin clearance is 100 min in infants younger than 6 months of age vs. 67 min in adults.

Body composition also differs with age. Total body water is 70 to 75% of body weight in neonates vs. 50 to 55% in adults. Extracellular fluid is 40% of body weight in newborns vs. 20% in adults. There is thus a greater fluid volume for dilution of water-soluble drugs in these infants. Their basal metabolic rate, moreover, is higher than in adults.

Other differences include greater permeability of skin, which has resulted in toxicity due to absorption of the hospital germicide hexachlorophene. Gastric pH is higher and gastric emptying is prolonged so that heavy metal absorption is increased. The intestinal flora will differ and biotransformation by microbes will be different as well. In later childhood, these situations may be reversed because systems are functioning at peak efficiency.

After age 75, these same systems may have slowed down significantly compared with those of a 30-year-old. Renal function and respiratory tidal volume are down, drug-metabolizing enzymes are less efficient, and even the number of drug receptors may be lower. Body composition also changes. The ratio of fat to lean body mass increases with age and total body water is down. Cardiac output is reduced and perfusion is lower. Plasma albumin content is down. The toxicity of water-soluble toxicants such as alcohol will be increased because they will be more concentrated. The concentration of the free component of albumin-bound agents will be increased because fewer albumin-binding sites are available. Biotransformation and excretion of xenobiotics will be impaired. The following are some examples of function at age 75 expressed as a percent of function at age 30.

Table 3 Function at Age 75 as a Percent of Function at Age 35

Nerve conduction velocity	90%
Basal metabolic rate	84%
Cardiac output	70%
Glomerular filtration	69%
Respiratory function	43%

Body composition

Body composition also varies considerably among normal individuals independent of age, and the factors discussed above with respect to fat and water content will be in play here as well.

Sex

Men and women may differ in response to xenobiotics. In part, this is due to differences in body size, fat content, and basal metabolic rate. In one study, the $t_{1/2}$ of antipyrine (an old and toxic analgesic) was 30% longer in young men than in young women. Differences in response to sex steroids obviously occur. Pregnancy is a special situation involving great changes in the metabolism, body composition, and fluid content of the mother. Placental transfer of many agents occurs and this may put the fetus at risk (see below). Sex-related differences also occur in experimental animals. Cessation of respiration has been shown to occur more frequently in female than in male rats after barbiturate anesthesia.

Genetic factors

For the majority of the population, biotransformations are controlled by multigenetic determinants. This is the basis of the continuous variation in response as reflected in the characteristic, normal population distribution curve. In some cases, however, a single gene locus may be responsible for altering the metabolism of a substance. This occurs in a subset of the population and affects a particular enzyme. The result is a discontinuous, bimodal distribution which reflects two overlapping normal distribution curves.

Figure 13 shows representative response distributions for acetylsalicylic acid (ASA, aspirin) as an example of the multigene type of control and isoniazid as the single gene type, in this case for an acetylating enzyme.

Pharmacogenetics is the sub-discipline of pharmacology that deals with this phenomenon. Over 100 examples have been identified of genetic differences in the biotransformation of drugs and chemicals. One of special concern for toxicology is the enzyme N-acetyltransferase. It acetylates and detoxifies many drugs and chemicals, including the aryl amines that are potential carcinogens, as well as many drugs including isoniazid, an anti-tubercular agent, and sulfa drugs. Slow acetylation is the dominant pattern in most Scandanavians, Jews, and North African Caucasians. Fast acetylation predominates in Inuit and Japanese. Similar patterns have been found in other species, including rabbits. Slow and fast oxidative metabolism has also been reported. About 9% of North Americans are slow metabolizers. In one study, a group of dye workers exposed to N-substituted aryl compounds were surveyed for the occurrence of *in situ* carcinoma of the bladder. Those with the disease were predominantly slow acetylators, suggesting that slow acetylators probably accumulated a carcinogenic agent normally detoxified by N-acetylation.

Thiopurine s-methyltransferase (TPMT) also is subject to genetic polymorphism. This enzyme is responsible for S-methylation of the antineoplastic drugs azathioprine and 6-mercaptopurine. About 88.6% of humans have high, 11.1% intermediate, and 0.3% low or undetectable levels of enzyme activity. Individuals with high levels may be poor responders to cancer chemotherapy unless dosage is adequate, while those with low or undetectable levels are in danger of developing complete suppression of bone marrow function unless, again, dosage is adjusted downward. Other aromatic and heterocyclic sulfhydryl compounds are methylated by this class of enzyme, and there is little doubt that toxicity can be affected by genetic differences in enzyme activity. Similar genetic polymorphism has been

Figure 13 Hypothetical construct of multigenetic (A) vs. single gene (B) influence on drug metabolism.

demonstrated for the drugs debrisoquine and metoprolol, which are detoxified by hydroxylation.

It is now technically possible through the use of modern techniques in molecular biology, notably DNA genotyping by the use of appropriate primers and the polymerase chain reaction (PCR) method, to identify individuals in advance who may pose therapeutic problems because of a lack or excess of detoxifying enzyme activity. Such technology can also be applied in the workplace to identify persons who are at excessive risk of a toxic reaction from a chemical in the work environment, or perhaps individuals who have already incurred DNA damage because of past exposure to a mutagen. These people could be excluded from high-risk areas (for them) or denied employment. Paradoxically, this possibility has already raised concerns in some union quarters that individuals might be denied their right to earn a living on biochemical grounds. In the future, employers might have to balance union concerns against the possibility of future litigation by workers made ill by their jobs. There is certainly no doubt that new ethical dilemmas will arise out of this technology.

Genetic factors also determine the emergence of strains of organisms resistant to normally toxic agents. There are many examples, including resistance of mosquitoes to DDT, rats to warfarin, malarial parasites to many drugs, cancer cells to anticancer drugs, and bacteria to antibiotics. In all cases, susceptible cells are killed off, leaving resistant mutants to proliferate. Mutations are occurring all the time. In bacterial populations, it has been estimated that a mutation imparting a degree of resistance to a drug occurs once in every 10^9 cell divisions. Unless that particular drug is present, however, the mutant strain will have no selection advantage and it will be overwhelmed by nonmutant cells. If the drug is present, however, it will have a distinct survival advantage and will become the dominant form.

Genetic factors can also influence the response to a toxicant at the target site as well as at the site of biotransformation. Inherited disorders that render individuals more susceptible to drug-induced hemolytic anemia include glucose-6 phosphate dehydrogenase (G-6-PD) deficiency and sickle cell anemia. A host of drugs, including antimalarials and sulfonamides, will induce a hemolytic attack in these people. There are also several inherited disorders of hemoglobin synthesis that act similarly. Many of the drugs involved are oxidizing, aromatic nitro compounds and nitrates. As noted above, altered gene coding plays a major role in carcinogenesis.

Presence of pathology

Given that the liver and the kidneys are the major organs of detoxication, it follows that any serious impairment of their function will have a significant impact on the toxicity of xenobiotics. This has been observed in fatty necrosis of the liver, hepatitis, and cirrhosis. These will impact mostly on highly lipid-soluble agents, requiring biotransformation to more water-soluble forms. Pharmacologically, this is seen with CNS depressants, including the tranquilizers Librium and Valium. Kidney dysfunction is reflected mainly in water-soluble agents and their elimination can be greatly impaired by renal disease. Water-soluble antibiotics such as gentamicin have a greatly prolonged $t_{1/2}$ in the presence of renal disease. Cardiovascular disease can also affect tissue perfusion and the delivery of the xenobiotic to, or conversely its removal from, its target site. Pulmonary disease will also affect the transfer of volatile agents across the alveolar membrane. In meningitis, the presence of inflammation compromises the integrity of the blood-brain barrier, and substances that would normally be excluded may reach significant concentrations in the spinal fluid. This fact makes some antibiotics (e.g., penicillin) useful to treat meningococcal meningitis although though they normally do not penetrate the barrier.

One of the most serious consequences of pre-existing pathology may simply be the fact that if an organ has already lost much of its function, further damage by a toxicant may destroy it completely and create a life-threatening situation. This is one of the main reasons why the elderly are more vulnerable to the toxic effects of drugs and chemicals.

Xenobiotic interactions

The effect that one drug may have on the action on another, collectively known as drug interactions (or drug-drug interactions), is an important aspect of clinical pharmacology. The effects of two drugs given together may be:

1. *Additive*, if they induce the same response (even through different mechanisms);
2. *Synergistic*, if the total response of their combined effect is greater than the predicted sum of their individual effects; or
3. *Antagonistic*, if one drug diminishes or prevents the effect of the other.

Mechanisms involved in drug interactions include altered absorption from the gastrointestinal tract, altered excretion (renal, biliary, respiratory), competition for receptors (antagonism) summation of pharmacological effects, and altered biotransformation. Drug interactions are usually associated with drug adverse reactions but they may also be exploited. All antidotal remedies are based on drug or chemical interactions. Examples of antidotes include:

1. The use of atropine to treat organophosphorus poisoning,
2. Metal chelators for treating lead and other heavy metal poisoning, and
3. The use of activated charcoal to bind toxicants in the gastrointestinal tract and prevent their absorption, emetics to induce vomiting and the removal of toxicants, naloxone to reverse the respiratory depression of opiates, and N-acetylcysteine to treat acetaminophen poisoning

When more than two substances are present, the possibility for a drug interaction is much greater. This is of special concern in the area of environmental toxicology because of the multiplicity of xenobiotics that may be present in water, food, and air. Most of the attention has centered on the possibility that the presence of one substance may increase the concentration of carcinogens from other sources by affecting their metabolism through enzyme induction. The effect of cigarette smoke on hepatic enzymes has been discussed. Many volatile solvents are enzyme inducers and chronic exposure to low levels of these, as in the industrial setting, can lead to increased enzyme activity. It has been shown experimentally that bedding rodents on softwood shavings induces hepatic microsomal enzyme activity because of the volatile terpenes emitted. Workers in softwood sawmills could experience similar effects. Literally hundreds of industrial chemicals have been shown to be enzyme inducers, including benzo[a]pyrene and 3-methylcholanthrene. Others include insecticides, (DDT, aldrin, dieldrin, lindane, chlordane), polychlorinated biphenyls (PCBs), polybrominated biphenyls (PBBs), dioxin, and drugs such as phenobarbital and other barbiturates, steroids, etc. Thus, although a substance is itself not toxic at a particular exposure level, it may influence the toxicity of other agents. The enzymes usually involved are the CYP-450 monooxygenases, but conjugating enzymes may also be induced. Food may contain natural enzyme inducers. The potato contains "a-solanin," the tomato "tomatin," both of which are steroidal alkaloids. The bioflavonoids are fairly potent inducers. They are found in species of *Brassica*, including Brussels sprouts where they have been shown to affect the metabolism of some drugs. Rutin is a bioflavonoid found in buckwheat in fairly large amounts. While it is difficult to identify situations in which these various inducers have influenced toxicity (with the exception of cigarette smoke), their ubiquity illustrates how different individuals or groups may display different sensitivities to the same toxic exposure on different occasions.

Organic solvents have been shown to influence ethanol toxicity. Toluene depresses alcohol dehydrogenase and prolongs the ethanol $t_{1/2}$. Ethanol itself may increase the liver and CNS toxicity of CCl_4, trichloroethylene, and others.

Diet can also affect the toxicity of substances in other ways. A toxicant may be adsorbed onto dietary components that reduce its absorption. The ability of dietary calcium to reduce lead toxicity is well-documented. Lead and calcium appear to compete for the same absorption site on the intestinal mucosa so that a diet high in calcium will reduce lead absorption. A diet high in fiber will shorten the transit time of the gastrointestinal tract so there is less time available for absorption. This may be one reason why a high fiber diet is associated with a lower incidence of colon cancer.

Some toxicological considerations

Acute vs. chronic toxicity

Acute and chronic toxicity for a single agent may be quite different and one is not a reliable predictor of the other. For example, acute benzene intoxication involves CNS disturbances such as excitation, confusion, stupor, and convulsions. Chronic toxicity includes depression of the bone marrow and a reduction of all circulating blood cells (pancytopenia), and benzene is carcinogenic in experimental animals. Chronic carbon monoxide (CO) poisoning is experienced by heavy smokers who may suffer from headache, dizziness, and shortness of breath. Acute CO poisoning affects firemen and fire victims who may become comatose. Numerous other examples exist. A major area illustrating this is that of tumor formation. A short-term exposure to a substance may elicit acute toxicity without significant risk of carcino-genesis, whereas long-term exposure to very low levels may not result in any toxic manifestation but could induce tumor formation. Dioxin (TCDD) is a prime example of this. Acute exposure causes the skin rash known as chloracne, whereas long-term exposure may be carcinogenic.

A detailed discussion of toxic mechanisms is beyond the scope of this chapter. More detailed descriptions are given with the discussions of specific chemicals and groups of chemicals. What follows is a brief overview to illustrate the role of target organs and systems in toxic reactions.

Acute toxicity

Acute toxicity generally refers to effects that occur following a 24- to 72-hr exposure to a single dose or multiple doses of a toxicant. Effects are usually observed within a few days. If the agent is rapidly absorbed, the effect may be immediate. The CNS is very vulnerable to acute toxicity from very lipid-soluble agents.

Peripheral neurotoxins

Organophosphorus and carbamate pesticides inhibit acetylcholinesterase (AChE) so that acetylcholine (ACh) accumulates and overstimulation of receptors occurs. ACh is a neurotransmitter in both the central and peripheral nervous systems. There are many naturally occurring neurotoxins, including tubocurarine, which blocks nerve transmission to voluntary muscle; botuli-num toxin (from *Clostridium botulinum*), which prevents the release of ACh from nerve endings; and tetrodotoxin (from the puffer fish), which paralyzes nerves by blocking sodium channels. Belladonna alkaloids (atropine and scopolamine) from nightshade are muscarinic blockers with peripheral and central effects, and muscarine from mushrooms is a muscarinic stimulant. Nerve gases are also irreversible AChE inhibitors. Many other examples exist (see Chapter 11).

Central neurotoxins

The inhalation of many volatile organic solvents and petroleum distillates can act like anesthetics and may cause unconsciousness. This has occurred in industrial accidents and in substance abuse (e.g., gasoline sniffing). Ethyl, methyl, and isopropyl alcohols are CNS depressants.

Inhibitors of oxidative phosphorylation

Cyanide (CN) in the form of cyanogenic glycosides (e.g., amygdalin) is present in many plant components, including almonds and the pits or seeds of cherries, apples, peaches, plums, apricots, and wild chokecherries. Human poisonings have occurred from consuming too many of these seeds, and livestock are often poisoned from eating chokecherry bushes. Cyanide binds to heme to prevent electron transfer. Cyanide may be present in metal ores, in some pesticides, and in metal polishes. The tragic accident at Bhopal, India, which killed over 2000 people, was due to the release of 40 tons of methyl isocyanate from an American Cyanamid plant. Azide and hydrogen sulfide act like CN. Carbon monoxide (CO) combines with hemoglobin and cellular cytochromes and prevents association with O_2. It thus causes cell hypoxia and also interferes with O_2 transport by red blood cells.

Uncoupling agents

Many agents act as uncouplers, preventing the phosphorylation of ADP to ATP, the high-energy phosphate. Uncouplers include the herbicide 2,4-D, halogenated phenols, nitrophenols, and arsenate. Most contain an aromatic ring structure. O_2 consumption and heat production are increased without an increase in available energy.

Inhibitors of intermediary metabolism

Certain fluoroacetate compounds of natural origin are used professionally as rat poisons. They inhibit the citric acid cycle to deplete available energy stores. The heart and the CNS are the organs of toxicity.

Chronic toxicity

Exposure to some toxicants must occur over days, weeks, or months before signs of toxicity appear. Heavy metals tend to act in this manner. Several outbreaks of methylmercury poisoning have followed this pattern. Mercury (Hg) from industrial discharges may be converted to methylmercury by microorganisms in the water. Monomethylmercury is CH_3Hg and dimethylmercury is $(CH_3)_2Hg$. These accumulate up the food chain to concentrate in fish and shellfish that may be consumed as food. As the toxicant accumulates in the tissues, severe neurological disorders occur. This happened at Minamata Bay in Japan and on the Grassy Narrows reserve in Northern Ontario. Infants born to exposed mothers may suffer from a cerebral palsy-like syndrome.

Cadmium (Cd), used in nickel-cadmium batteries, electroplating, and pigments, may accumulate in workers and cause kidney damage. Similar cases in the general population have been reported in Japan, from eating rice and other grains grown in soil contaminated with industrial wastes. Carbon tetrachloride (CCl_4) was used extensively in the dry-cleaning industry before it was discovered that it caused hepatic necrosis because it was activated to a free radical by CYP-450-dependant monooxygenase.

Mutagenesis and carcinogenesis

Sites of intracellular damage

Some toxic effects cannot be related to the level or frequency of exposure because the xenobiotic or its metabolite attaches irreversibly by binding to an essential component of the living cell to disrupt normal function. The binding sites are usually on macromolecules such as nucleotides, nucleosides, regulatory proteins, RNA, and DNA. The effects may be cumulative and they cannot be related to blood or tissue levels at the time of their manifestation. There may be a considerable delay from the initial exposure to the emergence of toxicity. This kind of reaction is involved in mutagenic and carcinogenic effects, and also causes a type of anemia (called aplastic anemia) in which the capacity of the bone marrow to produce blood cells is permanently destroyed. Some older drugs such as chloramphenicol and phenylbutazone have caused aplastic anemia in a very small number of patients — perhaps one in 40,000 exposed persons. The solvent benzene has also been associated with aplastic anemia. These low-frequency toxicities are difficult to identify and necessitate careful risk/benefit analysis when therapeutic interventions are contemplated.

A toxic effect of great concern in society today is, of course, the propensity of some xenobiotics to induce cancer. A wide range of natural and synthetic chemicals may induce alterations in DNA that, depending on the nature of the defect and the timing of its occurrence, can cause a neoplasm, a heritable change (*mutation*) or a developmental birth defect. It should be noted that birth defects (*teratogenesis*) may result from chemical interference with many other cell processes not involving altered DNA, such as interference with essential substrates and precursors, impaired mitosis, enzyme inhibition, and altered membrane characteristics. Most anti-cancer drugs are teratogenic for the same reason they are anti-neoplastic. Mutations are not always harmful and they provide the necessary genetic diversity for natural selection to occur, but they can also be responsible for fertility disorders, hereditary diseases, cancers, and malformations. Three types of genetic abnormalities can be induced.

1. Point mutation, also called gene-locus mutation, involves the alteration in some way of a small number of base pairs of nucleic acids. This may involve deletion, addition, or the substitution of an incorrect base pair. Strictly speaking, the term *mutagenesis* refers exclusively to this type of DNA alteration.
2. When the total number of chromosomes is altered, this is termed *aneuploidization*.
3. Chromosomal aberrations involving gaps, breaks, and translocations are referred to as *clastogenesis*.

Genetic diseases of eukaryotes can arise from all three types of chromosomal abnormalities. A characteristic of chemically induced carcinogenesis is that it is believed usually to involve decades-long exposure to very low levels of carcinogens, making predictions from animal studies very difficult. The induction of a cancer is thought to be a complex, multistage process involving interactions among the carcinogen and environmental and endogenous factors. Three stages are generally recognized: (1) initiation, (2) promotion, and (3) progression.

1. *Initiation.* The induction of a mutation by an electrophilic chemical or metabolite (a genotoxic carcinogen) that binds to DNA is believed to be the initial step. DNA thus altered is called an *adduct*. The heritable characteristic is not expressed. Chemicals that are capable of inducing tumors after a single exposure are sometimes called *complete carcinogens*. This is usually seen in animal studies. Polyaromatic hydrocarbons (PAHs) such as dimethylbenz[a]anthracene and 3-methylcholanthrene are initiators.
2. *Promotion.* Promoters are agents that increase the number of tumors, increase their growth rate, or decrease the latency period. They do not bind to DNA and are therefore called *epigenetic carcinogens*. Exposure to the initiator must occur first. Promoters do not themselves generally cause tumors, although there is evidence that some promoters are merely weak carcinogens that act synergistically with other carcinogens. Co-carcinogens are agents that, when present just before, or together with, a carcinogen, result in significantly higher tumor yields in experimental animals. Some substances may be both promoters and co-carcinogens (e.g., phorbol esters). Nutritional factors (e.g., saturated fat intake), hormones, trauma, and viruses may act as co-carcinogens.
3. *Progression.* This refers to the natural history of the disease. Figure 14 summarizes the steps involved in carcinogenesis.

An important advance in understanding carcinogenesis was the discovery of the existence of oncogenes. Proto-oncogenes are normal constituents of the cell that exist in every cell component (membrane, cytoplasm, nucleus, etc.) and that are involved in every signal transduction cascade. They modulate cell function, growth, and proliferation. When acted upon by external influences they are converted to oncogenes, resulting in uncontrolled cell growth.

A scientist named Peyton Rous first postulated in the 1920s that a sarcoma of chickens was virally induced but it was not until the 1970s when the rare T-cell leukemia was shown to be due to a retrovirus that the theory became accepted as fact. The Rous sarcoma virus is also a retrovirus. Retroviruses are a major factor in the conversion of proto-oncogenes to oncogenes. These viruses contain RNA rather than DNA and redirect host cells to produce viral DNA through the process of reverse transcription. This involves

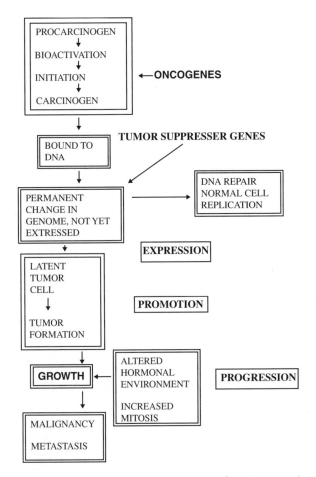

Figure 14 The steps in the progression of carcinogenesis.

an enzyme, "reverse transcriptase." Proto-oncogenes may be mutated to oncogenes during this process. Over 30 human oncogenes have been identified to
date. The oncogene may remain in the host cell but it may also be incorporated
into viral DNA (the process is called *transduction*) and passed on to another
cell. Oncogenes first identified in a cell are labeled with the prefix *c*. Those
first identified in a retrovirus are labeled by with prefix *v*. It may require the
participation of more than one oncogene to induce tumor formation. The Rous
sarcoma gene, *src*, is not a normal viral gene, but an oncogene of cellular
origin. Retroviruses may cause the excessive production of an oncogene,
induce its expression in an inappropriate cell, or mutate the coding of a gene
during transduction. Mutant versions of proto-oncogenes generally lead to
uncontrolled cell proliferation. Oncogenes may also be formed intracellularly
by chemical carcinogens or by chromosomal abnormalities. Over 100 oncogenes have now been identified. Human tumor oncogenes have been identified as alleles of the *ras* family of proto-oncogenes. Others include *fos, jun,*

and *myc*. The *myc* oncogene was first identified as a viral one and is important in human tumorigenesis as a growth factor stimulant.

An area of research that is currently attracting much attention relates to the existence of tumor suppressor genes. These have been called the "flip side" of the oncogene story and are sometimes called *anti-oncogenes*. Like proto-oncogenes, they also appear to modulate normal cell growth to prevent the emergence of neoplastic cells. Whereas oncogenes are dominant, tumor suppressor genes are recessive. They were discovered as a result of the observation that when normal cells are fused with tumor cells in culture, suppression of the tumor cells resulted. Inactivation of these suppressor genes may be an essential step in tumorigenesis, at least in some cases, and prevention of this inactivation by inserting normal copies of tumor suppresser genes, or by mimicking their function pharmacologically, may lead to new therapies. Tumor suppressor genes may be responsible for the existence of "cancer families." If an inherited defect in one copy acquires a matching defect in the other copy through retroviral or chemical damage, or from the other parent, cancer may result. Retinoblastoma, the most common eye tumor of children, may be heritable in this manner. Important suppressor genes include p53, involved in sarcomas and breast and brain cancer; Rb, involved in retinoblastoma and osteosarcoma; and APC, involved in colon cancer. E-cadherin, a Ca^{2+}-dependent trans-membrane protein involved in epithelial cell-cell interactions, has been suggested as a candidate tumor suppressor gene.

Genes are involved in cancer treatment as well. A multi-drug resistance gene, MDR-1, encodes for a P-glycoprotein that pumps cytotoxic, anti-neoplastic drugs out of tumor cells.

An even more recent development in the genetic manipulation of tumor cells is the synthesis of "anti-sense" sequences. These oligonucleotides are short (15 to 25 bases), single-stranded DNA sequences that have been altered so that they target a specific mRNA sequence, resulting in a defective DNA segment in a highly specific manner. Thus, if a specific segment (sense) that plays a key role in tumorigenesis can be identified, the appropriate anti-sense sequence can be formulated and inserted to interfere in the process. This advance holds great promise for the future treatment of cancer.

The World Health Organization's International Agency for Research on Cancer maintains an excellent Web site with evaluations on the current carcinogenicity rating of hundreds of chemicals (http://www.iarc.fr/).

DNA repair

DNA becomes damaged as a result of everyday exposure to toxic substances, the generation of free radicals, exposure to natural source radiation, etc. It is therefore not surprising that enzymatic DNA repair mechanisms exist. The characteristic of self-repair appears to be unique to the DNA molecule. Both constitutive (ongoing) and inducible (activated in response to an event) systems exist. The natural, endogenous repair system can be overwhelmed by massive or repeated exposure to genotoxic agents. In that case, primary DNA damage occurs (strand breaks, loss of bases, or adduct formation). The ability to remove and replace damaged segments of DNA is central to the repair process. If repair can be effected prior to the fixation of the mutation, there may be no adverse affect from the DNA damage. DNA damage may inactivate both error-prone and error-free processes. Error-prone processes may actually lead to new mutations through nucleotide mismatches. Error-free processes will correct the damaged DNA site with the correct nucleotide sequence. Which process is dominant depends on a number of factors, including species (processes tend to be species-specific), cell type, the nature of the chemical mutagen, and the nature of the lesion. 3'- and 5'-endonucleases cleave the DNA on either side of the adduct and an exonuclease cuts out the damaged region, including nucleotides on either side of it. A DNA polymerase inserts the correct bases into the patch and a DNA ligase seals it.

Even in error-free repair, the wrong bases are sometimes incorporated into the DNA patch. In this case, the repair process may be repeated to effect the correct pairing. This process is called *mismatch repair* and it is triggered by non-hydrogen-bonding base pairs. Evidence suggests that the repair process can correct damage resulting from low levels of exposure to genotoxic agents with a high frequency of success. This raises questions about the degree of risk associated with such exposures. The processes of mutation and repair are illustrated in Figure 14.

Genetic predisposition to cancer

Just how much cancer is due to environmental factors and how much is due to a genetic predisposition has been a matter of considerable debate. Recently, a study conducted by the Department of Medical Epidemiology at the Karolinska Institute has shed some light on these questions. Researchers combined data from 44,788 pairs of twins listed in the Swedish, Danish, and Finnish twin registries to examine the risk of cancer for persons with a twin with cancer at one of 28 anatomical sites. Statistically significant evidence of effects of heritable factors was observed for several cancers. Analysis indicated that genetic predisposition accounted for 27% of the risk for breast cancer, 35% of the risk for colorectal cancer, and 42% of the risk for prostate cancer. Overall, the risk of being diagnosed with any of these cancers by age 75 ranged from 11 to 18% for a person whose identical twin had that cancer and from 3 to 9% for a person whose fraternal twin had that cancer. The authors concluded that environmental factors were by far the greater contributors to the risk of getting cancer.

Epigenetic mechanisms of carcinogenesis

Carcinogens that function through non-DNA-reactive processes have been identified in recent years. They do not induce mutation in cell assays nor do they induce direct DNA damage in target organs. Precise mechanisms by which these agents operate have not been established, but changes in gene expression and cell growth (proliferation) seem to be central to the process. Cytotoxic agents, by killing large number of cells, trigger the production and release of growth factors and the frequency of mutations is thus also increased. Some of these may be malignant cell types. Chloroform, by inducing liver necrosis and hyperplasia, can induce tumors in this way. Phenobarbital, conversely, can induce mitosis without cytotoxicity by mechanisms that remain unclear. Rapid cell division can lead to error-prone repair and hence mutation. Under the influence of mitogenic factors, a mutant cell may proliferate and become the dominant cell type.

Another mechanism that may operate for non-genotoxic carcinogens is interference with normal apoptosis. Apoptosis, or programmed cell death, is a normal process by which senescent cells or cells with massive DNA damage are removed. Cancer results from an imbalance between cell growth and cell death with resulting unchecked proliferation. Several tumor promoters have been shown to inhibit normal apoptosis. The process of apoptosis involves oncogenes, tumor suppressor genes, and cell cycle regulatory genes; interference with these can lead to cell proliferation and neoplasia.

Many of these processes can be facilitated by interference with cell-cell communication through gap junctions. Gap junctions are areas of cell-cell close contact-containing channels that allow the passage of ions and small water-soluble molecules. Signal-transducing substances such as calcium, inositol triphosphate, and cyclic AMP are transmitted in this way. Gap junction communication has been shown to be decreased in the presence of some epigenetic carcinogens.

Induction of cytochrome P450 may result in the formation of reactive intermediates that can act as promoters, often through the formation of free radicals.

The role of cell repair and regeneration in toxic reactions

The ability of a tissue to repair itself after chemical insult is dependent on the percentage of cells damaged, the nature of the damage, and the rate of cell division. Cells of the epithelium (skin, hair, nails) and mucous membranes (gastrointestinal tract, lungs, urogenital tract) are capable of rapid cell division and tissue regeneration. Sloughing of these tissues is protective because it carries away any toxicants accumulated in the cells from low-level exposures. Bone marrow also has a rapid rate of cell division, as do the germ cells of the developing fetus. Damaged bone marrow may regenerate if enough cells survive.

Conversely, such rapidly dividing tissues are vulnerable to destruction when exposed to higher levels of cytotoxic agents. It is for this reason that hair loss, anemia, diarrhea, and gastrointestinal hemorrhage may occur with the use of anti-neoplastic drugs and following exposure to radiation. These adverse reactions occur because the interval between cell damage and cell division is too short for repair to occur. Cell division cannot proceed or is so impaired that second-generation cells cannot themselves reproduce.

Cells of the liver, kidney, exocrine and endocrine glands, and connective tissue have less rapid rates of cell division but are capable of proliferation and repair. Heart cells, peripheral nerve cells, and voluntary muscle cells regenerate poorly but recent evidence suggests that these may be replaced to a greater extent than previously thought. Increased muscle mass from conditioning is due to increased actin and myosin content of the cells.

Response of tissues to chemical insult

1. *Tissue Regeneration.* The destruction of some cells may lead to the release of growth factors, which stimulate cell division and proliferation, followed by maturation of the cells to a functional state. Function is thus restored.
2. *Cell regeneration.* If a portion of a cell is destroyed but the nucleus remains intact, the cell may regenerate. Thus, the axon of a neuron may eventually regenerate if the cell body is undamaged.
3. *Hyperplasia.* This refers to an increase in the size of an organ or tissue in response to increased demands placed upon it. This is a normal adaptive phenomenon and involves an increased number of normal cells. It tends to occur in organs of intermediate specialization such as the liver, pancreas, thyroid, adrenal cortex, and ovary. Hyperplasia of the liver can occur in response to an increased demand to detoxify a xenobiotic presented in high concentration or for a long period.

Fetal toxicology

Teratogenesis

The existence of malformed infants and animals has been recorded for thousands of years. Early explanations tended to be supernatural. Deformed infants were often regarded as warnings from the gods. The word *monster* comes from the Latin *monstrum*, meaning "portent." Malformed animals were thought to be the result of copulation with humans. Those suspected of such an act were often put to death painfully.

Although it was recognized as early as 1932 that dietary deficiencies could result in birth defects, it was not until 1962 that the possibility of drug-induced teratogenesis was recognized as a result of the phocomelia observed in the thalidomide infants in Germany, Japan, the United Kingdom, and other countries (including Canada). This tragedy resulted in the introduction of regulatory requirements for tests for teratogenic effects of new drugs in all industrialized countries.

The placenta is a semipermeable membrane that will pass many xenobiotics and their metabolites. Whereas most nutrients cross the placental membrane by energy-consuming, active transport systems, toxicants cross mainly by passive diffusion; thus, lipid solubility is an important determinant of toxicity to the fetus. The exceptions are anti-metabolites (anti-cancer drugs), which are analogs of natural substrates and can use their transport pathways. Highly lipid-soluble agents will establish equilibrium very quickly between the maternal plasma and the fetus. More polar agents will take longer to accumulate, but there is no such thing as a "placental barrier." The placenta possesses biotransforming enzymes so that some detoxication may occur, but it is not enough to protect against any but very low exposures. The human placenta is capable of Phase I and II biotransformations, but the enzymes are not inducible. Biological functions in the fetus, even near term, are poorly developed. The blood-brain barrier is imperfect, and biotransforming enzymes are undeveloped, as is the excretory function of the kidney. The earlier in gestation, the more this is so. The consequences of fetal exposure to toxicants are several, including:

1. If the exposure is high enough, embryonic death will occur and, possibly, reabsorption of the embryo. This tends to occur early in gestation.
2. If the exposure occurs during structural development, teratogenesis may occur and an anatomical defect may result, the nature of which will reflect the stage of development when the exposure occurred. There is, thus, a critical period during which a target site exists which does not exist before or afterward. This is what occurred when pregnant women took the drug thalidomide to control morning sickness.
3. If structural development is more or less complete, a functional defect can occur that may not become evident until later in life when the affected function would normally come into play. The newborn infant may thus be vulnerable to developmental toxicity for several weeks after birth until organ systems become fully matured.

Examples of known teratogenicity resulting from fetal exposure to toxicants include the following (others doubtless exist):

1. Folic acid antagonists (aminopterin) are used as anti-cancer agents.
2. Androgenic hormones (natural and synthetic progesterones) used to treat breast cancers (and by athletes to increase muscle mass) will masculinize female offspring.
3. Thalidomide, a very potent teratogen, use for a single day during days 20 to 50 of pregnancy was associated with phocomelia (flipper-like limbs). Most animal species are much less sensitive than humans.
4. Alcohol: fetal alcohol syndrome (FAS) involves impaired growth, impaired mentation, and distinct facial characteristics (small head, small eye openings, thin upper lip). Fetal alcohol effects (FAE) involve only one or two of these criteria.

5. Methylmercury: infants exposed in utero may develop severe neurological disturbances similar to cerebral palsy. This occurred during the Minamata exposure in Japan.

Teratogenesis may occur from causes other than drugs and environmental chemicals. In fact, in 65 to 70% of cases, the cause is never established. Developmental toxicity can occur in the absence of any maternal toxicity. Of the remainder, the breakdown is shown in Table 4.

Table 4   Causes of Teratogenesis

Known genetic transmission	11–18%
Chromosomal aberration	3–5%
Environmental causes	
Radiation (therapeutic, nuclear)	<1%
Infections (rubella, herpes, toxoplasma, cytomegalovirus, syphilis)	2–3%
Maternal metabolic imbalance (endemic cretinism, diabetes, phenylketonuria, rilizing tumours)	1–2%
Drugs and environmental chemicals	2–3%

Transplacental carcinogenesis

The development of cancer as a result of fetal exposure is a possibility if the reproductive system is involved. The best-known example is the occurrence of carcinoma of the vagina and cervix in young women exposed *in utero* to diethylstilbestrol (DES). DES was given to prevent impending abortion, primarily between 1950 and 1970.

Carcinogens abound as natural substances in the environment. Table 5 compares some relative risks of these with synthetic ones. The problems associated with attempting to predict the carcinogenic potential of new chemicals from animal studies is discussed in Chapter 2. The problem is critical, due to the high cost and protracted period it takes to evaluate a new chemical.

It is generally accepted that the root of most anthropogenic environmental problems is overpopulation. The doubling time of the Earth's population is down to about 30 years, and this has created incredible pressures on our ability to provide food, living space, energy, and manufactured goods, and particularly on our ability to deal with the waste products of our society (see Table 6). Yet there has been little official recognition of this in the West, nor are there specific actions to control population explosion. This is changing, but fear of reaction from religious groups and right-to-life activists, who tend to link population control to such issues as abortion and birth control, has made the issue unattractive to politicians. The lessons of nature should convince us of the need to address this issue. When a population exceeds the ability of its ecosystem to support it, it dies out.

Table 5   Relative Carcinogenic Hazards in the Environment

Source	Cancers/100,000 Population
U.S. average lifetime risk (all types)	25,000–30,000
Foods	
Four tablespoons peanut butter/day (aflatoxin)	60
One pint milk/day (aflatoxin)	14
8 oz. broiled steak/week (nitrosamines, polycyclic aromatic hydrocarbons, PAHs)	3
One diet soda/day (saccharin, U.S.)	70
Average U.S. fish consumption/day	33
Lake Michigan sport fish (based on median consumption levels and EPA potency values)	480–3300
U.S. sport fish (based on Kim & Stone potency values)	77–340
Drinking water	
Average U.S. groundwater 2 L/day	1
Niagara River water 2 L/day (EPA potency value)	0.3
Air	
Various estimates, U.S. urban	10–560

Table 6   Some Potential Carcinogens Encountered in Industry

1. PAHs (e.g., benzo[a]pyrene); produced by any combustion process, refining and distilling of petroleum, breakdown of lubricants, in welding and foundry processes
2. Benzene (32 million tons/year produced); chemical and petrochemical industries, solvents (paint industry), as an impurity, etc.
3. Carbon tetrachloride (CCl_4); solvent in industrial processes
4. Halogenated hydrocarbons (PCBs, PBBs, and TCDD); in lubricants, transformer insulation, pesticides, etc.
5. Asbestos from mining, in insulation
6. Vinyl chloride in plastics industry
7. Formaldehyde (embalmers, pathologists)

Further reading

Aldridge, W.N., *Mechanisms and Concepts in Toxicology,* Taylor & Francis Ltd., London, 1996.

American Medical Association Council on Scientific Affairs. Formaldehyde report. *JAMA,* 261, 1183–1187, 1989.

Assennato, G., Cervino, D., Emmett, E.A., Longo, G., and Merlo, F., Followup of subjects who developed chloracne following TCDD exposure at Seveso, *Am. J. Indest. Med.,* 16, 119–125, 1989

Benet, L.Z., Mitchell, J.R., and Sheiner, L.B., Pharmacokinetics: the dynamics of drug absorption, distribution and elimination, in *Molecular Oncology,* Bishop, J.M. and Weinberg, R.A., Eds., Scientific American, New York, 1996, chap. 1.

Cartwright, R.A., Glashan, R.W. et al., Role of N-acetyltransferase phenotypes in bladder carcinogenesis: a pharmacogenetic epidemiological to bladder cancer, *Lancet*, 2, 842–846, 1982

Hardman, J.G. and Limbird, L., Eds. in Chief, *Goodman and Gilman's The Pharmacological Basis of Therapeutics*, 9th ed., McGraw-Hill, New York, 1996 .

Klassen, C.D., Amdur, M.O., and Doull, J., Eds., *Casarett and Doull's Toxicology: The Basic Science of Poisons*, 5th ed., McGraw-Hill, New York, 1996.

Klaunig, J.E., Kamendulis, L.M., and Yong, X., Epigenetic mechanisms of chemical carcinogenesis, *Belle News Lett.*, 9, 2–21, 2000.

Lichtenstein, P., Holm, N.V., Verkasalo, P.K. et al., Environmental and heritable factors in the causation of cancer-analyses of cohorts of twins from Sweden, Denmark and Finland, *New Engl. J. Med.*, 343, 78–85, 2000 (comments 135–136).

Marx, J., Research news: learning how to suppress cancer, *Science*, 261: 1385–1387, 1993.

News and Comment. Animal carcinogen testing challenged, *Science*, 250, 743–745, 1990.

News and Comment. Experts clash over cancer data, *Science*, 250, 900–902, 1990.

Niesink, R.J.M., Hollinger, M.A., and de Vries, J., Eds., *Toxicology: Principles and Applications*, CRC Press, Boca Raton, FL, 1996.

Oberbauer, R., Not nonsense but antisense–applications of antisense oligonucleotides in different fields of medicine, *Wein. Klin. Wochenschr.*, 109, 40–46, 1997.

Putnam, D.A., Antisense strategies and therapeutic applications, *Am. J. Health Syst. Pharm.*, 53, 151–160, 1996.

Research News. Dioxin revisited, *Science*, 251, 624–626, 1990.

Ross, E.M., Pharmacodynamics: mechanisms of drug action and the relationship between drug concentration and effect, in *Toxicology*, Marquardt, H., Schafer, S.G., McLellan, R., and Welsch, F., Eds., Academic Press, New York, 1999.

Van Loon, J. and Weinshilboum, R.M., Thiopurine methyltransferase isoenzymes in human renal tissue, *Drug Metab. Dispos.*, 18, 632–638, 1990.

Review questions

1. Match the following terms with the appropriate definition:
    a. Pharmacodynamic
    b. Probit ∨
    c. Economic toxicology iii
    d. Xenobiotic i
    e. Anthropogenic IV
        i. A substance foreign to living systems. ▷
        ii. A chemical process in biological systems in which neither of two agents, which interact, is permanently altered. A
        iii. The application of toxicology to achieve an advantage for humankind. c
        iv. Resulting from human activity. E
        v. A probability unit used for making comparisons of potency or toxicity. B

2. Select the correct statement.
    T   a. A quantal response is one that is "all or nothing."
    F   b. The response to increasing doses of a drug or toxin continues to increase indefinitely.
    F* c. The minimum lethal dose value is applicable to all species.
        d. A quantal response can never be converted to a graded one.
        e. Semilogarithmic dose-response plots are used to clean up bad data.   *Nothing can clean up bad data.*

Answer the following questions True or False.

3. If drug A has a Therapeutic Index (TI) of 500 and drug B has a TI of 10,000, drug A is safer than drug B. F
4. An acute toxicity reaction is defined as one that occurs within 48 hr of exposure. T
5. Parallel dose response curves suggest that the two agents in question probably work through the same mechanism. T
6. If the $t_{1/2}$ of an agent exceeds the dosage/exposure interval, it will never accumulate in the body. F
7. Insecticides, which inhibit the enzyme acetylcholinesterase, cause rapid loss of consciousness. T
8. Nerve toxins may work by blocking axonal conduction, blocking the enzymatic destruction of a neurotransmitter, blocking the attachment of a neurotransmitter to its receptor, or blocking its release from the nerve terminal. T
9. Methylmercury is less toxic than elemental mercury. F
10. Cadmium is a nontoxic heavy metal.   *↳mercury in the soil.*

For questions 11 to 16, use the following code:

    Answer A if statements a, b, and c are correct.
    Answer B if statements a and c are correct.
    Answer C if statements b and d are correct.
    Answer D if only statement d is correct.
    Answer E if all statements (a, b, c, d) are correct.

11. a. Most anticancer drugs are teratogenic. T
    b. Genetic abnormalities may involve point mutations or chromosomal breaks. T
    A
    c. Cancer induction is likely a multistage process.
    d. Viruses have nothing to do with cancer. F
12. a. Promoters increase tumorigenesis in response to other agents.
    b. Exposure to the promoter must occur before exposure to the initiator.
    c. Promoters do not bind to DNA.
    d. Promoters always cause cancer in their own right.

13. a. Selective toxicity is usually absolute; that is, a substance is completely toxic for the target species and completely harmless for other ones.
    b. The portal of entry can significantly affect the toxicity and carcinogenicity of a xenobiotic.
    c. Cancer studies in animals are highly predictable for cancer risk in humans.
    d. Oncogenes can be activated to convert a cell to a malignant form.
14. Highly water-soluble chemicals are:
    a. Excreted by the kidney without biotransformation.
    b. Poorly absorbed from the gastrointestinal tract.
    c. Very polar.
    d. Cross the blood-brain barrier very poorly.
15. Highly lipid-soluble chemicals:
    a. Are well absorbed from the gastrointestinal tract.
    b. Do not enter the cerebrospinal fluid.
    c. Require biotransformation by the liver before being eliminated by the kidneys.
    d. Generally have a very short $t_{1/2}$. F
16. Which of the following statements is/are true?
    a. Serum albumin has numerous binding sites for weakly acidic xenobiotics.
    b. Substances with epoxide bonds tend to bind irreversibly to macromolecules within cells.
    c. Mixed-function oxidase enzymes in the smooth endoplasmic reticulum of the liver cells are responsible for much drug metabolism.
    d. Conjugation of a xenobiotic with glucuronide is defined as a Phase II reaction.

Answer the following questions True or False.

17. The kidney is the sole route of elimination of xenobiotics.
18. Water-soluble agents of small molecular size pass through the glomerulus.
19. Body composition has no influence on the fate of chemicals in the body. F
20. Some agents may be secreted by an active process across the wall of the renal tubule.
21. The presence of one chemical in the body may influence the metabolism of another. T

For Questions 22 to 24, use the following code:

Answer A if statements a, b, and c are correct.
Answer B if statements a and c are correct.

Answer C if statements b and d are correct.

Answer D if only statement d is correct.

Answer E if all statements (a, b, c, d) are correct.

22. a. Proto-oncogenes are involved in regulating cell-signaling cascades.
   b. Proto-oncogenes may be converted to oncogenes by retroviruses.
   c. Oncogenes generally cause uncontrolled cell proliferation.
   d. Retroviruses may become carriers of oncogenes.

23. a. Tumor suppressor genes display dominant characteristics.
   b. Oncogenes are recessive in nature.
   c. Tumor suppressor genes have no function in the normal cell.
   d. Tumor suppressor genes normally prevent uncontrolled cell growth.

24. a. "Sense" refers to short segments of DNA coding for some important aspect of cell function.
   b. "Anti-sense" is an altered sequence of oligonucleotides matched specifically to a "sense" sequence.
   c. Anti-sense sequences inhibit tumor growth *in vitro*.
   d. Anti-sense sequences generally contain 15 to 25 bases.

25. List five chemicals that can induce hepatic microsomal enzymes.

26. List five known industrial carcinogens.Ry 38

27. Define the following:
   a. Teratogenesis
   b. Carcinogenesis
   c. Mutagenesis

28. List three biological variables that can affect the body's response to xenobiotics.

Answers

1. i = d, ii = a, iii = c, iv = e, v = b
2. a
3. False
4. True
5. True
6. False
7. False
8. True
9. False
10. False
11. A
12. B
13. C
14. E
15. B

16. E
17. False
18. True
19. False
20. True
21. True
22. E
23. D
24. E

For Questions 25 through 28, find the answers in the text.

chapter two

Risk analysis and public perceptions of risk

(Risky Business)

Introduction

It is estimated that between 60,000 and 70,000 industrial and commercial chemicals are currently in use in North America, with the possibility of more coming on-stream every day. Only about 3500 of these have been studied sufficiently to conduct any sort of risk assessment regarding human health, and such studies characteristically use only one route of administration (portal of entry). Approximately 600 chemicals are currently judged to constitute a significant potential risk to human health, either because of their toxicity or because they are manufactured in such quantities that there is likely to be a high level in the environment. The public seems unwilling to give up the advantages accruing from such chemicals (plastics, pesticides, petroleum fuels, etc.) but is also increasingly vociferous in its demands to be protected from any adverse effects arising from their use. The environmental damage caused by some of these agents is becoming more and more evident and indeed this may be the real danger facing humankind. Nevertheless, legislators and regulators are faced with the task of making decisions regarding safe limits for thousands of chemicals, often on the basis of very limited data and in the face of pressure from consumer groups, environmental activists, and industry lobbies.

Assessment of toxicity vs. risk

Toxicity assessment is the determination of the potential of a substance to act as a poison, the conditions under which this potential will be realized, and the characterization of its action. Conversely, the assessment of risk involves the quantitative assessment of the likelihood of these deleterious effects occurring in a given set of conditions. This subtle difference is not always appreciated by the public — and especially not by the news media. Thus, statements frequently appear to the effect that dioxin is the "most potent poison known to man." In fact, *botulinum* toxin is 100× more potent in mice and the toxicity of dioxins in man has not been fully established. Moreover, the real question of risk must consider such factors as:

1. What is the biological half-life of the substance? Dioxins are very stable.
2. What is the partition coefficient? Dioxins are very lipid-soluble and are therefore sequestered in the body.
3. Does the toxin concentrate up the food chain? Yes, because of the partition coefficient.
4. What are the long-term effects? Is the substance carcinogenic? Yes, in experimental animals. In humans, the evidence is much less conclusive.
5. What are the predicted risks to humans and the environment based on known levels of contamination? This is the area that causes the most controversy because it is highly speculative.
6. What are the costs of avoiding these risks? This is very difficult to estimate and therefore also controversial. While risk to the general public is difficult to assess and usually of a very minor nature, risks encountered by industrial workers may be much greater because of the higher exposures and because of the risk of accidental contamination. Populations in some regions, however, may be exposed to similar risks from industrial accidents or from uncontained dump sites.

Predicting risk: workplace vs. the environment

Acute exposures

Information from industrial accidents and from preregulation exposures is very valuable because it eliminates the need to make extrapolations from test animals. Prediction of risk following defined exposures is thus fairly accurate as, for example, in the case of cholinesterase-inhibiting insecticides. Animal data are still useful, however, because they also deal with acute exposure.

Chronic exposures

Predictions are less reliable due to biological variations in susceptibility to chronic, lower levels of exposure. Individual susceptibility to lung damage from paraquat, for example, may vary considerably.

Very low-level, long-term exposures

It is more difficult to predict organ toxicity from animal studies with this type of exposure but they are still useful. Epidemiological data from human exposures are most useful if available. For example, extensive data have accumulated over many decades regarding pneumoconiosis (black lung disease in miners).

Carcinogenesis

At best, predictions from animal data can only provide a rough approximation due to the need to extrapolate from very high to extremely low exposures and the possibility of species differences. Differences in the nature of the exposure can further complicate extrapolations from animal data to the human situation. Moreover, predictions of risk due to low-level exposures are complicated by the presence of other risk factors, many of them from natural sources. For example, volcanic eruptions can pour huge volumes of gases and particulates into the atmosphere, equal to years of industrial pollution. After the Mount St. Helen volcanic explosion, the word *pneumoultramicroscopicsilicovolcanopneumoconiosis* was coined as the longest word in the English language. It refers to pneumoconiosis from inhaling volcanic ash. Smoking would be an example of an "anthropogenic" risk factor (i.e., of human origin).

Risk assessment and carcinogenesis

As already noted, this is the most complicated and least reliable area regarding the prediction of risk to human health in the general population from exposure to very low levels of environmental pollutants. There are several mathematical models for predicting carcinogenic risk, either by extrapolation from animal data or from human industrial exposures. Regarding animal studies, there is general agreement among these models for extrapolation to human exposures at high doses. At very low exposure levels, predictions of cancer risk can vary by several orders of magnitude and this is the very type of exposure that creates the greatest concern in the public's mind. These differences arise because of the application of different theories of carcinogenesis to the development of models for calculating risk (see also Chapter 1). Examples of these models include:

- Distribution models (log probit, logit) assume that every individual has a threshold below which no adverse effect will occur (a No Observable Adverse Effect Level or NOAEL).
- Mechanistic models are based on presumed mechanisms of tumorigenesis and assume that a cancer can arise from a single mutated cell. The single-hit model assumes that the exposure of DNA to a single molecule of a carcinogen is sufficient to induce carcinogenesis. The gamma multihit model assumes that more than one "hit" is required. Multistage models assume that carcinogenesis is a process requiring several stages (a series of mutations, biotransformations) involving carcinogens, co-carcinogens, and promoters that can best be

modeled by a series of multiplicative mathematical functions. Predicted dose responses are linear at very low exposure levels and
assume that there is no NOAEL.

All of these methods differ in the nature and shape of the dose response
curve at the low-exposure end. Figure 15 illustrates how these differences
affect predictions.

The U.S. Environmental Protection Agency (EPA) uses the "Linearized,
Multi-Stage Assessment Technique," which assumes that there is no NOAEL
and which involves the following steps (see Figure 16):

1. Evidence of carcinogenesis is obtained from animal studies in rabbits,
   rats, and mice, with dose response data for oral, inhalation, or dermal
   portals of entry (routes of administration).
2. From this dose response data, the dose is calculated that would
   theoretically cause one cancer per million animals. The assumption
   is made that the dose response curve is linear all the way to zero;
   that is, that there is no "no effect" level for the carcinogen.
3. An equivalent human dose is calculated that would cause the same
   incidence of cancer. This stage employs arbitrary factors to adjust for
   differences in absorption, metabolism, and excretion based on what
   data are available for humans, or simply uses a safety margin if no
   data are available. The 1/1,000,000 risk level is the "red line" that the

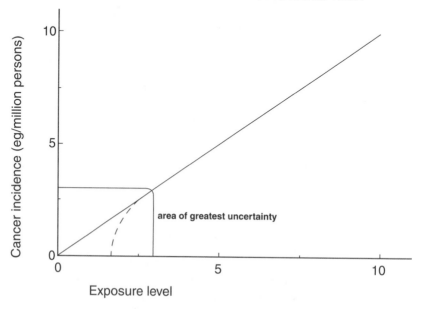

Figure 15 Area of greatest inaccuracy (threshold vs. no threshold) in predicting
cancer risk.

EPA has set for acceptable risk and it is used to determine safe limits in the environment.

4. Using knowledge of the average human intake orally or by inhalation, maximum allowable limits are set for the toxicant that would keep daily intake below the level that would induce one additional cancer per million people. An additional safety margin can be introduced, based on the lowest levels that can be achieved at an acceptable cost. In Canada, the Canada Environmental Protection Act (CEPA) defines the Tolerable Daily Intake (TDI) as the maximum to be permitted. It

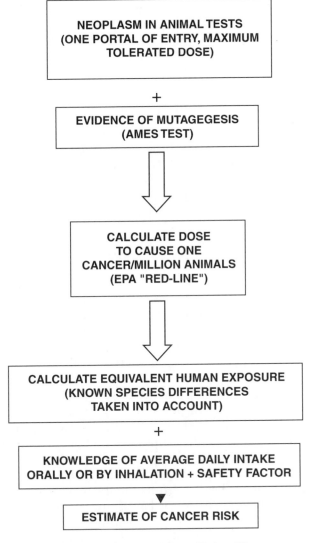

Figure 16 Stages in the process of cancer risk prediction. There are several points of uncertainty.

uses a safety factor of 100 times the threshold obtained from animal studies. It also uses the Exposure/Potency Index (EPI), a value that takes into account the level of environmental exposure as well as the known toxicity of a substance, to rank chemicals as to degree of risk. Thus, Canada has identified some 44 "priority" chemicals that are felt to be significant risks. The United States has 128 on a similar list.

The linearized, multistage model assumes that there is no threshold for carcinogenesis, a reasonable assumption for electrophilic carcinogens affecting DNA, but this may not be true for epigenetic carcinogens such as dioxin. Canada and some European countries set dioxin limits 170 to 1700 times higher than EPA limits because they do not apply the linear approach to dioxin risk analysis. The CEPA defines such "threshold" chemicals where possible and treats them separately from those where no threshold exists or where none has been demonstrated.

Sources of error in predicting cancer risks

Obviously, there are several points in this method that require estimations and therefore there may be wide variations in resulting predictions. This is the greatest source of contention between governments and various special-interest groups. Environmentalists generally press for reductions in allowable levels, whereas industry may lobby for higher levels if lower ones involve significant cost factors. Some specific sources of contention in risk analysis are discussed below.

Portal-of-entry effects

1. The method may not be reliable when exposure of humans involves multiple portals of entry. Volatile chemicals, for example, may be inhaled, ingested, or absorbed through the skin.
2. Toxicity may be affected by differences in absorption or biotransformation that occur at the portal of entry, so that data obtained from one type of exposure may not be applicable to others. As an extreme example of portal-of-entry effects, the purest air can be fatal if injected intravenously, as can the purest water if inhaled. Ethyl acrylate produces a 77% incidence of tumors in rats at 200 mg/day orally. The same dose applied to the skin causes no tumors. Cadmium (Cd) is carcinogenic by inhalation, but not orally or dermally. Conversely, epichlorhydrin will cause tumors at the point of contact with any epithelium. It has been stated that of the more than 500 risk assessments that have been completed, nearly all involve a single route. This applies both to carcinogenic and noncarcinogenic effects. Numerous examples of route-specific effects exist; for example, trichloroethylene causes central nervous system (CNS) depression at 7 ppm

if inhaled, but the same concentration taken orally has no effect because of incomplete absorption.

3. The area of contact may affect uptake, even for the same portal of entry. Thus, if a large area of skin is exposed to a toxicant, more will be absorbed. Moreover, the skin of the forehead absorbs 20 times, and that of the scrotum 40 times, more effectively than the skin of the forearm. Transit time for ingested material in the intestinal tract can vary from 10 to 80 hr, depending on age, diet, and other factors; thus, the time available for absorption will vary as well. The relatively rapid transit time through the small bowel may partly explain the rarity of cancer in this area.

Figure 17 summarizes the possible fate of xenobiotics (literally "foreign to life") that can occur at various portals of entry and thereafter. The mammalian

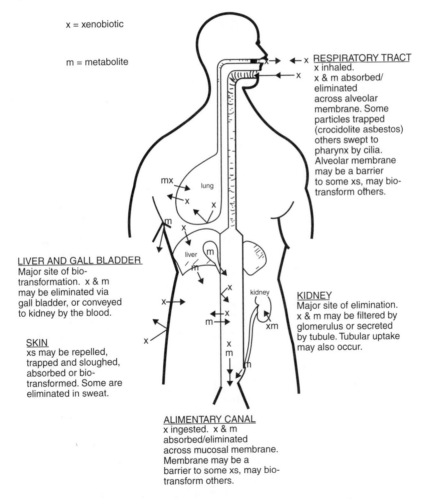

Figure 17 The possible fate of xenobiotics in the body.

body can be visualized as a thick-walled tube with the outer surface (skin) and inner surface (gastrointestinal tract) in contact with the environment.

Excretion of toxicants and waste products back to the environment takes place in sweat, expired air, feces, urine, and the sloughing of cells in contact with the environment.

Age effects

The age of the population at risk may affect the degree of risk. Infants — especially premature ones — absorb chemicals through the skin much more efficiently than adults. Infants have died from absorbing pentachlorophenol used as an antibacterial agent in hospital bedding before the practice was abandoned. Even data from human industrial exposures usually deal with adult males and may not be applicable to the elderly or to females.

Exposure to co-carcinogens and promoters

It is often difficult to control for the presence of co-carcinogens and promoters, even in animal studies. Regarding human data, such factors as smoking, alcohol consumption, and intake of nitrites, nitrates, and saturated fats may differ considerably from an exposed, industrial population to the public at large.

Species differences

These are the focus of considerable attention both from the scientific community and from animal rights activists who use them to trivialize the value of animal data. It must be stated at the outset that pharmacokinetic differences can be far greater among human beings than between them and experimental animals. Biological variation is a governing force in all living things. Nevertheless, there are important differences that are known and others that are only now being identified, including:

1. Human skin, for example, is much more impervious than that of laboratory animals, being more similar to that of the pig.
2. The rat forestomach is devoid of secretory cells and is a better model of squamous epithelium than of secretory tissue.
3. Moreover, the rat forestomach contains an active microflora that can alter chemicals, whereas the stomach and upper bowel of the human are virtually sterile because of the acidity.
4. This same acid medium can serve to denature and detoxify potentially harmful chemicals.
5. Anatomical differences in the branching patterns of bronchi exist in the lungs of rodents vs. primates. This can result in vastly different deliveries of inhaled volatile toxins. The pattern in humans is described as dichotomous-asymmetric, whereas that in the rat is

monopodial-symmetric. In the latter case, the primary bronchi penetrate deep into the lungs and have secondary bronchi branching off their length. The distance to the terminal bronchiole may vary greatly and hence also the target cell exposure (see Figure 18).

6. The rat has no gall bladder; thus, bile flow tends to be continuous and unaffected by food. Stasis of the bile, which can affect contact time, is rare.

7. There are numerous differences in the nature and location of biotransforming enzymes. Knowledge of these differences (e.g., for cytochrome P450) can be exploited to select the most appropriate model for study. Chapter 10 further examines species differences and how they affect toxicity.

Despite the problems with extrapolation from animals to humans, it should be remembered that DNA varies from the human array by only 5% in mice, by less than 2% in most primates, and by less than 1% in chimpanzees. The similarities are far greater than the differences. Moreover, the extrapolation of risk to the general public from data acquired from industrial exposures, including accidents, has its own problems. Numerous differences usually exist between workers and the populace. The former tend to be

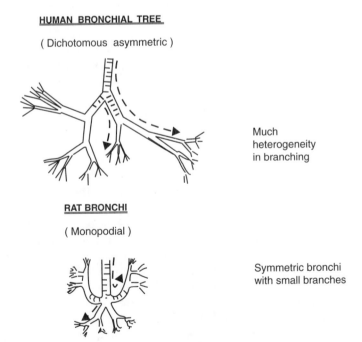

HUMAN BRONCHIAL TREE

(Dichotomous asymmetric)

Much heterogeneity in branching

RAT BRONCHI

(Monopodial)

Symmetric bronchi with small branches

Figure 18 Comparative anatomy of human and rat bronchial trees. Dotted lines represent the differences in distances to the terminal bronchioles.

predominantly males, 18 to 65 years old, from the lower end of the socio-economic scale, and possibly with different habits regarding such health factors as smoking, alcohol consumption, and diet. There is also the need to extrapolate from moderate to high exposures (and possibly from very high single exposures, as in an industrial accident) in the workplace to very low ones in the environment.

Extrapolation of animal data to humans

One source of continuing dispute is the reliability of animal data in extrapolating cancer risk to humans. One critic of the current system is Bruce Ames, inventor of the Ames test for mutagenicity. He now feels that it is too sensitive and thus it is predicting cancer risks that are artificial for many chemicals. One basis for his argument is that many chemicals are cytotoxic at the high concentrations in standard tests for carcinogenicity and therefore they induce a high rate of cell proliferation for repair. This in turn increases the likelihood of mutations that could lead to malignancy. Critics claim that any substance, at high enough doses, can be carcinogenic. The debate revolves around the use of the Estimated Maximum Tolerated Dose (or EMTD) as the high dose level in cancer bioassays. This is defined as the highest dose in chronic studies that can be predicted not to alter the animals' longevity from effects other than cancer. According to the proliferation-mutagenicity theory, lower doses should not be carcinogenic if they do not induce cell proliferation. Defenders of the current (U.S.) National Toxicology Program, however, point out that approximately 90% of chemicals defined as carcinogens induced tumors at doses well below the EMTD, and that, of 33 proven human carcinogens, 91% were shown to be carcinogenic in the animal tests.

Hormesis

The term *hormesis* refers to a U-shaped dose response where the effects at the low end of the dosage/exposure scale are markedly different from those at the high end. The arms of the dose response curve are separated by a flat area of no observable effect. Such low dose effects have often been shown to be beneficial. The phenomenon of hormesis has been demonstrated in both experimental animals and humans for a wide range of toxic substances. Lest it seem bizarre that toxic substances should have beneficial effects at low doses, it should be remembered that we exploit this fact every day, as, for example, in the use of chlorine in drinking water and fluoride in tooth-paste. A publication put out by the University of Massachusetts School of Public Health, *The BELLE Newsletter* (BELLE for Biological Effects of Low Dose Exposures) has devoted considerable attention to this subject. Exposure to low doses of ionizing radiation has long been held to impart some beneficial effects. Both experimental and epidemiological evidence suggests this.

Experimentally, it has been shown that exposure to very low doses of uranium imparted resistance of cultured cells to higher, carcinogenic doses. Moreover, increased resistance to oxidizing agents (e.g., hydrogen peroxide) and to anticancer drugs also occurred. This adaptive response is thought to explain several instances of hormesis — but not all of them. The fluoride effect, for example, is due to hardening of dental enamel and increased resistance to caries. It is unlikely that a single, mechanistic explanation can be applied to all examples of hormesis.

The incorporation of hormetic effects into the risk assessment process has yet to occur and remains fraught with political pitfalls. The implications for the risk assessment process are enormous and likely to create considerable public controversy and resistance. Imagine the consequences of stating that low doses of dioxins might actually be beneficial to health! The concept that no dose of a carcinogen is safe would be radically altered. Little change is likely to occur until hard evidence of a hormetic effect is obtained for each specific agent.

Natural vs. anthropogenic carcinogens

A more valid criticism perhaps, also raised by Bruce Ames, is the fact that there are hundreds of natural carcinogens in foods to which we are exposed daily; and of 77 that have been tested by the standard methods, approximately half (37) were carcinogenic. People are thus likely exposed to many more natural carcinogens than synthetic ones. Our natural defenses probably take care of many of these. Humans slough our epithelial layer regularly (skin, gastrointestinal tract, and lungs) and with it, its accumulation of toxins. Humans have killer lymphocytes that destroy abnormal cells and detoxifying mechanisms that render many toxins harmless unless these defenses are overwhelmed by high doses. The natural decline in these defenses with age is a major factor in the increasing incidence of cancer in the elderly. In view of the abundance of natural carcinogens, elimination of all synthetic ones may not reduce cancer incidences as much as one might expect. The counterargument is that humans have had several million years to evolve defenses against natural carcinogens and these may not work as efficiently against synthetic ones. One indisputable statistic, however, is that life expectancy has been steadily increasing for several decades. In fact, a great concern in all developed countries is that the ratio of working to retired persons has been declining steadily with the eventual consequence that there may not be enough revenues contributed to government pension and Medicare plans to support all those who need them.

Reliability of tests of carcinogenesis

Another concern has been raised recently about the reliability of tests of carcinogenicity. A particular strain of mice, $B_6C_3F_1$, has been widely used in such tests because of its known tendency to readily develop tumors in response to a wide variety of chemicals. Information has been accumulating concerning significant differences in the metabolic processes of rodents and people, and it is likely that using this cancer-sensitive strain of mice may yield too many false-positive findings of carcinogenicity. For example, the volatile solvent butadiene, used in the production of synthetic rubber, is exhaled unchanged in most animals and thus does not display carcinogenicity. In mice, however, much more is retained in the lungs and absorbed (33 times as much as in monkeys). It is then oxidized to a mutagenic epoxide. The mouse has a much lower activity of the detoxifying enzyme "epoxide hydrolase" than do humans, so the carcinogenic risk is many times greater in mice. This strain of mice also harbors a murine leukemia virus that has been shown to enhance carcinogenicity. The (U.S.) National Institute of Occupational Safety and Health (NIOSH) based its estimate of butadiene cancer risks entirely on studies of this strain of mice. The NIOSH model predicted that exposure to 2 ppm butadiene for 45 years would cause 597 excess cancers in 10,000 workers. In fact, 1066 workers who had been exposed to levels as high as 1000 ppm since the industry began in the 1940s, had only 75% of the cancer incidence in the overall population.

The cost of overestimating cancer risks can be horrendous as unnecessary and expensive protective measures are legislated. This can drive industry to seek homes in countries with less restrictive legislation, which may then lead to loss of control over other, truly hazardous industrial chemicals.

Environmental monitoring

Environmental monitoring occurs in two ways. Ambient monitoring refers to measurements in water or air downstream or downwind from the source and is primarily a measure of the state of the environment. So-called "end-of-pipe" or point of emission monitoring refers to the measure of effluent levels from drains and stacks and is used to ensure compliance with legislative regulations.

Bioassays are used to look for effects rather than to identify specific chemicals. For water, the water flea, *Salmonid* fingerlings, and the opossum shrimp are used. Earthworms and germinating plants are used for testing soil. Sensitive bacteria are used to detect mutagens. The Ames test can detect a few mutations in several million cells and a newer test, the Microtox assay, measures reduced bacterial luminescence resulting from inhibited cell division. Genetically engineered species such as nematodes are being developed to detect contamination in indoor air — the modern version of the canary in the coal mine. EC_{50}, LD_{50}, and TLV values (see below) can be calculated for these.

The author's own research has found that marine sponges accumulate some metals, notably cadmium, to a greater extent than other benthic marine species. This is due, at least in part, to the large quantities of seawater they are capable of filtering. The author's research has also shown that such high levels of cadmium (10 to 30 µg/g dry weight) may also interfere with cell function. Others have shown such interference in several marine species. Species such as sponges may constitute useful biomarkers of environmental damage.

Setting safe limits in the workplace

As noted above, because of data collected after industrial accidents and from preregulation exposures, setting acceptable limits for toxic substances in the workplace can be done with somewhat greater confidence. Exposures are generally higher but of relatively short duration as compared to those in the environment. It should be noted that tolerance limits apply only to the particular type of exposure stated (inhalation, skin contact, etc.). Jurisdictions over worker safety vary widely from country to country. Most Western, industrialized countries have similar legislation. In Canada, provincial ministries are responsible for occupational health and safety. In Ontario, this comes under the Occupational Health and Safety Act, Revised Statutes of Ontario, 1980. Regulations made under this act deal specifically with biological, chemical, and physical agents in the industrial, construction, and mining settings and, most recently, regulations governing the Workplace Hazardous Materials Information System (WHMIS). Under federal guidelines, each province has enacted comparable legislation.

Historically, the development of safety legislation in Ontario dates to the Ontario Factories Act of 1884. The formation of the Royal Commission on the Health and Safety of Workers in Mines in 1976 led to significant additions to legislation dealing with workplace safety. In 1979, the Occupational Health and Safety Act was proclaimed, together with regulations for the industrial, mining, and construction settings. In 1984, a Royal Commission on Asbestos resulted in a further regulation under the Act in 1985. In 1981, a list of "designated substances" was begun. These are chemicals that are considered to be especially hazardous and therefore to require special controls, restrictions, or even prohibition. The regulations apply to all workplaces and other projects (except construction sites) where designated substances are likely to be inhaled, ingested, or absorbed. The 1990 list included acrylonitrile, arsenic, asbestos, benzene, coke oven emissions, ethylene oxide, isocyanates, lead, mercury, silica, and vinylchloride.

In 1988, the WHMIS regulations were brought into effect. These define the information that must be on labels of chemical containers in the workplace and the information that must be readily available to the worker in the form of Material Safety Data Sheets. Ontario Bill 208 expands on the act. Some important features of Bill 208 are as follows:

1. It is designed to work on the "Internal Responsibility System," the key aspect of which is that workers have:
   a. The right to be informed about the nature of the hazards they might be exposed to in the workplace

b. The right to participate in the decision process concerning job safety
c. The right to refuse to work in conditions they consider to be unsafe without fear of reprisal from the employer

2. Workplaces (as defined in the act) must form a Joint Health and Safety Committee with representation from both the employer and the employees (union) who must have at least numerically equal membership. This committee may make recommendations concerning health and safety but the employer is not bound to accept them. This is a weakness in the current system.
3. Responsibilities for safety in the workplace must be shared by the employer and the employees.
4. The definition of a worker is "a person who performs work or provides services for monetary compensation" but does not include inmates of a correctional institution, owners or occupants of a private residence or their servants, farmers, or hospital patients. Persons performing work in their homes for monetary compensation are considered to be workers.
5. The act contains a blanket clause to the effect that "the employer must take every reasonable precaution to ensure the health and safety of the worker" where specific regulations do not exist (they do for such things as protective clothing and equipment).

Some important definitions include:

TWAEV: Time-Weighted Average Exposure Value. The average concentration in air of a biological or chemical agent to which a worker may be exposed in a workday (8 hr) or workweek (40 hr).
STEV: Short-Term Exposure Value. The maximum concentration in air of a biological or chemical agent to which a worker may be exposed in any 15-min period. If not specifically defined in the regulations, this is taken as 3 times the TWAEV for up to 30 min.
CEV: Ceiling Exposure Value. The maximum concentration in air of a biological or chemical agent to which a worker may be exposed at any time. If not specifically defined in the regulations, this is taken as 5 times the TWAEV. Levels in air are expressed as ppm or mg/m^3 of air (sometimes called an excursion limit).

The NOEL and the ADI are values that are employed in animal tests.

NOEL: No Observable Effect Level. The highest level at which no effect is observed in experimental animals.
NOAEL: No Observable Adverse Effect Level. The exposure level at which no toxic effect is observed.
ADI: Acceptable Daily Intake. This is the NOEL divided by an arbitrary number, at least 100.

Reliability depends on the use of a large number of animals to determine the NOEL. The ADI is not a predictor for carcinogenic risk. ADIs are also calculated for human exposures to determine allowable levels in air, water, and food. The act also provides regulations governing noise exposures. Standards are different for the mining and industrial sectors.

In the United States, somewhat different abbreviations are employed, but with similar definitions. For example:

TLV: Threshold Limit Values. These apply mostly to vapors.
TLV-C (Ceiling): That concentration that should never be exceeded (= CEV).
TLV-STEL (Short-Term Exposure Limit): The exposure that can be tolerated for up to 15 min without irritation, tissue damage, or sedation (= STEV).
TLV-TWA (Time-Weighted Average): That level to which a worker may be exposed continuously for up to 40 hr/week without adverse effects (= TWAEV).

Environmental risks: problems with assessment and public perceptions

Establishing acceptable levels of toxic substances in the environment is difficult and inexact because:

1. Exposure may be for a lifetime and no data usually exist regarding long-term, very low-level exposures.
2. Toxic effects may be difficult to identify unless they are unusual. Angiosarcoma of the liver was readily associated with vinylchloride because it is otherwise very rare. For the same reason, Kaposi's sarcoma was associated with AIDS. If harelip were the teratogenic effect of a chemical pollutant, however, the association would be difficult because it is a fairly common congenital defect.
3. Public perceptions of risk may be very exaggerated, forcing the legislation of much lower maximum levels than necessary.
4. The concept of risk-benefit analysis may seem callous and calculating to segments of the populace. In their view, there may be no such thing as an acceptable level of risk and they frequently have an imperfect understanding of the meaning of statistical probability. Moreover, the possibility that avoiding one risk may increase another is often overlooked. The debate over the expansion of nuclear vs. fossil fuel power plants is a good example of this. The elimination of nuclear power plants would force greater reliance on coal-fired generators with increases in acid rain production, release of greenhouse gases, and a greater risk of fatal mining accidents (see Chapter 12).
5. Risk factors may be additive or even synergistic, making analysis even more difficult.

The psychological impact of potential environmental risks

One of the greatest adverse effects of environmental pollutants is undoubtedly psychological in nature. Human beings have a natural fear of the unknown and incomprehensible, and the field of environmental toxicology brings both of these into play. The human mind is a highly impressionable instrument. Many years ago, when one city's drinking water was to be first fluoridated, it was announced that fluoride would be added on a certain date. On that date, the switchboard of the municipal offices was flooded with calls complaining about the taste of the water. It was then announced that the fluoride had actually been added a week earlier! Only then did the phone calls subside. It is well-established that after every major environmental accident there is a rash of vague medical problems such as headache, nausea, and dizziness. Cancerphobia (fear of cancer) leads to attributing every new case of cancer in the area of exposure to the accident. Such beliefs may persist even after extensive studies have failed to reveal any difference in incidence between exposed and non-exposed populations. A sheep farmer living near a nuclear generating station may attribute every abnormal birth and fetal deformity (not uncommon in lambs) to radiation from the power plant despite evidence from extensive testing indicating that there were no radiation abnormalities on the farm.

Voluntary risk acceptance vs. imposed risks

Public pressure may force the expenditure of vast sums of money to avoid risks that are virtually nonexistent. Conversely, a large segment of the public may steadfastly refuse to take steps that are inexpensive and proven to reduce premature deaths. The use of seat belts in cars is a prime example of this. In the United States, motor vehicle accidents affect one of every 50 to 60 persons each year. The statistical probability of incurring such an injury during one's lifetime is thus quite high, but the risk of being injured on any given trip is very low, and this is what influences the public's perception of risk. Recent surveys indicate that, in the absence of state laws requiring their use, only about 15% of American drivers routinely use passenger restraints despite extensive efforts to educate the public. This pattern has not changed much in the last decade. Users differed from non-users in a number of ways. Non-users tended to be less well-educated, were more often smokers, rated seat belts as uncomfortable or inconvenient more frequently, often considered seat belt legislation to be an infringement of personal liberty, and less frequently knew someone who had been injured in a car accident. This pattern appears to be being repeated when the mandatory use of helmets by bicyclists is newly introduced to a jurisdiction, with indignant letters to newspaper editors protesting this latest infringement on personal freedom of choice. A recent report evaluated similar measures for horseback riders and found that the rate of serious injury per number of riding hours was greater than for either motorcyclists or automobile racers. In a recent 2-year period, there were nearly 93,000 emergency room visits in the United States for riding-related injuries with over 17,000 head and neck injuries. Competition riders are required to wear safety equipment. In North America, the National Hockey League recently chose to make helmets for players optional, illustrating the degree of risk elements of the public will accept in some circumstances.

Costs of risk avoidance

Because it is not possible to truly "save" a life, but only to postpone death, it is common in this field to refer to "premature deaths avoided." This requires an arbitrary decision regarding normal life expectancy and what constitutes a premature death; this definition will keep changing as life expectancy is extended. Approximately one in three people will develop cancer by age 70 as a result of the gradual deterioration in immune and cellular defenses. If a carcinogen in the environment significantly increases the incidence of cancer but mostly in the >70 segment of the population, are these premature deaths?

The following are some calculated costs, in U.S. dollars, of avoiding a premature death by instituting some simple safety and health measures:

1. Screening and education for cervical cancer: $25,000
2. Installation of smoke alarms in homes: $40,000
3. Installation of seat belts in autos at time of manufacture: $30,000
4. Stopping smoking: actually a saving of $1000/year plus medical costs avoided. Ironically, a Canadian group calling itself the "Smokers Freedom Society," using data from 1986, recently claimed that smokers actually save society money by dying prematurely and costing less in pension benefits and custodial care. They failed to take into account such factors
   a. The damage done to the health of others by sidestream (second-hand) smoke
   b. The cost in workdays lost because of generally poorer health (a higher incidence of respiratory infections for example)
   c. The cost in lives and property from fires started accidentally by smokers
5. Installing fire detection and control systems in commercial aircraft cabins: $200,000

In contrast, banning of diethylstilbestrol, a synthetic estrogen used as a weight-gain promoter in cattle and suspected of being carcinogenic: $132 million.

A recent example of this cost-effectiveness problem comes from the experience of some American states that introduced compulsory AIDS testing for couples applying for a marriage license. While this approach may seem somewhat draconian in the current social climate, it is by no means a new concept. Not many decades ago, most jurisdictions required a Wasserman test for syphilis before a marriage license would be granted. As a result of the mandatory AIDS tests, 160,000 tests were performed at a cost of $5.5 million: 23 subjects were identified as HIV-positive, for a cost of $239,130 each.

Cost-benefit analyses frequently involve conclusions that may be repugnant to a segment of the public. The cost of detecting one breast cancer through annual mammography in 40- to 50-year-old women has been estimated at $144,000. In the 55- to 65-year-old group, it drops to $90,000. Legislators are required to make such difficult choices because of fiscal restraints, often in the face of severe criticism.

Some examples of major industrial accidents and environmental chemical exposures with human health implications

Radiation

In 1979, a serious accident occurred at the Three Mile Island Nuclear Generating Station near Middletown, Pennsylvania. This was a disaster without noise, smoke, or visible evidence of damage. The information the public received came from the press, which, early on, labeled the event a manifestation of the "nuclear disease." Over 10 years later, the clean-up was still in progress. It produced 2.3 million U.S. gallons of weakly radioactive water, mostly from tritium. This could have been discharged into the river without exceeding federal standards, but public pressure forced the installation of a special evaporator at a cost of $5.5 million. The calculated cost of avoiding one premature death is as follows. The maximum exposure from the river discharge would have been 2 microrems (μrems). This is equivalent to a 4-min exposure to natural-source radiation such as radon or cosmic radiation. The collective dose (mean exposure × number of persons exposed) would have been about 1 person-rem (prem). The calculated incidence of cancer for radiation exposure is 1/5000 prem. The computed cost of avoiding one premature death is about $25 billion. The Chernobyl disaster is discussed in Chapter 12.

Formaldehyde

In the late 1970s, a report was published indicating that formaldehyde fumes caused nasal cancer (squamous cell carcinoma of the mucosa) in rats and in one mouse strain. No evidence of carcinogenesis in hamsters could be shown and there were no human studies suggesting a carcinogenic potential for formaldehyde. This type of tumor is rare in people and no evidence of increased incidence in workers exposed to high levels of formaldehyde could be found. The EPA used the linear extrapolation multistage method to evaluate risk. A federal court overturned the initial evaluation (1981) on the grounds that there was not enough data to determine risk. Legal wrangles between the EPA and the chemical industry continued for the next few years. Urea formaldehyde foam insulation was banned in the United States and Canada and many homes were ripped apart and the insulation removed. Homeowners agitated for government subsidies to finance this. Over the next few years, additional epidemiological data accumulated on populations deemed to be exposed to higher than average levels of formaldehyde. Whereas people exposed to 0.4 ppm (medical, dental, nursing, and science students) had cancer incidences little different from the general population, pathologists and funeral service workers exposed to 3.0 ppm had almost 2000 additional cases of cancer per 100,000 over the expected frequency. In 1987, the EPA defined formaldehyde as a probable human carcinogen. The EPA set the time-weighted average exposure value (TWAEV) at 1 ppm, the short-term exposure value (STEV) at 2 ppm, and an action level of 0.5 ppm. Above this, regulations come into force governing such things as medical surveillance, protective equipment, and training. Most people can detect formaldehyde by odor at 0.5 ppm. For example, the characteristic smell of new carpeting is due to formaldehyde used in the adhesive. Particleboard also contains formaldehyde in the binding adhesive. The smell of formaldehyde is rarely if ever detectable in homes insulated with urea formaldehyde insulation, which releases it as the foam slowly breaks down. Levels are unlikely to reach the "action level" set by the EPA for industry, especially if ventilation is adequate, except in new homes where levels up to 1 ppm have been measured due to its release from building materials and carpeting. The Canadian government, however, subsidized removal of such insulation to the tune of $10,000 per home. An important source of toxic aldehydes as a potential health hazard is cigarette smoke. Acrolein is much more irritating than formaldehyde and the industrial TLV is set at 0.1 ppm. It is a major contributor to the irritant properties of cigarette smoke and photochemical smog.

Dioxin (TCDD)

On July 10, 1976, a serious explosion and fire occurred at a chemical plant near Milan, Italy. As a result, over 1 kg TCDD (tetrachlorodibenzo-*p*-dioxin, the most toxic of the dioxins) was spread over the adjacent countryside. The chemical fallout was heaviest in the town of Seveso, where concentrations in some parts reached 20,000 $\mu g/m^2$ of surface area. By the end of July, 753 people were evacuated from the area; 3300 animals died and 77,000 were killed; 500 people were treated for acute skin irritation and 192 eventually developed chloracne. There also was evidence of liver damage, as indicated by increases in serum enzyme levels. Several follow-up studies were reported, the most recent in 1985. The chloracne was completely reversible; all but one person had recovered by 1983. Enzyme levels also returned to normal and no other abnormalities were reported. To date, no conclusive evidence of excess cancer incidence has been detected. Although recent findings suggested increases in the incidence of leukemia and lymphoma in contaminated areas, the overall incidence of cancer was not elevated. Moreover, it was not possible to relate cancer incidences to exposure levels. In the area thought to be most heavily exposed, none of the cancer rates was significantly elevated. One suggestion is that other carcinogens, such as 4-aminobiphenyl, may have played a role in cancer generation. (For a more complete discussion of TCDD carcinogenicity, refer to Chapter 5.)

Some legal aspects of risk

De minimis *concept*

Recently, courts in the United States and Great Britain have begun to apply an old legal tradition to the question of risk. This is the concept of *de minimis non curat lex*, which means "the law does not concern itself with trifles." Applied to risk analysis, the *de minimis* concept means that in some cases the computed risk is so small that it does not justify regulation. For example, the U.S. Supreme Court ruled in 1980 that the Occupational Safety and Health Administration (OSHA) could not further limit levels of benzene fumes in the workplace unless it could demonstrate significant risk to workers from existing exposure levels. Similarly, the appeals court of Washington, D.C. ruled that regulations governing plastic containers could not be introduced on the purely theoretical grounds that toxic substances might leach into the contents.

Delaney Amendment

The Delaney Amendment to the U.S. Food and Drug Act has had a significant impact on worldwide public attitudes regarding cancer risks and on the responses of politicians to those attitudes. It states that no substance that has been shown to induce cancer in animal experiments can be permitted in any foodstuff. Strictly interpreted, it would prohibit the addition of even one molecule of a suspected carcinogen to any food. The U.S. Food and Drug Administration (FDA) sought to apply the *de mimimis* concept to two food colors that had been shown, in very high doses, to induce cancer in animal tests. A review court ruled against the FDA on the grounds that Congress intended the Delaney clause as an absolute prohibition against carcinogenic dyes in foods. Other authorities, however, are stating that it is time to reassess the Delaney clause in light of the extreme sensitivity of current methods of detecting impurities in food. In all areas of toxicology, the sensitivity of test methods has outstripped our knowledge of the significance of such low exposure levels for human health.

Cancerphobia, an unreasonable fear of developing cancer, has spawned a new phenomenon: suing out of fear of developing the disease in the absence of any signs or symptoms. Some litigants are winning. In 1989, residents of Toone, Tennessee, accepted an out-of-court settlement from the Velsicol Chemical Co. of $9.8 million because they had been exposed to contaminated drinking water as a result of corroding barrels in a toxic dumpsite owned by the company.

A recent controversy concerning a chemical hazard in the environment centered on the use of daminozide (Alar) sprayed on red apples to prevent premature windfall. Early in 1989, the Natural Resources Defense Council (NRDC), an American environmental activist group, published a report claiming that children who consume large amounts of apples and apple products could be at increased risk of cancer. A breakdown product of daminozide, unsymmetrical dimethylhydrazine (UDMH), has been shown to be carcinogenic in animals. The maximum, allowable level of Alar in apples is set by the FDA at 20 ppm. In Canada, it is 30 ppm. The NRDC claims that these limits were set before information regarding the cancer risk was available and that the increased risk of cancer could be 45/1,000,000, which exceeds the "socially acceptable" level of 1/1,000,000 that the EPA uses as a guideline. Moreover, this group claims that the acceptable level (1/1,000,000) is exceeded when levels of Alar exceed 0.01 ppm. This contention has been vehemently disputed by the FDA and by many scientists, including Bruce Ames, developer of the Ames test for mutagenesis. Thomas Jukes, a prominent microbiologist at Berkeley, pointed out that "organic" apple juice, recommended by the NRDC, may contain up to 45 ppm of patulin, a carcinogen produced by a *Penicillium* mold. A study undertaken by Health and Welfare Canada in response to this debate found that 13% of 30 samples of apples tested had daminozide levels of from 0.06 to 3.0 ppm, well below the legal limit but above the acceptable level claimed by the NRDC. A dilution effect could well lower it further, even to their limit. How real is the risk? The question is now moot because the manufacturer has withdrawn the product as a result of adverse publicity.

Statistical problems with risk assessment

In toxicity studies, to reach the 95% confidence limit, one must have 12% responders in a group of 50 subjects, 30% in 20 subjects, and 50% in 10 subjects. To have a chance of seeing one single case in a population at risk, one must test a population three times as large. In other words, to detect an incidence of 1/100, one must test 300 subjects; to detect 1/1000, one must test 3000; etc. This is an extremely important concept when dealing with toxic reactions that are not dose dependent. Aplastic anemia from bone marrow depression occurs very rarely in response to some chemicals. For example, the antibiotic chloramphenicol is estimated to cause this in one of 35,000 to 50,000 treated patients. Even assuming that the same genetic predisposition existed in a test animal as it appears to do in humans, one would have to test 150,000 animals to see one case. The effects of high doses, moreover, cannot always be extrapolated to low-dose situations, even if the effect is dose dependent. The time available for the repair of chemically damaged DNA may be adequate at very low doses, so that the defect is not expressed.

Risk assessment is thus a very inexact process when applied to low-risk situations relating to environmental pollutants. The public tends to overestimate risk and underestimate or ignore avoidance costs. Vested interests tend to do the reverse, and politicians are sometimes influenced by the most powerful lobby, which may be either an industry or an environmental organization. Before the Challenger disaster, NASA estimated the risk of a shuttle accident to be 1/100,000. Empirical data now suggest that the real risk was 1/25. Unconscious bias, small sample size, lack of human data, failure to consider interactive factors such as food chain biomagnification, and public pressure resulting from unjustified fears — all may affect the decision process. It is a characteristic of human nature that people will accept significant levels of risk if they can exert some personal choice in the situation, but they will almost universally reject even the slightest degree of a risk if it appears to be imposed upon them by government or industry. Examples of voluntary risks discussed above include smoking, not wearing seat belts, not wearing protective helmets, and, of course, a whole range of hazardous sports involving considerable risk to life and limb. The question of relative risk assumes great significance when comparing various sources of energy. Is nuclear energy inherently more dangerous than that from coal-fired generators? This subject is considered in Chapter 13, "Radiation hazards."

Risk management

The Royal Society of Canada and the Canadian Academy of Engineering have formed a Joint Committee on Health and Safety. In 1993, it released a report entitled "Health and Safety Policies: Guiding Principles for Risk Management." These principles can be summarized as follows.

1. Risk should be managed so as to maximize the net benefit to society as a whole. The reference to net benefit recognizes the need to evaluate any negative aspects of an action, including cost; that is, to conduct a cost-benefit analysis.
2. The desired benefit is quality-adjusted life expectancy (QALE). In other words, the determination of benefit must include an evaluation of how the quality of life has been improved — not just its length.
3. Health and safety decisions affecting the public must be open to public scrutiny and applied across the complete range of risks.

A number of problems are recognized in the report, some of which were discussed above. Some individuals may be adversely affected by an action that benefits the majority of society. Jobs lost as a result of the closure of a polluting industrial site is an obvious example. Measures taken to reduce one risk may increase another risk. (The debate over nuclear vs. fossil fuel power generators is ongoing.)

A significant problem — the focus of this entire chapter — is how the public perceives a risk and how it may influence politicians to allocate huge sums to deal with minuscule risks. This too has been dealt with above but the report makes an important, additional point. Namely, that prevention of accidental death and injury is one of the most cost-effective risk management actions that can be taken, yet much regulatory effort has been directed at controlling occupational, environmental, and dietary cancer risks and such regulations are extremely costly. It is evident that, in keeping with principle 3 above, much work needs to be done to educate the public regarding the differences between real and perceived risks.

Efforts have been made to devise systems that allow comparisons of risks. In 1993, the EPA introduced the IRIS method (Integrated Risk Information System). IRIS is an EPA database, updated monthly, that contains the EPA's consensus positions on potential adverse effects of 500 substances. It contains information on hazard identification and dose response evaluation, the first two steps in risk assessment (followed by exposure assessment and risk characterization). An IRIS chemical file may contain any or all of the following: an oral reference dose, an inhalation reference concentration, risk estimates for carcinogens, drinking water health advisories, U.S. EPA regulatory action summaries, and supplementary data.

$$\text{Oral Reference Dose (RFD)} = \frac{\text{NOAEL (mg/kg/day)}}{\text{UF} \times \text{MF}}$$

where UF is an uncertainty factor that is used to allow for the variability in species and in the extrapolation to humans, and MF (modifying factor) is essentially a fudge factor used to cover the possibility of unknown variables.

The Inhalation Reference Concentration is the same except that the NOAEL is multiplied by the Human Equivalent Concentration taken from industrial exposures where data exist. *The Federal Register Notice 58 FR 11490* was first published on February 25, 1993.

Information regarding IRIS and its database can be found on the internet at http://toxnet.nlm.nih.gov.

The precautionary principle

The precautionary principle advocates the "better safe than sorry" attitude toward risk management. The principle has been stated in many ways. Principle 15 of the 1992 Rio Declaration on Environment and Development, a heavily negotiated example of it, states that "where there are threats of serious or irreversible damage, lack of full scientific certainty shall not be used as a reason for postponing cost-effective measures to prevent environmental degradation." Although this expression relates to the environment, similar statements have been applied to potential threats to human health.

Implicit in all such statements, although not always expressed, is the concept that the cost of corrective or preventative measures must be commensurate with the estimated degree of risk associated with the potential hazard. Two uncertainties thus attend every attempt to apply the principle: (1) How great is the risk? (2) How much will it cost to reduce it? There have been calls to reform or refine the existing expression of the Precautionary Principle (PP). In the past, legislative action has been taken based on the PP, and subsequent research has shown the action to be unnecessary. Saccharin, for example, is no longer considered to be carcinogenic for humans because of newer and better scientific data. A particular risk may have attending benefits as well as risks. Moderate consumption of alcohol reduces the risk of heart attack but slightly increases the risk of hemorrhagic stroke. Consumer and environmental advocates may tend to take a one-sided view of such situations. It appears that the PP should be applied to the introduction of precautionary regulation.

Further reading

Abelson, P.H., Testing for carcinogenicity with rodents (Editorial), *Science*, 249, 1357, 1990.

Abelson, P.H., Exaggerated carcinogenicity of chemicals (Editorial), *Science*, 256: 1609, 1992.

Ames, B.N. and Gold, L.S., Pesticides, risk, and applesauce, *Science*, 244, 755–757, 1988.

Ames, B.N. and Gold, L.S., Too many rodent carcinogens: mitogenesis increases mutagenesis, *Science*, 249, 970–971, 1990.

Assennato, G., Cervino, D., Emmett, E.A., Longo, G., and Merlo, F., Followup of subjects who developed chloracne following TCDD exposure at Seveso, *Am. J. Indust. Med.*, 16, 119–125, 1989.

Cogliano, V.J., Farland, W.H., Preuss, P.W. et al., Carcinogens and human health: part 3 (letter), *Science*, 251, 607–608, 1991.

Covello, T., Flamm, W.G., Rodericks, W.V., and Tardiff, R.G., Eds., *The Analysis of Actual vs. Perceived Risks*, Plenum Press, New York, 1981.

Flamm, W.G., Pros and cons of quantatitive risk analysis, in *Food Toxicology: A Perspective on the Relative Risks*, Taylor, S.L. and Scanlon, R.A., Eds., Marcel Dekker, New York, 1989, chap. 15, 429–446.

Formaldehyde-Council on Scientific Affairs (AMA) Report, *JAMA*, 261, 1183–1187, 1989.

Graham, J.D., Making sense of the precautionary principle, *Risk in Perspect.*, 7(6), 1–6, 1999.

Goldfarb, B., Beyond reasonable fear, *Health Watch*, Sept./Oct., 14–18, 1991.

Infante, P.F., Prevention versus chemophobia: a defence of rodent carcinogenicity tests, *Lancet*, 337, 538–540, 1991.

Joint Committee on Health and Safety (Royal Soc. Can. & Can. Acad. Eng.). Health and Safety Policies: Guiding Principles for Risk Management. Inst. for Risk Research, Waterloo, Ontario, Canada, 1993.

Marshall, E., A is for apple, Alar, and … alarmist? News and comment, *Science*, 254, 20–22, 1991.

Marx, J., Animal carcinogen testing challenged, *Science*, 250, 743–745, 1990.

Perera, F.P., Carcinogens and human health: part 1 (letter), *Science*, 250, 1644, 1990.

Rall, D.P., Carcinogens and human health: part 2 (letter), *Science*, 251, 10–11, 1991.

Stone, R., New Seveso findings point to cancer, *Science*, 261, 1383, 1993.

Weinstein, I.B., Mitogenesis is only one factor in carcinogenesis, *Science*, 251, 387–388, 1991.

Weinstein, N.D., Optimistic biases about personal risks, *Science*, 246, 1232, 1989.

Zeckhauser, R.J. and Viscusi, W.K., Risk without reason, *Science*, 248, 559–564, 1990.

Review questions

For Questions 1 to 6, use the following code:

Risk = Hazards × time.
How much   How long.

Answer A if statements a, b, and c are correct.
Answer B if statements a and c are correct.
Answer C if statements b and d are correct.
Answer D if only statement d is correct.
Answer E if all statements (a, b, c, d) are correct.

1. Which of the following factors contributes to the degree of risk of a toxicant in the environment?

   a. The biological half-life of the substance.

   E   b. The partition coefficient of the substance.

   c. The toxicity of the substance.

   d. The level of exposure likely to occur.

2. Which of the following statements is/are true?

   a. Individual susceptibility to a toxicant may vary considerably.

   b. The prediction of risk associated with very low levels of exposure can be done with reasonable accuracy.

   A   c. Mechanistic models assume that cancer can arise from a single, mutated cell.

   d. Distribution models assume that there is no threshold below which a cancer-causing agent will induce tumor formation.

3. Which of the following statements is/are true?

   a. The carcinogenicity of a substance is not affected by the portal of entry.

   b. Toxicity studies are generally conducted using a single portal of entry.

   c. The age of the population likely to be exposed to a cancer-causing agent does not affect the degree of risk.

   no extrapolation   d. Cancer incidence data acquired from accidental or industrial human exposures are of greater use for predicting risk in the population at large, if exposed to the same chemical, than animal data.

4. Which of the following statements is/are true?

   a. A population of workers may not be representative of the population at large.

   b. The DNA of other mammals differs by 50% compared to human beings.

    c. Infants generally absorb chemicals through the skin much more efficiently than adults.

    d. Smoking habits have no bearing on cancer incidence due to other carcinogens.

5. Safety in the workplace requires:
   a. A means of measuring levels of potentially hazardous substances in the work environment.
   b. A knowledge of the toxicity of the substance.
   c. A knowledge of the influence of level and duration of exposure on risk.
   d. A knowledge of the effectiveness of appropriate safety measures such as respirators.

6. Hormesis:
   a. Involves a U-shaped dose response curve.
   b. Always refers to beneficial effects at the low dose.
   c. Is not yet incorporated into risk assessment models.
   d. Has been clearly proven in human exposures.

7. Match the appropriate acronym to the definitions given below:

    a. STEV   b. TWAEV   c. NOAEL or NOEL   d. CEV

    i. The average concentration, in air, of a toxicant to which a worker may be exposed during an 8-hr day or a 40-hr week

    ii. The level of exposure at which no (adverse) effect is observed

    iii. The maximum concentration, in air, of a toxicant to which a worker may be exposed at any time

    iv. The maximum level of a toxicant in air to which a worker may be exposed in any 15-min period

Answer the following questions True or False.

8. Industrial exposure to vinylchloride has been associated with an increased incidence of angiosarcoma.

9. About one in three North Americans will develop some form of cancer by age 70.

10. The *de minimis* concept means that no level of risk from industrial pollutants is acceptable.

11. The cost of one "premature death avoided" from installing smoke alarms in homes is about $1000.00.

12. Most people can detect formaldehyde by odor at a concentration in air of 0.5 ppm.

13. The main toxic effect related to the dioxin accident at Seveso has been chloracne.

14. The rat stomach is a good model of the human one.

15. The skin of the forehead absorbs chemicals much more efficiently than the skin of the forearm.

16. The acceptable level of risk from environmental anthropogenic chemicals is defined by the EPA as one additional cancer per 1,000,000 population.
17. There are no natural carcinogens of great concern.
18. Patulin is a carcinogenic mycotoxin.
19. Highly lipid-soluble toxicants with long $t_{1/2}$ values tend to concentrate up the food chain.
20. Skin does not have any biotransforming properties.
21. By sloughing our skin cells, we can also eliminate some carcinogens that they have accumulated.

Answers

1. E
2. A
3. C
4. B
5. E
6. B
7. i = b, ii = c, iii = d, iv = a
8. True
9. True
10. False
11. False
12. True
13. True
14. False
15. True
16. True
17. False
18. True
19. True
20. False
21. True

Case study 1

A population of miners has been identified as having a high incidence of a rare form of cancer. It accounted for 18% of all deaths in workers who had been exposed 20 to 30 years previously. Those living close to the mine and family members of miners regardless of the location of their home, also had an elevated incidence of this same tumor (2 to 3% of all deaths). The incidence of death from this cancer in the general population is only 0.0066% of all deaths. The incidence of lung cancer in miners who smoke is 60 times the incidence in miners who do not smoke. In the general population, smoking increases the risk of lung cancer by about 20 times.

Question 1. What factors, including biological and social, could account for this distribution of the rare cancer?

Question 2. Why is the incidence of lung cancer so much higher in the smoking miners?

Question 3. Why should family members of miners who live some distance from the mine have an elevated incidence of the rare cancer?

Case study 2

Five men were employed in plugging leaks in, and waterproofing, an underground storage tank using epoxy resin paint. The tank contained several inches of water, so the ventilating fans that were available were not used (for fear of electrocution). The tank measures 20 yards long, 6 yards wide, and 3 yards deep. It was divided into three connecting compartments and a single ladder and hatch provided the only means of entry and egress. At 10:00 a.m., one worker left the tank because he was drowsy and nauseated. On reaching the surface, he vomited. At 11:30 a.m., another man left the tank to get coffee. When both men returned, they found the remaining three men dead in the tank.

Question 1. What violations of safe working procedures occurred here?

Question 2. What safety measures could have been instituted to prevent these deaths?

Question 3. What can you say about the source and nature of the toxic substance involved in these deaths?

For both of the above case studies, apply common sense and some of the information provided in this chapter.

chapter three

Water and soil pollution

(Hang your clothes on a hickory limb but don't go near the water)

Introduction

Three components of the biosphere can serve as toxicological sinks: soil, air, and water. These are often considered separately, but it should be obvious that they function as an integrated system. Thus, rain will transfer toxicants to soil and water; evaporated surface water and soil as airborne dust can move them back into the air where they may be transported over great distances by wind. Moreover, runoff from the soil, sewage, and industrial discharge are the main sources of water contamination. Seepage into deep aquifers from soil and surface water can also occur, and freshwater reservoirs are connected to the sea by rivers and estuaries. Thus, while this chapter tends to focus on water, both as an essential resource for human consumption and as marine and aquatic ecosystems, this should not detract from an understanding of the integrated nature of the biosphere. Soil often becomes the repository for our most toxic waste products and the consequences of this are touched upon in this chapter. Chemicals may also enter foodstuffs grown in contaminated soil, and the spraying of crops with pesticides has been a matter of considerable public concern.

Water pollution is of considerable importance for several reasons. The most obvious is the possibility that xenobiotics may enter drinking water supplies and constitute a direct threat to human health. The contamination of fish and shellfish obtained both from the sea (marine organisms) and freshwater lakes and rivers (aquatic organisms) may further threaten human health when these foods are consumed. Larger (and older) fish often have higher levels of lipid-soluble toxicants, but younger ones have higher metabolic rates and can concentrate them more quickly.

Many toxicants are taken up initially by unicellular organisms that serve as a food source for larger (but still microscopic) ones, which in turn are food for bigger ones, etc. This process can lead to increasingly higher concentrations progressing up the food chain, and this is called *biomagnification*.

Freshwater and marine organisms are themselves vulnerable to toxicants, which may threaten their survival. Toxicants can shift the selection advantage for a species, such that hardier ones may proliferate to the detriment of others. A classical example of this is the process known as *eutrophication*, which results when excessive phosphate and nitrate levels in water develop from fertilizer runoff from farmlands and from sewage effluent containing detergents. The high phosphate and nitrate levels favor the growth of certain algae and bacteria that bloom extensively and consume available oxygen until there is not enough to support other life forms. Sunlight will also be blocked out, further altering the nature of the ecosystem.

Factors affecting toxicants in water

All natural water contains soil and all soil contains water, but there is considerable variation in the mix. In fact, it is necessary to distinguish among the various types because the behavior of pollutants differs in them. Moreover, the nature of the water itself can vary with regard to hardness, pH, temperature, and light penetration with consequences for the fate of pollutants. These modifying factors are considered in more detail below.

Exchange of toxicants in an ecosystem

Figure 19 is a schematic representation of a body of water showing sources of contamination (rain, runoff, effluent discharge, and percolation through soil) and some of the means of transferring toxicants to aquatic organisms. Of particular note is the layer of soil/water mix at the bottom interface. This is described as the "active sediment" and it contains, at the surface, a layer of colloidal particles suspended in pore water. The sediment contains organic carbon that tends to take up lipophilic substances. An equilibrium state is thus established with the pore water, which is in equilibrium with the body of water itself. The active sediment is a rich environment for many forms of aquatic life; particle feeders can concentrate toxicants from the suspended particles, whereas filter feeders will do so from the pore water. Dilution of the toxicant in the principal body of water will shift the equilibrium and release more from the bound state. Thus, removing a source of contamination may not be reflected in improved water quality for some time. The active sediment can thus be both a sink and a source of toxicants (see below).

Limnologists accustomed to working in streams and small lakes encounter the floccular, low-density active sediment in a depth of only a few centimeters. Commercial divers on the Great Lakes, however, describe the phenomenon of sinking up to their helmet top in soft, bottom sediment — a disturbing sensation when first experienced.

Factors (modifiers) affecting uptake of toxicants from the environment

Modifiers are classified as *abiotic* (not related to the activity of life forms) or *biotic* (related to the activity of life forms).

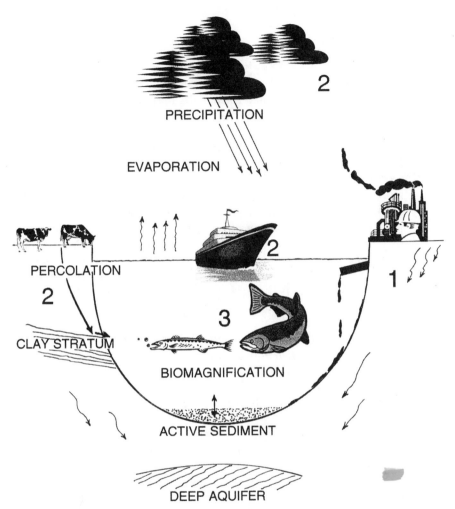

Figure 19 Sources and distribution of toxicants within the ecosystem. Point sources of pollution (1) include chemical spills, industrial effluents from stacks and out-falls, sanitary and storm sewer out-falls, runoff from mine tailings, leakage and seepage from dump sites, treatment lagoons, and old sludge ponds. Non-point sources of pollution (2) include precipitation of dissolved chemicals, downwind fallout of particulates, runoff and tile drainage from agricultural lands, wind drift from sprayed chemicals, discharge of ship's ballast and sewage, and runoff of road chemicals (salt, calcium chloride, etc.) into soil and waterways. Evaporation and precipitation may exchange pollutants among air, water, and soil. Pollutants may travel along a clay belt and enter a body of water, and they may eventually percolate to deep aquifers.

Pollutants may enter the food web through bioaccumulation and biomagnification in marine aquatic species (3) with active sediment acting as a sink. Pollutants may be absorbed into food crops (or accumulated on their surfaces). Humans and animals may accumulate toxicants through the consumption of contaminated plants and animals (including fish and shellfish), through the inhalation of polluted air, through the drinking of polluted water, and even by transcutaneous absorption.

Abiotic modifiers

Abiotic modifiers include:

1. *pH.* As is the case in any solvent–solute interaction, the pH of the solvent will affect the degree of ionization (dissociation) of the solute. Because the nondissociated form is the more lipophilic one, this will influence uptake by organisms. The wood preservative pentachlorophenol, for example, dissociates in an alkaline medium so that, in theory at least, acid rain would increase the bioavailability of the toxicant by favoring a shift to the lipophilic form. Copper, which is very toxic to fish and other aquatic life forms, exists in the elemental cuprous (Cu^{2+}) form at more acidic pH, but as less toxic carbonates at about pH 7. Toxicity to rainbow trout decreases around neutral pH. An important aspect of pH concerns the methylation of mercury by sediment microorganisms. This occurs over a narrow pH range and is a detoxication mechanism that allows the microorganisms to eliminate the mercury as a small complexed molecule. Approximately 1.5 % per month is thought to be converted under optimal conditions (pH 7) (see also Chapter 6).

2. *Water hardness.* Carbonates can bind metals such as cadmium (Cd), zinc (Zn), and chromium (Cr), rendering them unavailable to aquatic organisms. Of course, equilibrium will be established between the bound and the free forms so that removal of the dissolved copper will cause the carbonate to give up some of its copper. There also is an intimate interaction between hardness and pH, so that the lethality curve for rainbow trout will be bimodal at a given degree of hardness, with dramatic increases in the LC_{50} (Lethal Concentration 50%) at pH 5 and 8. It should be noted that Canadian Shield lakes tend to be soft because they do not receive drainage from limestone. Figure 20 illustrates this relationship for copper toxicity at a single degree of hardness.

3. *Temperature.* Apart from a few mammals, aquatic and marine species are poikilotherms, so that water temperature greatly affects their metabolic rate, which in turn will be reflected in the circulation time of blood through gills, the activity of transport processes, and, hence, the rate of uptake of xenobiotics. Rates of biotransformation and excretion may also be affected. Temperature will also affect the rate of conversion of mercury to methylmercury.

4. *Dissolved organic carbon.* These will complex with a variety of lipophilic toxicants and serve as a sink for contaminants in sediment and suspended particles. Again, an equilibrium state will exist and if dissolved toxicants are removed or diluted, more will be released from the sink. Sediment typically consists of inorganic material (silt, sand, clay) coated and admixed with organic matter, both animal and vegetable, living and dead.

EFFECTS OF CALCIUM CARBONATE AND pH ON COPPER LETHALITY IN TROUT

Figure 20 A hypothetical bimodal lethality curve for copper in rainbow trout showing the influence of pH and water hardness.

5. *Oxygen.* As noted, oxygen depletion by algal blooms will compromise other life forms that may be involved in processes of toxification or detoxification, including the microbes that form methylmercury.

6. *Light stress* (photochemical transformations). Ultraviolet radiation can induce chemical changes in contaminants that may result in more toxic forms of a chemical. Thus, photooxidation can increase the toxicity of polycyclic aromatic hydrocarbons (PAHs) through the formation of highly reactive free radicals. In clear water, this effect can be significant at a depth of 6 meters and it can have a marked impact on levels of toxicants.

Biotic modifiers

Biotic modifiers are similar to the factors that can affect a patient's response to a drug and include:

1. *Age.* Old trout are less sensitive than fry to some toxicants; larval forms of aquatic organisms usually differ metabolically from adult forms and may concentrate or metabolize toxicants differently.

2. *Species.* Many differences exist regarding species sensitivity to toxicants. Salmonid species are generally more vulnerable than carp, which can exist under a wider variety of environmental conditions. Disturbances of the natural ecosystem by the introduction of foreign species can have drastic consequences. The Great Lakes are especially

vulnerable to the effects of introduced species because of the introduction of the system of locks connecting them with the sea. Introduced marine species have included:

a. Lamprey eel (still a major problem for sport and commercial fisheries),

b. Alewife (a small coarse fish which died by the thousands and fouled the beaches until the introduction of the coho salmon controlled them),

c. Zebra mussel that clog water intakes and foul ship's hulls,

d. Deep-dwelling quaga mussel, and

e. Some species of gobies; aggressive, highly territorial fish.

Natural transfer of mollusks from marine to aquatic environments is rare because the larvae are not strong enough to swim against river currents. When adults or larvae hitchhike in the hold of a ship, however, it is quite another matter. The current invasion of zebra mussels is thought to have occurred as a result of a ship emptying, in the Great Lakes, its hold of ballast water taken on in a European port, a practice that is in fact prohibited by law.

3. *Overcrowding.* This constitutes an additional stress factor that can influence responses to toxicants.

4. *Nutrition.* The level of nutrition will affect such factors as depot fat (an important storage site for lipophilic toxicants) and the efficiency of detoxifying mechanisms. The nutritional state in turn may be affected by abiotic factors.

5. *Genetic variables.* Unidentified genetic variables are undoubtedly at work, influencing the response of individuals to xenobiotics.

Some important definitions

Acclimation: This refers to the process of adaptation to a single environmental factor under laboratory conditions. Acclimation to heavy metals such as cadmium occurs because of an increase in the levels of metallothionein, a metal-binding protein.

Acclimatization: This refers to the adaptation of an organism to multiple environmental factors under field conditions.

Anthropogenic: This refers to the process of arising from human activity.

Bioaccumulation: This refers to the uptake of the dissolved plus the ingested phases of a toxicant; for example, gill breathers absorb lipophilic substances through the gills and consume them in food.

Bioconcentration: This refers to the uptake of the dissolved phase of a toxicant to achieve total body concentrations that exceed that of the dissolved phase in the water (i.e., against a concentration gradient).

Biomagnification: This refers to the concentration of a toxicant up the food chain so that the higher, predatory species contain the highest levels; for example, polycyclic aromatic hydrocarbons (PAHs) such as benzo[a]pyrene (BaP) (complex ring structures, implicated as carcinogens, formed from incomplete combustion during forest fires or coming from oil spills) are concentrated but not metabolized by mollusks. These may bioaccumulate; BaP has been detected in the brains of beluga whales taken from a polluted area of the St. Lawrence River.

Toxicity testing in marine and aquatic species

A wide variety of marine and aquatic organisms is employed for toxicity testing. This is important because of the biomagnification factor discussed above. Testing species only at the top of the food chain would not provide any information regarding the likelihood that those species might bioconcentrate and bioaccumulate the toxicant. Nor would it give any indication of how the toxicant might distribute in the aquatic or marine environments. Species commonly employed include the organisms *Daphnia magna* (water flea, an aquatic crustacean 2 to 3 mm in length), *Selenastrum* (duckweed), rainbow trout, and fathead minnows. Fish species are important because the gills are an important mechanism for uptake of toxicants. The gills will pass molecules less than 500 Daltons. Large molecules may clog the gills and suffocate the fish. A marine species gaining importance is the opossum shrimp, *Mysidopsis bahia*. This is a tiny, live-bearing estuarine species with a rapid life cycle and adaptability to laboratory culture conditions. It is being used as a bioassay for sewage effluent and petroleum spill toxicity.

Water quality

Liquid freshwater (as opposed to water vapor) exists on Earth either as surface water (lakes, rivers, streams, ponds, etc.) or as groundwater. Groundwater may be in the form of a shallow water table that rather quickly reflects changing levels of xenobiotics at the surface, or as much deeper aquifers that acquire surface contaminants more slowly, but just as surely nonetheless. An aquifer is a layer of rock or soil capable of holding large amounts of water. Subterranean streams and pools also exist. A significant difference between surface water and groundwater is the accumulation of sediments by the former. It is estimated that 50% of croplands in the United States lose 3 to 8 tons topsoil/acre/year and another 20% lose more than 8 tons/acre/year. Soil erosion contributes more than 700 times as much sedimentary material to surface water as does sewage discharge. Both surface and groundwater can serve as a source for drinking, household, and industrial use. Groundwater, however, provides a supply for 50% of all of

North America, 97% of all rural populations, 35% of all municipalities, and 40% of all agricultural irrigation.

Sources of pollution

Sources of pollution include:

1. *Agricultural runoff.* Drainage systems conduct any soil contaminants to surface water and, by seepage, to groundwater. This includes agricultural chemicals (pesticides, chemical fertilizers), heavy metals leached from the soil by acid rain, atmospheric pollutants carried to the soil in rainfall, bacteria from organic fertilizers, seepage from farm dumpsites (old batteries, used engine oil), etc.

2. *Rain.* Rain will transfer atmospheric pollutants directly to surface water. Gases may be dissolved directly in water.

3. *Drainage.* Drainage from municipal and industrial waste disposal sites and effluent from industrial discharge is an important potential source of contamination if not controlled.

4. *Runoff.* Runoff from mine tailings, which may be rich in heavy metals, can contaminate both surface and groundwaters. In northern Ontario, a small town named Wawa recently launched a suit against a mining company that had operated a mine, now defunct, in the area for many years. Arsenic contamination of soil from mine tailings has been detected to a depth of 10 cm. Heavy fall rains in 1999 contaminated the local water supply with arsenic to levels many times the maximum allowable level, forcing residents to use water trucked in tank trucks or purchased bottled water. This single incident clearly illustrates the close relationship between soil and water. In India, arsenic leached out of mountain soil and rock by rivers, a natural phenomenon, has made arsenic poisoning an epidemic problem.

5. *Municipal sewage discharge.* Even if treated, this discharge may carry phosphate detergents, chlorine, and other dissolved xenobiotics into water courses. *The Globe and Mail* (August 18, 1999) reported that major Canadian cities annually dump more than 1 trillion L of poorly treated sewage into water courses. *The Globe* was quoting a study conducted by the Sierra Legal Defense Fund. Five cities actually dump raw sewage into rivers. This is illegal under the Federal Fisheries Act but some municipalities are chronic offenders. The average Canadian generates about 63,000 L of wastewater each year.

6. *Municipal storm drains.* These constitute another source of pollution through runoff. In the Great Lakes basin, salt is used extensively on roads to melt ice and improve traction for vehicles. The salinity of rivers and lakes is increasing as a result. Used engine oil from home oil changes in automobiles may be dumped down storm drains. In Canada, as estimated 30,000,000 L from such usage is not recycled annually. Calcium chloride also may be conducted to lakes, along with residues from vehicle exhaust.

7. *Natural sources.* Although the primary concern of many people is toxicants of anthropogenic origin, it must be remembered that natural toxicants such as methylmercury can form as a result of bacterial action on mercury leached from rock, and of special concern is the level of natural nitrates in drinking water. Nitrates form from nitrogenous organic materials derived from decaying vegetation. Natural levels are not usually a source of concern, but the addition of nitrates from agricultural activity (nitrate fertilizers, animal wastes) may increase the content to dangerous levels. Nitrates are converted by intestinal flora to nitrites that oxidize ferrous hemoglobin to ferric methemoglobin, which cannot transport oxygen. Infants are especially sensitive and cases of poisoning numbering in the thousands have been reported, with a significant mortality. Adults and older children possess an enzyme, methemoglobin reductase, that can reform normal ferrous hemoglobin. Normal nitrate levels in water are about 1.3 mg/L, contributing about 2 mg/day to the total intake of 75 mg per person per day. Levels as high as 160 mg/L have been reported in some rural areas where wells serve as the source of water (see also section on food additives in Chapter 8). Both the EPA and the Environmental Health Directorate of Health and Welfare Canada have set maximum acceptable limits for toxicants in drinking water. For example, the EPA limit for nitrates is 10 mg/L measured as nitrogen.

Water pollutants can be described as oxygen-depleting (contributing to eutrophication), synthetic organic chemicals (detergents, paints, plastics, petroleum products, solvents) that may be very persistent in the environment, inorganic chemicals (salts, heavy metals, acids), and radioactive wastes from nuclear generating plants. Low-level radioactive liquid wastes are produced in the primary coolant.

Some major water pollutants

Specific classes of xenobiotics will be dealt with in detail later in this text as they may serve to contaminate soil, water, or air. The more important groups in water are reviewed here, including:

- *Detergents.* A wide variety of substances is employed as wetting agents, solubilizers, emulsifiers, and anti-foaming agents in industry and in the home. They have in common the ability to lower the surface tension of water (surfactant effect) and, as cleaning agents, this increases the interaction of water with soil, solubilizing the latter. Chemically, they possess discrete polar and nonpolar regions in the same molecule. The nonpolar region is usually a long aliphatic chain. Sodium dodecylbenzenesulfonate (an anionic detergent) and polyphosphates such as sodium tripolyphosphate are in this group. The latter, $Na_5O_{10}P_3$, is commonly known as STP, the engine oil additive.

In sewage, it is readily hydrolyzed to form orthophosphate. Removal efficiencies for sewage treatment are typically 50 to 60%, so that significant amounts can enter surface water to contribute to the process of eutrophication (discussed above). Despite a ban on phosphate detergents by most states and provinces bordering the Great Lakes, water phosphate levels have not dropped significantly. The ban has apparently been offset by the use of phosphate fertilizers. The average North American uses about 23 kg of soaps and detergents yearly. The biochemical, or biological, oxygen demand (BOD) is a measure of the organic material dissolved in the water column and hence of the oxygen requirement for its decomposition. It includes natural sources such as phytoplankton, zooplankton, and organic material from vegetation, as well as nitrates. Pure water has a defined BOD of 1 ppm. BODs above 5 ppm suggest significant pollution. Pulp mill effluents may have levels greater than 200 ppm and agricultural animal wastes may approach 2000 ppm.

- *Pesticides.* This class of chemicals has generated great public concern, sometimes in the absence of any hard evidence of toxicity for humans at the level of exposure likely to be encountered. For example, the European Economic Community, in its "Drinking Water Directive" of 1980, set limits of 0.1 µg/L for any single pesticide and 0.5 µg/L for all pesticides combined, without regard for their toxicity or their economic importance to agriculture. Included in this group are insecticides, herbicides, fungicides, rodenticides, and specific agents to kill snails (molluskicides) and nematodes (nematocides or nemacides). Nematodes (roundworms), from the Greek *nema* meaning thread, are a huge class of parasites that infect humans and animals as well as many plants. The galls that one sometimes sees on leaves of trees are usually due to nematode infestation. Although not strictly pesticides, the public tends to include other agricultural chemicals used to improve growth or ripening in this category. Alar, for example, holds red apples on the tree to allow for even color development. It was recently withdrawn voluntarily by the manufacturer because of concern about carcinogenicity.

Chemical classification of pesticides

Pesticides can be classified as:

1. Chlorinated hydrocarbons such as DDT, lindane, aldrin, dieldrin, and heptachlor (also called organochlorine insecticides). PCBs are also chlorinated hydrocarbons but are not insecticides.
2. Chlorphenoxy acids including the herbicides 2,4-D and 2,4,5-T, which contains dioxins as impurities.
3. Organophosphorus insecticides such as parathion, malathion, DDVP, and TEPP.

4. Carbamate insecticides, which act like organophosphorus compounds (cholinesterase inhibitors) but are derivatives of carbamic acid. There are also carbamate herbicides that lack significant anti-cholinesterase activity.
5. Paraquat, a bipyridyl herbicidal agent that is not considered to be an important environmental contaminant but which is extremely toxic to handlers if used recklessly.

Health hazards of pesticides and related chemicals

Chlorinated hydrocarbons

Chlorinated hydrocarbons are very persistent in the environment and are slowly degraded by bacteria and other microbes. In addition, they are very lipid soluble and thus have very long biological half-lives. Although this group is considered to have low acute toxicity, the combination of lipophilicity and long $t_{1/2}$ leads to biomagnification up the food chain and greater potential for chronic toxicity. This is not easily demonstrable in humans; but in nature, DDT (dichlorodiphenyltrichloroethane) and its metabolites have been shown to interfere with calcium metabolism, causing softening of the eggshell in many species of birds (e.g., gulls, peregrine falcon, bald eagle, brown pelican) with consequent loss of reproductive efficiency. Human fat may contain up to 10 ppm, with a clearance of about 1% content/day. Acutely, DDT is a neurotoxin, causing tremors and convulsions. The oral LD_{50} for humans is estimated at 400 mg/kg. Polychlorinated biphenyls have been used for many years for their insulating properties and the fact that they can withstand temperatures up to 800°C. These properties make them ideal for use in electrical transformers, hydraulic fluids, brake fluids, etc. Their stability means that they are very persistent in the environment when contamination occurs through accident or improper disposal. In the United States, the Environmental Protection Agency (EPA) has set a maximum allowable level of 0.01 ppb in streams. In the Baltic Sea, PCBs have been incriminated in reproductive failure in seals (pinnipeds).

Considerable concern has been generated over seepage of PCBs into groundwater and streams from deteriorating containers in dumpsites. Levels of 5 to 20 ppm have been detected in Lake Ontario fish. Toxic effects in the environment include reproductive defects in phytoplankton and, in mammals and birds, microsomal enzyme induction, tumor promotion, estrogenic effects, and immunosuppression. The potential for human toxicity is therefore high, but existing data are somewhat controversial (see below).

Chlorphenoxy acid herbicides

The chlorphenoxy acid herbicides 2,4-D and 2,4,5-T have been widely used on lawns and along road and railway rights-of-way. They mimic plant growth hormones so that accelerated growth exceeds the energy supply.

2,4,5-Trichlorophenoxyacetic acid (2,4,5-T) is weakly teratogenic but the main concern is the presence of the contaminant 2,3,7,8-tetrachlorodibenzo-*p*-dioxin (TCDD, dioxin) a by-product of synthesis. 2,4,5-T is teratogenic and very toxic to some animals. The LD_{50} for rats is 0.6 to 115 µg/kg. It causes degenerative changes in the liver and thymus, weight loss, changes in serum enzymes, porphyria, chloracne, and cancer. In humans, the only confirmed toxic effect is chloracne (see Chapter 2 on the Seveso accident). Although Vietnam veterans have been very concerned about the use of Agent Orange (which contains equal parts 2,4-D and 2,4,5-T) as a defoliant, several epidemiological studies have failed to confirm long-term effects. A study released by the U.S. Air Force in March 2000 demonstrated a modest but statistically significant association between exposure to Agent Orange and an increased incidence of diabetes. A cause-and-effect relationship has not yet been established. Some recent evidence suggests that industrial workers with prolonged exposure to high levels of these agents may have a slightly increased incidence of cancer (see also Chapter 5).

Organophosphates (organophosphorus insecticides)

Organophosphates irreversibly inhibit acetylcholinesterase; the symptoms of acute toxicity are those of massive cholinergic discharge and include profuse sweating and diarrhea, tremors, mental disturbances, convulsions, and death. Although parathion is fairly toxic to humans, it does not persist in the environment and thus is not a significant environmental hazard. These agents are water soluble but are hydrolyzed to nontoxic by-products.

Carbamates

Carbamates act generally like the organophosphates. The exception is aldicarb, which is not hydrolyzed and which has entered groundwater in several locations, including Long Island, New York, where it is predicted to exceed maximum allowable levels of 7 ppb for up to 20 years. It is highly toxic but it is a reversible inhibitor and is rapidly degraded and excreted. Pesticides are considered further in Chapter 9.

Acidity and toxic metals

Acidity, largely from acid rain (the causes of which are discussed in Chapter 5), can have several deleterious effects on water quality. This subject has already been introduced (see "Abiotic Factors" above). Acidity can leach toxic metals such as lead (Pb), cadmium (Cd), and aluminum (Al) from soil into groundwater. It can contribute to the formation of more toxic methylmercury from mercury. It may also leach lead from solder in the plumbing of houses and cottages as has been shown in a study by Health and Welfare Canada. Water at pH 4.5 to 5.2 was allowed to stand in plumbing systems and reached maximum leaching rates after 2 hr. After 10 days, the water showed levels of 4560 μg/L for copper (Cu), 3610 μg/L for zinc (Zn), 478 μg/L for Pb, and 1.2 μg/L for Cd. The upper limits recommended for Canadian drinking water are 100 μg/L for Cu and 50 μg/L for Pb. Arsenic (As) and selenium (Se) have also been detected. It is highly advisable to flush plumbing systems in houses and cottages that have been standing vacant for any length of time.

At pH 5 or less, aluminum can be leached from soil and transported as complexes with bicarbonate, organic materials, and in the ionic form. Acid surface water may contain 4 to 8 μmol/L, which can be toxic to fish. In humans, high concentrations of Al may be deposited in bones and brain tissue to cause osteomalacia and symptoms of dementia. Microcytic-hypo-chromic (i.e., small, pale cells) anemia can also occur. These problems have been encountered in hemodialysis patients due to the leaching of Al from the dialyzer into the blood of the patient and from oral aluminum hydroxide given them in antacids. Bauxite miners suffer respiratory problems from inhaling ore dust. Al also appears in drinking water because of the use of alum $[Al_2(SO_4)_3 \cdot 12H_2O]$ to precipitate suspended organic material in the third (tertiary) stage of water treatment. The first stage involves the removal of large solids by screens and the second stage removes most of the organic material with filters.

In 1980, the U.S. Congress commissioned the National Precipitation Assessment Panel (NAPAP), consisting of over 2000 scientists from virtually all of the major universities, to study the acid rain problem. In 1987, it issued a highly controversial interim report that concluded that the situation was not as bad as had been previously suspected. Of several thousand lakes studied in upper New York state, only 75 were seriously affected. Sulfur emissions were found to have declined by 25% in the previous decade. The final 6000-page report was released in 1990 (at a total cost of over $570 million) and concluded that 10% of eastern lakes and streams were adversely affected, that acid rain had contributed to the decline of red spruce at higher altitudes and to the corrosion of buildings and materials. A more controversial statement was that there was no evidence of widespread decline of forests in the United States or Canada. Acid precipitation, how-ever, can be deposited hundreds of miles from its site of formation. More-over, lakes that drain limestone bedrock areas are much more resistant to acidification because of their buffering capacity. Lakes that drain granite bedrock, however, are very susceptible because they have virtually no buff-ering capacity. This includes all of the lakes in the Canadian Shield area. Again, aluminum plays an important role. Scientists have discovered that fish in a laboratory setting can withstand a pH of 4.5 or less while in the natural setting, such a low pH is frequently fatal.

Aluminum silicates are a major soil component, and soft-water lakes that drain soil areas acquire significant levels. A suspension of fine aluminum precipitate forms in water, blocking the sodium and oxygen exchange sys-tems in the gills of fish, which then expire. Freshwater fish must take up sodium across the gills to compensate for that lost in urine. Freshwater fairy shrimp have "chloride cells" that also regulate sodium levels and accumulate toxic levels of aluminum. Some success has been achieved in selectively breeding aluminum-resistant strains of aquatic plants such as duckweed that may be used to revitalize dead lakes. Selective breeding of plant species was developed early in the twentieth century to combat the effects of acidic soils that poison plants by interfering, through metal solubilization, with calcium and phosphorus. Phosphorus, especially, binds to aluminum. Because sodium regulation in nearly all cells involves sodium/potassium ATPase (the sodium pump), the aluminum-bound phosphate can attach to, and disable, the sodium pump. This phosphorus link may be involved in the massive diebacks of forests exposed to high levels of acid rain, and selective breeding of resistant species may provide a partial solution.

Another hazardous chemical introduced as a result of water treatment is chloroform. It is introduced as a contaminant in the process of chlorination and it is a known carcinogen. Others, such as benzene and carbon tetrachloride, may enter groundwater from industrial sources.

Chemical hazards from waste disposal

In addition to the types of hazardous contaminants discussed above, numerous substances may enter water from industrial, agricultural, institutional, and domestic sources. They may be solids, liquids, sludge, or gases and may be corrosive, flammable, explosive, radioactive, or biologically toxic. Risks range from minimal to extreme and there may be short- or long-term effects on human health. Usual disposal methods for these substances include surface impoundments used in industry (45–55%), landfill sites (domestic and other, 25–35%), burning (10–15%), and other means, e.g., disposal at sea (2–5%).

An idea of the extent of the problem of buried toxic substances can be gleaned from the experiences of workers in tunnel construction projects. Traditionally, compressed air has been used in tunnel construction for the purpose of excluding water from the tunnel and also for stabilizing the surrounding ground. A new use is emerging, however. Increasingly, tunnel projects in urban areas (for sewer mains, rapid transit systems, auto tunnels under rivers, etc.) are encountering pockets of toxic materials such as gasoline, probably leaked from old service station storage tanks, chlorinated hydrocarbons, and other dumpsite toxins. The use of compressed air in the tunnel prevents the seepage of toxic fumes and liquids into the tunnel and provides a safer working environment.

Water from drinking wells continues to be a source of concern with regard to chemical contamination. In 1987, a study of the Tutu well fields in St. Thomas, U.S. Virgin Islands, showed that 22 wells were contaminated with benzene, *trans*-1,2-dichloroethylene, trichloroethylene, and tetrachloroethylene originating from several sources. Although levels were low, an estimated 11,000 people were exposed for about 20 years. In Minnesota in 1981–1982, 41 of 137 private and commercial wells located downhill from an industrial complex were found to be contaminated with trichloroethylene and trichloroethane. Such wells generally should be sealed with concrete or clay and abandoned.

The Love Canal story

The Hooker Chemical Co., between 1942 and 1953, disposed of about 420,000 metric tonnes of approximately 300 organic chemicals by burying them in steel drums in the abandoned Love Canal near Niagara Falls, New York. The site was sold to the Niagara Falls Board of Education for $1.00 in 1953 and subsequently a subdivision was built over it. As the barrels rusted out, chemicals seeped into the groundwater. There were some early warning signs, including chemical odors in people's basements, sinking areas over collapsing barrels, some exposed pools of waste, and some minor health problems such as rashes and eye irritation after contact with exposed chemicals. Overall, the residents of Love Canal seemed unaware of any unusual health problems or circumstances.

In 1976, however, the International Joint Commission on the Great Lakes undertook a study to find the source of the banned pesticide Mirex in Great Lakes fish and the chemical contamination was discovered. The New York State Department of Environmental Conservation studied the situation over the next 2 years amid great controversy and in the face of emerging anecdotal claims of serious health problems. In August of 1978, a series of dramatic events occurred. The state health commissioner declared a health emergency and recommended that pregnant women and children under 2 years of age be evacuated. President Jimmy Carter declared the area a federal disaster zone, thereby creating a mechanism for federal assistance to be provided. The governor of New York state announced that 239 families would be relocated at state expense. The immediate consequence was that Love Canal became a ghost town and considerable anxiety was created about long-term health effects — not only in the evacuees, but also in those who lived near the edge of the arbitrary demarcation line.

In 1988, a federal district court found Hooker Chemical (by then Occidental Petroleum) to be liable for the cost of the clean-up, estimated at $250 million. The state health commissioner declared that some areas were safe to resettle, but new information challenged this decision (see below). Numerous health studies had been conducted in the intervening decade. They generally failed to reveal any significant evidence of an increased incidence of illnesses. Cancer registries are relatively new, and some states still do not have one. An analysis of the New York state cancer registry found that the incidence of lung cancers was generally higher throughout the Niagara Falls region, but no differences in the occurrence of liver cancer, lymphomas, or leukemias were noted in the Love Canal region. The increased incidence of lung cancer is intriguing in light of a study by the University of Toronto and Pollution Probe, which found that the mist from Niagara Falls contains PCBs, benzene, chloroform, methylene chloride, and toluene. Although the levels were not themselves high enough to pose a risk, they could be additive with other carcinogens in cigarette smoke, auto exhausts, etc., and the mist could settle out on crops to enter the food chain.

Statistics compiled in the United States and by Health and Welfare Canada showed statistically significant differences in mortality from different types of cancer among and within counties bordering the Great Lakes but did not indicate that these populations suffered substantially higher cancer mortality than non-basin counties. Nor were there consistent differences among municipalities within Niagara County. Currently, attempts are being made to track the some 15,000 persons who lived in the Love Canal area prior to the evacuation in order to identify increased incidences of cancer and other diseases. If this can be accomplished, some answers about the nature of the Love Canal risks may finally be obtained.

One fairly convincing bit of data is that the families living closest to the canal had a higher occurrence of miscarriages, abortions, and low birth weight babies. Health and Welfare Canada also attempted to uncover evidence of an increase in birth defects by surveying the occurrence of cleft palate in Ontario. There was definitely a higher incidence in the Niagara peninsula, but the area of higher incidence extended all the way north to Lake Simcoe, and other areas of concentration were discovered. There is a small one at the southern tip of Lake Huron (but upstream from the "chemical valley" of the St. Clair River) and another very large one running from the north shore of Lake Superior north to Hudson's Bay. These areas are highly diverse, both demographically and geographically, so it is difficult to identify a common factor to explain this distribution.

There is no question that a major component of the health impact of exposure to such toxic disasters is psychological in nature, which is not to say that it is not real. People tend to accept natural disasters with much greater equanimity. Flood victims do not appear to suffer the same prolonged mental stress as victims of a toxic disaster, nor do they attach blame as readily, even when there may be legitimate questions about the adequacy of flood control measures. Such things tend to be accepted as acts of God, and there may even be some essence of pioneer spirit to be taken from battling the forces of nature. In contrast, victims of toxic disasters tend to feel at the mercy of forces that are man-made and out of control. They are not just victims; they feel victimized by greed and callousness. Their sense of helplessness may be exacerbated by contradictory statements from government officials and by sensational reporting by the news media. An understanding of this phenomenon may be an important step toward minimizing the human price of such disasters. Following the Love Canal disaster, there was a high frequency of insomnia and severe depression, an increase in suicide attempts, lowered performance in schoolchildren, and other manifestations of emotional illness, including symptoms of malaise for which no organic cause could be found. The PCB fire at St. Basile-le-Grand, Quebec, and the tire-dump fire at Hagersville, Ontario are two more recent incidents in Canada with the potential for similar costs in mental anguish.

In April 1989, a peer review panel consisting of ten scientists familiar with the Love Canal situation met to examine charges that the area was not safe for resettlement. The charges claimed that the site had not been adequately cleaned up, and that an area in a zone designated as a control area for purposes of comparison had been shown to contain a "hotspot" with high levels of chlorinated organics under a church parking lot. The charges were brought by the Environmental Defense Fund, which also was responsible for the Alar controversy (see Chapter 2). The EPA, however, concluded that the area comparison method was still valid for the Love Canal and that the habitability decision should not be reconsidered.

Problems with Love Canal studies

Problems with the Love Canal studies include the following:

1. Initial studies were incomplete and not well organized. Most of the chemicals involved were very volatile and would not show up in human blood or tissues except for a short time after exposure. Only lindane and dioxins were persistent enough to be detected later, and lindane was not found in the subjects who were monitored. Dioxin assays are expensive and time-consuming.
2. When the Centers for Disease Control was asked to conduct a study in 1980, those people with the highest risk of exposure had been evacuated. Little information was available on previous exposures.
3. The episode was highly emotionalized. These were the homes of working-class people who had invested their life savings in their properties.
4. The dispersion of the evacuated families has made it difficult to collect valuable data on health effects.

The story of Love Canal has been repeated in many other communities in smaller ways. In London, Ontario, a playground in St. Julien Park was built over a waste disposal site. There were anecdotal reports of local clusters of brain tumors and other problems, but a study conducted by local health officials failed to confirm increased health risks. Clean-up and preventive measures have been instituted. A very definite hazard of landfill sites is the production and accumulation of methane gas from decaying organic material. Seepage into basements has resulted in explosions with deaths and injuries. Landfill sites that are covered over and landscaped or built upon should always have vents installed to prevent the accumulation of methane.

Chemical spills in the St. Clair River have on several occasions threatened the water serving Wallaceburg, Ontario and the Wallpole Island Indian Reserve.

Toxicants in the Great Lakes: implications for human health and wildlife

While the Great Lakes are not threatened by acid rain because of their large size and buffering capacity, they are definitely threatened by toxic chemicals. In 1991, the Joint Commission issued a report on toxicants present in Great Lakes water which they felt posed a potential threat to the populations of provinces and states in the Great Lakes basin by virtue of their ability to interfere with reproduction. Table 7 provides a partial listing of these toxicants.

Table 7   Some Great Lakes Contaminants with Potential Reproductive Effects

PCBs: linked to embryolethality, deformities, development
DDT: now banned; disrupts hormone balance
Dieldrin and aldrin (pesticides): linked to the death of adult bald eagles
Chlorinated dibenzofurans: carcinogenic
Toxaphene: an insecticide used in the cotton belt; has been found as far afield as the high Arctic
Dioxin (TCDD): probably carcinogenic for humans exposed to high levels.
Polycyclic aromatic hydrocarbons (PAHs): carcinogens
Hexachlorobenzene fungicide: causes organ toxicity, is carcinogenic, and may cause infertility
Mirex: carcinogenic insecticide that causes reproductive problems
Mercury: toxic to the CNS, liver, and kidney
Lead: toxic to the CNS

It is important to emphasize that these threats are largely potential, but there is an obvious need to improve the situation dramatically. Recent studies of white suckers from Lake Ontario have shown that, despite the fact that these bottom-feeding fish contain large amounts of PAHs, only about 5% showed cancer, and these all had parasitic liver disease. A study at the University of Guelph suggested that glutathione-S-transferase (GST) enzymes in the liver normally protect the fish by detoxifying the PAHs, but that the efficiency of these enzymes is destroyed by the liver parasites. This illustrates that there are many factors, including genetic ones, which likely must combine before a tumor will develop. Nevertheless, there is abundant evidence that numerous pollutants present in the Great Lakes are toxic to wildlife. These include lead, mercury, hexachlorobenzene, polychlorinated biphenyls (PCBs), polycyclic aromatic hydrocarbons (PAHs), dioxin (TCDD), Mirex, DDT, dieldrin, and toxaphene. Embryos and chicks of fish-eating herring gulls have been shown to suffer from a disease called expanded chick edema disease. This is very similar to a disease that occurs in domestic poultry exposed accidentally to high levels of polychlorinated dibenzodioxins (PCDDs) and PCBs. TCDD (2,3,7,8-tetrachlorodibenzodioxin) has been shown to have an LD_{50} in lake trout eggs of 80 parts per trillion (ppt), and in hamsters, 58% fetal mortality at the 9th gestation day has been shown at a dose of 18 µg/kg. Neurobehavioral and neurochemical abnormalities have been detected in fish-eating birds that consumed contaminated fish from Lake Michigan. Abnormalities have also been shown in laboratory rats fed a diet containing 30% salmon from Lake Ontario.

One of the difficulties in attempting to estimate the risk to humans of exposure to toxicants in the environment is the fact that little is known about the combined effects of several of these. An attempt has been made to deal with this problem as it relates to the polyhalogenated aromatic hydrocarbons that are structural analogs of TCDD. Recent evidence indicates that all of these are inducers of a hepatic microsomal monooxygenase enzyme, aryl hydrocarbon hydroxylase (AHH). The potency of these agents in inducing this enzyme in cultured hepatocytes correlates well with their experimental toxicity and this has led to the development of a set of Toxicity Equivalency Factors, or TEFs. The EPA has adopted the TEF method as a means of determining the toxicity of mixtures of these agents, as their effects on AHH seem to be additive. This approach, however, has limitations. The use of an *in vitro* culture system cannot take into account differences in pharmacokinetic parameters among different compounds, nor the extent to which they bioaccumulate. For example, one of the hexachlorobiphenyls has been shown to be a significant contaminant of mother's milk although it is one of the less common environmental contaminants.

Evidence of adverse effects on human health

It is axiomatic that toxicity to humans resulting from exposure to PAHs is related to total body burden. Portals of entry include the lungs (contaminated air) and the gastrointestinal tract (water and diet). Polychlorinated dibenzodioxins (PCDDs) and polychlorinated dibenzofurans (PCDFs) are known products of municipal incinerators. The Ontario Ministry of the Environment has calculated a worst-case scenario of 126 pg TEQ (picograms toxic equivalents)/day for people living near a municipal incinerator with inadequate pollution controls. In contrast, because of their poor solubility in water, drinking water is estimated to contain less than 2 pg TEQ/L. Diet is probably the most significant source of exposure and may account for 95% of all human intake. Levels of TCDD in human fat have been calculated to average 10 ppt worldwide, with little variation from place to place (evidence of the ubiquity of the substance). The EPA has set an acceptable daily intake of TCDD of 1.0 pg/kg/day.

Accidental poisonings in Yusho, Japan and Yucheng, Taiwan resulted in total body burdens of PCDDs and PCDFs that were 200 to 300 times the North American average. These levels were associated with nausea and anorexia, increased frequency of premature births, low birth weights, impaired growth, impairment of neuromuscular and intellectual development, and a higher frequency of health problems. The Michigan State Department of Health has assembled three large cohorts of individuals exposed to increased levels of organohalogens and it maintains a registry of these individuals.

Included in their records are farmers who were exposed accidentally to high levels of PBBs. This fire retardant was accidentally labeled as a mineral supplement for cattle. Farmers who were exposed to chlorinated biphenyls ("silo farmers") and a population of Lake Michigan anglers and their families, who consumed large amounts of fish from the lake and who were exposed to a mixture of contaminants, are also included. These cohorts have been followed for over 20 years. Mothers in the "angler" families consumed an average of three meals of contaminated fish per month from the Great Lakes while they were pregnant. Levels of PCBs in the fish were low and believed to be nontoxic, and the degree of impairment of the infants was much lower than in the Yusho and Yucheng incidents. The differences were nonetheless significant, even when a number of confounding variables such as smoking and alcohol consumption were taken into account. The effects were greatest in infants whose umbilical vein serum PCB levels were greater than 3.5 ng/mL. It must be emphasized that the effects seen in the Michigan infants were largely subclinical and, therefore, it is difficult to determine the true degree of risk to the population at large.

In 1989, a workshop was held at which findings from these studies were reviewed and compared to a similar study from North Carolina in which 800 infants believed to have had moderate *in utero* exposure to PCBs (estimated from maternal breast milk levels) were followed for over a year. Most studies showed a delay in cognitive performance as measured by the Brazelton and Bayley scales. These decrements in performance were detectable in neonates and in children at age 4, as were deficits in body size.

The Michigan studies have been criticized for flaws that include:

1. Using anecdotal reports (of the type and amount of fish consumption going back over 6 years);
2. Having differences between control and test groups (e.g., in alcohol consumption, use of medications, caffeine consumption, and the frequency of assisted births); and
3. Limiting the study's statistical power by restricting participants to one third of those exposed.

If nothing else, this illustrates the difficulties of conducting epidemiological studies when looking for subtle, subclinical findings. Nevertheless, the potential for toxicity and the significance of these warning signs cannot be ignored. The large body of evidence from laboratory studies and from the examination of numerous wildlife species provides ample indication that many of the halogenated hydrocarbons (organohalogens) have serious reproductive consequences if sufficient amounts are consumed. The questions of extrapolation to humans and of the effects of much lower exposures are as troublesome here as they are for carcinogenesis (see Chapter 5 for a further discussion of these accidental exposures).

Evidence of adverse effects on wildlife

There is certainly increasing concern about the consequences of environmental, estrogen-like chemicals for reproduction in aquatic species. Salmon with mixed male/female sex organs have been observed in the Great Lakes and abnormal sexual behavior has been seen in other fish. The subject of hormone modulators in the environment will be expanded upon in Chapter 12.

Global warming and water levels in the Great Lakes

In 1998, The International Joint Commission on The Great Lakes issued a dire warning. Unless the population of North America took water conservation more seriously, water loss from the system due to drought and evaporation could reduce water levels to the extent that the shoreline could move 50 to 100 meters out from its present position within a decade. In the summer of 1999, levels in the lower lakes were down nearly a meter, causing financial hardship for many marina operators. Pressure to export water from the system continues as world shortages of freshwater worsen. Concerns were expressed in strong terms at an October 1999 conference in Atlanta. It was pointed out that population growth of 30% is being predicted for the Great Lakes basin and that, even if the 3-year drought does not continue, this growth may well outstrip the available water supplies.

An article in the November 1999 issue of *Equinox* further elaborates the problem. It points out that 99% of the water in the Great Lakes is melt-water from the last Ice Age. It is therefore a nonrenewable resource. Only 1% comes from runoff, and the combined effects of global warming, which increases loss from evaporation, and several years of drought, which reduces runoff, could produce a crisis within a few years.

North Americans are the world's most profligate users of water. Americans use 5150.7 L/person/day and Canadians, 4383.6. In contrast, Sweden uses only 849.3 and the United Kingdom, 493.2 L/person/day. Water has also become a political and economic issue. Several companies, mostly American, have applied for permits to export water from Canada to the United States and elsewhere. Thus, Canada has banned bulk water exports but this policy is being challenged under the North American Free Trade Agreement (NAFTA). The court's decision will have significant consequences for the Great Lakes and other waters.

The marine environment

The marine environment is not exempt from the effects of pollutants. There is an area in the Gulf of Mexico called the "dead zone" that is virtually devoid of oxygen and hence of marine life, and it is increasing in size. It appears each summer off the coast of Louisiana and Texas, and in 1998 it was estimated to be 18,000 square kilometers in size. It is widely attributed to the flow of agricultural fertilizers, especially nitrates and phosphates, into the Gulf from the Mississippi River, which drains 41% of the continental United States and a small bit of western Canada. Algal blooms deplete the water of dissolved oxygen, making the environment uninhabitable for other organisms. There could be serious consequences for commercial fishing. Catches fell dramatically from 1983 to 1993, and the Gulf accounts for about 40% of the annual American commercial catch.

When 1998 was declared "The Year of The Oceans," the event went largely unnoticed, at least as far as the popular press was concerned. An exception was a feature article in the October 5 edition of *Maclean's* magazine. Appropriately titled "The Dying Seas," it documented the declining fish stocks worldwide, largely the result of over-fishing, and listed several species that are already on the endangered list or are threatened. Included are the bluefin tuna, barn-door skate, Atlantic haddock, Pacific salmon, anchovy, abalone, grouper, and orange roughy. The disastrous effects of bottom trawling, likened to strip mining, are also noted. The impact of over-fishing on the Atlantic cod is well-documented. Oil pollution, drift nets, and chemical pollutants are taking a heavy toll on cetaceans.

Not all toxicants in water are anthropogenic. Many microorganisms such as algae and diatoms are capable of producing lethal toxins that can concentrate up the food chain. There is presently concern that algal blooms in the ocean are creating a hazard for both marine life and people. This subject is discussed in Chapter 11.

Changing climatic conditions, whether of human origin or not, can have a significant impact on marine and aquatic environments. A predicted (by some models) increase of 10°C by the year 2050 would cause a one-meter rise in sea level. Salmon have already been detected in the Beaufort Sea and may be colonizing the Mackenzie River. An increase in the seal population is contributing to shoreline erosion of this river, along with associated permafrost melting. The 1992 El Niño caused a significant decline in the fish catch of the western coast of South America.

Aquatic toxicology

Until fairly recently, most research on aquatic pollutants has concentrated on their potential to cause problems for human health. More recently, the realization that they also pose a threat to aquatic and marine species has led to more research in this area. Aquatic organisms share many cell regulatory and signaling systems with mammalian cells. Toxicants may gain entry to these organisms through the integument, across the gills (a major uptake site), and, in the case of metal ions, by means of various ion transport channels. Experimentally, heavy metals such as cadmium have been shown to:

1. Disturb immune function in bivalves,
2. Reduce magnesium-ATPase in the gills of eels,
3. Inhibit metallothionein mRNA and the production of reactive oxygen species in oysters,

4. Disturb pigmentation in fiddler crabs, and
5. In our own studies, inhibit the aggregation of marine sponge cells.

Freshwater species may also be affected. Exposure of the freshwater sponge *Ephidatia fluviatilis* caused malformed gemmoscleres, and prey attacks by juvenile bluegills were inhibited by exposure to as little as 30 μg/L (see Philp, *Comp. Biochem. Physiol.*, 124C, 41–49, 1999 for further information).

Noted above was the importance of the active sediment as a sink for toxicants and a site of exchange with organic carbon and water. In the sea, a microlayer at the surface, approximately 50 microns thick, concentrates certain toxicants. This sea-surface microlayer may have metal concentrations 10 to 1000 times those of subsurface water. Organisms that spend a few hours daily at the surface, often responding to sunlight, are at risk of severe adverse effects, including growth deformities and cancer, often not manifested until later in their development, and even death. Phytoplankton and zooplankton including krill, larval stages of many organisms, and many other small crustaceans may be at risk. This has consequences for the entire food web as these are its foundation.

Biological hazards in drinking water

There is an oft-told story of a British physician who, in the nineteenth century, stopped a cholera epidemic in London by padlocking a communal pump. Cholera, caused by the bacterium *Vibrio cholerae*, and typhoid fever, caused by another bacterium *Salmonella typhi*, are just two of many water-borne intestinal infections spread when drinking water becomes contaminated by human feces, usually from untreated sewage. In the developed world, these have largely become historical diseases but they can resurface any time that water treatment facilities become overwhelmed and sewage contamination occurs. This is a serious problem during extensive flooding. Seafood contaminated with human sewage can also spread these infections. This is believed to have been responsible for an outbreak of cholera in Peru in 1991. There were 55,000 confirmed infections and 258 deaths. Another group of bacteria, *Escherichia coli*, also can cause enteritis, the severity of which depends on the particular strain involved. Contamination of drinking water can result from sewage, runoff from manure piles (many strains infect both animals and humans), and even bird droppings. The frequent summer closures of beaches can be caused by agricultural runoff, but also by concentrations of gulls fouling the bathing beaches. A period of high wave activity can stir up bottom sediment and temporarily increase bacteria counts in the water. *Aeromonas hydrophila* is a lesser known but potentially serious bacterial contaminant of water supplies and a cause of enteritis. Bird feces can spread infection, and the population explosion of ring-billed gulls has become a significant source of water contamination. These birds are highly adaptable and can survive on virtually any source of protein, including garbage.

Intestinal, protozoan parasites can also be spread by water. One of these, *Giardia lamblia*, is the cause of the erroneously named "beaver fever." It can infest wildlife such as deer as well as domestic animals. In 1997, there was an outbreak in the Kitchener area of Ontario, and in 1998, massive contamination of the drinking water of Sydney, Australia forced nearly 4 million inhabitants to boil their water for 2 weeks. Also found in their water was another parasite, *Cryptosporidium parvum.* This parasite caused an outbreak in Milwaukee in 1992–1993 in which over 400,000 people were infected and there were several deaths. Small outbreaks have occurred in Ontario as well. It is especially troublesome because it survives standard chlorination and filtration procedures and requires the installation of special filters.

Standard tests for water quality require that bacterial counts not exceed 100 cells/mL of water and there must be no *E. coli*. This organism is used as a marker for fecal contamination.

In 1991–1992, an extensive survey of groundwater quality was conducted in rural Ontario. The survey monitored 1300 farm wells for fecal coliform organisms, nitrate-N, several herbicides, and petroleum-based derivatives. The results indicated that 37% of all wells tested contained one or more of the target contaminants at levels above Provincial recommended limits: 31% had coliform levels above maximum acceptable limits and 20% had fecal coliforms. The incidence was higher in wells located on farms with manure systems: 8% had detectable levels of herbicides and 13% had nitrate-N levels in excess of the maximum acceptable concentration. Nitrates have been a cause of poisoning and deaths in infants fed formula made with contaminated water. The source is usually chemical fertilizers (see also Chapter 8). It is obvious that special hazards attend the use of water from farm wells, especially shallow ones that collect surface water.

Seafood, notably bivalves, may also be a source of infection with *Salmonella, E. coli*, and even hepatitis viruses. Contamination is almost always the result of the release of untreated sewage into the sea.

Anatomy of a small town disaster

Walkerton is an idyllic town of 5000 souls located in southwestern Ontario about 40 km from Lake Huron and 50 km from Owen Sound. In early May of 2000, citizens, including children, began showing up at the local medical clinic with severe, sometimes bloody, diarrhea, vomiting, fever, sweating, weakness, and other signs and symptoms of a severe gastrointestinal infection. There was no common event such as a public dinner or a meal at any particular restaurant that pointed to food as the source of the infection. Samples were sent for culture and identification as the number of affected people continued to mount. Infants were also being affected and soon the first death occurred. When the lab results came back, they showed that the offending organism was strain O157:H7 of *Escherichia coli.* Nonvirulent strains of *E. coli* are common inhabitants of the gastrointestinal tracts of animals and humans, but this strain can be a killer and is the organism responsible for so-called "hamburger disease." The outbreak continued for several weeks, eventually affecting at least 2000 people and causing 7 confirmed deaths, including the infant daughter of a local physician. Other deaths were suspected of being related to the outbreak. Testing soon revealed that the local water supply was the source of the infection. A "boil water" order was issued immediately and was predicted to be in effect until November. All water mains had to be scoured and disinfected right to the kitchen taps and thousands of liters of bottled water were shipped into the town over the summer. There followed charges and counter charges of improperly maintained and operated chlorination equipment, of inadequate testing, and of lack of communication by the provincial government. At the time of this writing, several investigations were underway and class action suits have been launched.

The source of contaminated water was traced to one of several deep drilled wells (well #5) serving the community. This well was in fairly close proximity to land grazed by beef cattle that were shown to harbor the organism. Manure had been spread on adjacent fields and the problem was compounded by a period of very heavy rain that facilitated seepage of surface water into the contaminated well. Another pathogenic bacterium, *Campylobacter*, was also found in the water. Over the course of the summer, other rural communities experienced high coliform counts in their water supplies, and public health authorities issued several "boil water" orders. These episodes call into question the generally assumed safety of deep drilled wells and indicate that standard chlorination procedures may not protect against an overwhelming influx of contaminated surface water with a high sediment content and the presence of pathogenic organisms. The mounting reliance on intensive livestock farming, with its massive production of animal wastes, is an increasing cause for concern in rural communities. A hog operation of 7000 animals will produce as much sewage as a town with the same number of inhabitants, but the treatment of the sewage is much more rudimentary. The installation of drainage tiles in cultivated fields may also provide a conduit for manure runoff to enter ponds and waterways. Many jurisdictions are contemplating more rigorous regulations, with the predictable conflict between farmers and rural residents. Complicating this picture is the fact that rural areas of North America are littered with abandoned, shallow, dug wells that have not been filled in. These wells, often the first source of water for a farm before a deep drilled well was installed, serve as conduits for surface water to enter the water table directly, instead of being filtered through layers of sand and gravel. Another potential source of contamination is private shallow wells that are cross-connected to municipal water systems. Farms and rural residents may draw from such wells for watering livestock, irrigation, or other outside uses. If common plumbing is employed, there is a chance that backflow may contaminate the general water supply.

E. coli O157:H7 is a normal inhabitant of the gastrointestinal tract of ruminants, where it does no harm. It is capable of producing a witches' brew of toxins. The relative amounts of each will vary with the strain. Strains are typed serologically and for the virulent strain O157:H7, the letters O and H represent specific antigens. In enteropathogenic strains, a plasmid-encoded (see Chapter 8) adherence factor, K88, promotes attachment of the bacterial cell to mammalian gastrointestinal epithelial cells, causing diarrhea. Entero-invasive forms such as O157:H7 cause a severe, hemorrhagic diarrhea. Hemolytic uremic syndrome (HUS) with kidney damage may also occur, and thrombotic thrombocytopenic purpura (TPP) can be a further compli-cation. These strains produce plasmid-encoded outer membrane polypep-tides similar to those produced by *Shigella* and are part of a family of toxins called Shiga toxins. A heat-labile exotoxin causes an increase in intracellular cyclic AMP levels with resulting increase in electrolyte and water secretion into the lumen of the gut. A heat-stable toxin, ST-I, acts similarly but on cyclic GMP. ST-II is thought to work via prostaglandin E2. Verotoxins 1 and 2, in O157:H7, inhibit protein synthesis after binding to specific glycoprotein receptors on the cell surface. Death from infection frequently involves kidney failure following massive fluid loss and hemorrhage. As few as ten ingested bacteria may be sufficient to cause disease. The toxin may be carried in blood attached to neutrophils and persist for a long time. Antibiotic therapy may actually make matters worse by lysing bacteria and releasing more toxin. It may also activate bacteriophages that carry the genes for the toxins.

Wildlife is not exempt from the effects of bacterial organisms. In Novem-ber 1999, thousands of loons and Merganser ducks were found littering the beaches of Lake Huron and Lake Erie. Four that I saw were lying in a perfectly straight line all facing inland, as if they had just fallen out of the sky, which indeed they had. The cause of this epizootic was the bacterium *Clostridium botulinum* type E. This is a different strain of the organism that usually causes botulinum food poisoning (Type C) and the toxin has the same effect of paralyzing voluntary muscles by blocking acetylcholine recep-tors. It is significant that all of the affected birds were fish eaters (piscivores). Type E is prevalent in aquatic environments in cold and temperate regions of North America and Europe. Several outbreaks of Type E botulism have occurred in people as a result of eating warm-smoked, contaminated fish. In contrast, Type C causes western duck fever and more often affects dabblers that feed in warm, shallow ponds on vegetation. Whether the November 1999 outbreak was connected to human activity is uncertain because out-breaks occur naturally. The possibility cannot be dismissed.

Further reading

Bart, A., They can't do it alone: some carcinogens in pollutants need help to cause cancer, *The (University of Guelph) Crest*, 2 (July), 9, 1991.

Boerlin, P. et al., Associations between virulence factors of Shiga toxin-producing *Escherichia coli* and disease in humans, *Clin. Microbiol.*, 37, 497–503, 1999.

Colbourn, T.E., Davidson, A., Green, S.N., Hodge, R.A., Jackson, C.I., and Liroff, R.A., *Great Lakes Great Legacy?* The Conservation Foundation, Washington, D.C. and The Institute for Research on Public Policy, Ottawa, Canada, 1990.

Continued use of drinking water wells contaminated with hazardous chemical substances — Virgin Islands and Minnesota, 1981–1993, *Morbid. Mortal. Wk. Rep.*, 43, 89–91, 1994.

Doyle, M.P., Pathogenic *Escherichia coli, Yersinia enterocolitica* and *Vibrio parahaemolyticus*, *Lancet*, 336, 1111–1115, 1990.

Edelstein, M.R., *Contaminated Communities: The Social and Psychological Impacts of Residential Toxic Exposure*, Westview Press, Boulder, CO, 1988.

Fujii, Y., Hayashi, M., Hitotsubashi, S., Fuke, Y., Yamanaka, H., and Okamoto, K., Purification and characterization of *Escherichia coli* heat-stable endotoxin II. *J. Bacteriol.*, 173, 5516–5522, 1991.

Gotfryd, A., Aluminum and acid: a sinister synergy, *Canad. Res.*, June/July, 10–11, 1989.

Habitability of Love Canal questioned after new discoveries, *Pesticide and Toxic Chemical News*, 17(19), 17(25), 7–8; 17(30), 21, 1990.

Henderson-Sellers, B. and Markland, H.R., *Decaying Lakes: The Origin and Control of Cultural Eutrophication*, John Wiley & Sons, Toronto, 1987.

Humphrey, H., Environmental contaminants and reproductive outcomes, *Health and Environ. Digest*, 5, 1–4, 1991.

Kimbrough, R.D., Health impact of toxic wastes: estimation of risk, in *The Analysis of Actual Versus Perceived Risks*, Covello, V.T., Flamm, W.G., Rodricks, J.V., and Tardiff, R.G., Eds., Plenum Press, New York, 1981, 259–265.

Korkeala, H., Stengal, G., Hyytia, E., Vogelsang, B., Bohl, A., Wihlman, H., Pakkala, P., and Hielm, S., Type E botulism associated with vacuum-packed hot-smoked whitefish, *Int. J. Food Microbiol.*, 43, 1–5, 1998.

Law, D., Virulence factors of *Escherichia coli* O157 and other Shiga toxin-producing *E. coli*. *J. Appl. Microbiol.*, 88, 729–745, 2000.

Malins, D.C. and Ostrander, G.K., Perspectives in aquatic toxicology, *Annu. Rev. Pharmacol. Toxicol.*, 31, 371–99, 1991.

Olsson, M. et al., Contaminants and diseases in seals from Swedish waters, *Ambio*, 21, 561–562, 1992.

Paneth, N., Human reproduction after eating PCB-contaminated fish, *Health and Environ. Digest*, 5, 4–6, 1991.

Philp, R.B., Cadmium content of the marine sponge *Microciona prolifera*, other sponges, water and sediment from the eastern Florida panhandle: possible effects on *Microciona* cell aggregation and potential roles of low pH and low salinity, *Comp. Biochem. Physiol.*, 124C, 41–49, 1999.

Roberts, L., News and comment. Learning from an acid rain program, *Science*, 251, 1302–1305, 1991.

Steward, R., Olson, J. et al., Toxicology and environmental chemistry of exposure to toxic chemicals, in *Human Health Risks from Chemical Exposure: The Great Lakes Ecosystem*, Flint, R.W. and Vena, J., Eds., Lewis Publ., Chelsea, MI, 1991, chap. 3.

Swain, W.R., Human health consequences of consumption of fish contaminated with organochlorine compounds, *Aquatic Toxicol.*, 11, 357–377, 1988.

Walker, C.H., Hopkin, S.P., Sibly, R.M., and Peakall, D.B., *Principles of Ecotoxicology*, Taylor & Francis Ltd., London, 1996.

Review questions

For Questions 1 to 10, define each of the following terms (answers can be found in the text).

1. Abiotic
2. Biotic
3. Eutrophication
4. Bioconcentration
5. Biomagnification
6. Acclimatization
7. Acclimation
8. Bioaccumulation
9. Aquatic
10. Marine

For Questions 11 to 20, answer True or False.

11. Species is a biotic modifier.
12. Pore water is bottom water in which sedimentary particles are suspended.
13. pH has no effect on the methylation of mercury by microorganisms.
F 14. Metals bound to carbonates in water become more toxic.
T 15. Concentration equilibria are established between bound and un-bound forms of toxicants in water.
F 16. Methylmercury never forms from natural sources.
17. DDT is an example of a chlorinated hydrocarbon.
18. Chlorinated hydrocarbons have a short biological half-life.
T 19. Human fat may contain up to 10 ppm DDT.
20. Organophosphate insecticides are reversible inhibitors of acetylcholinesterase.
F 21. All infectious organisms can be removed from drinking water by standard chlorination and filtration techniques.
T 22. Giardia and Cryptosporidium are protozoan parasites that can cause intestinal infections when present in drinking water.
23. The microlayer at the surface of the ocean concentrates many toxicants.
T 24. Heavy metals may interfere with the function of many marine and freshwater species.
F 25. Contamination of bathing beaches with *E. coli* and other coliform organisms does not pose a threat to human health.
F 26. Farm well water is always safer than municipal water supplies.

For Questions 27 to 333, use the following code:

> Answer A if statements a, b, and c are correct.
> Answer B if statements a and c are correct.
> Answer C if statements b and d are correct.
> Answer D if statement d only is correct.
> Answer E if all statements are correct.

27. Which of the following statements is/are true?
    a. Dioxin (TCDD) is a proven carcinogen for humans at levels wide-ly encountered in the environment.
    b. Chloracne (a skin rash) is the most common toxic manifestation of TCDD.
    c. TCDD is used as a herbicide.
    d. TCDD is a contaminant of the herbicide 2,4,5-trichlorophenoxy-acetic acid.
28. Which of the following symptoms is/are *not* characteristic of organo-phosphorus poisoning.
    a. Profuse sweating
    b. Tremors
    c. Confusion and other mental disturbances
    d. Constipation
29. Acidity in water can contribute to:
    a. Leaching of lead from solder joints in plumbing
    b. Acceleration of the transfer of metals from soil to groundwater
    c. A shift of copper to the elemental cuprous form (Cu^{2+}) from the carbonate form
    d. The existence of pentachlorophenol in the undissociated, more lipid-soluble state
30. Polychlorinated dibenzodioxins (PCDDs):
    a. Have been associated with accidental poisonings
    b. Are used industrially as pure chemicals
    c. Are products of municipal and industrial incinerators
    d. Are not themselves toxic
31. Which of the following statements is/are true regarding polycyclic aromatic hydrocarbons (PAHs)?
    a. Liver parasites may increase the toxicity of PAHs in fish by im-pairing their detoxication.
    b. Behavioral abnormalities have been observed in experimental an-imals exposed to PAHs.
    c. Herring gull embryos have shown signs of expanded chick edema disease, which has been associated with exposure to PAHs.
    d. Acid rain is a threat to the Great Lakes by increasing the toxicity of PAHs.

32. Which of the following statements is/are true about the Great Lakes?
    a. They constitute the largest, single supply of freshwater in the world.
    b. Toxic effects of chemical pollutants have been detected in wildlife.
    c. The introduction of foreign aquatic species has been a major problem.
    d. Government regulations have effectively controlled the dumping of poorly treated sewage into the system.
33. Which of the following is/are true regarding soil contaminated with toxic chemicals?
    a. Those with long biological half-lives cause the greatest concern.
    b. Transfer of contaminants to groundwater seldom occurs.
    c. Foodstuffs grown on contaminated soil may themselves become contaminated.
    d. Human activity such as mining is the only source of metal contamination of soil and water.

Answers

11. True
12. True
13. False
14. False
15. True
16. False
17. True
18. False
19. True
20. False
21. False
22. True
23. True
24. True
25. False
26. False
27. C
28. D
29. E
30. B
31. A
32. A
33. B

chapter four

Airborne hazards

> "The work is going well, but it looks like the end of the world."
>
> **— S. Rowland, co-discoverer of the CFC effect,
> to his wife.**

Introduction

When potentially noxious substances are discharged into the atmosphere at a rate that exceeds its capacity to disperse them by dilution and air currents, the resulting accumulation is *air pollution*. It may take the form of haze, dust, mist (which may be corrosive), or smoke and may contain oxides of sulfur and nitrogen and other gases that may irritate the eyes, respiratory tract, or skin and other substances that may be harmful to the environment or to human health. Absorption may occur in amounts sufficient to cause acute or chronic systemic toxicity. Air pollution has been greatly underestimated as a cause of illness and death. In May 2000, acting Canadian Environment Commissioner Richard Smith quoted government statistics indicating that smog adversely affected the health of 20,000,000 Canadians and caused 5000 premature deaths annually in 11 major population centers. This is in comparison to 4936 deaths from breast cancer, 3622 from prostate cancer, 3064 from motor vehicle accidents, and 665 from malignant melanoma. Air pollution obviously is an important health hazard.

Types of air pollution

Air pollutants may be gaseous or particulate in nature, and particulates may be either solid or liquid. Smog is a combination of air pollutants.

Gaseous pollutants

These are derived from materials that have entered into chemical reactions or combustion processes. They include carbon-based compounds such as hydrocarbons; oxides and acids; sulfur compounds such as dioxide, trioxide, and sulfides; nitrogen compounds (ammonia, amines, oxides); and halogenated substances (organic and inorganic halides).

Particulates

Particle or droplet size may range from 0.01 to 100 microns in diameter. The smaller particles are referred to as aerosols and can remain suspended, scattering light and behaving much like a gas. Below 10 microns, particles are capable of penetrating to all sites in the respiratory tract. Industrial particulates are usually solid and are carbonaceous, metallic oxides, salts, or acids and their porosity is such that they will absorb other gases and liquids.

Smog

The word is a combination of *smoke* and *fog* and is a popular term for a fairly uniform mixture of gaseous and particulate pollutants that accumulate over urban centers and persist for a prolonged period. Smog is a brown or yellow haze and usually occurs during the phenomenon of temperature inversion when a high-level mass of cold air traps warmer air beneath it to prevent mixing and dispersion. An especially bad "killer smog" occurred in London, England in 1952. It persisted for over a week and was responsible for about 4000 deaths, mostly from respiratory diseases. As a result, the Clean Air Act was passed in 1956, banning the use of soft coal for home heating.

Sources of air pollution

Air pollution may arise from natural sources and human activities. Volcanic eruptions, forest fires, and dust storms are natural sources, the importance of which should not be underestimated. The 1980 Mount St. Helen's explosion in Washington state pulverized half of a mountain and released millions of tons of dust. It affected weather patterns as far east as the Great Lakes. In 1912, a similar explosion of a volcano in Alaska released about 30 times the amount of dust as Mount St. Helen. The recent eruptions of Mount Pinatubo in the Philippines, together with smoke from the Gulf oil fires, have been blamed for unusually cool summers and excessive rainfall throughout most of North America in 1991–1992. Additional major eruptions in the "ring-of-fire" are predicted for the near future.

Human sources include discharge from coal-fired electrical generating stations, nuclear generating stations, industrial emissions, and domestic heating. Transportation sources include passenger autos, trucks, diesel locomotives, etc. Pollution may arise from all sources of combustion, industrial fuming and volatilizations, dust-making processes, photochemical reactions, biological sources (including microorganisms such as viruses, bacteria, and fungi), pollen, and chemicals from decaying organic matter. The breakdown of pollution sources in industrial countries is approximately as follows: transportation 50–60%, industry 15–20%, electric generating 10–15%, heating

15–20%, and waste disposal 3–5%. Considerable concern is arising over the problem of indoor air pollution. The hazards of side-stream cigarette smoke seem firmly established and this has led to increased restrictions on smoking in the workplace and in public buildings. Recent studies have shown that 4-aminobiphenyl, a potent human bladder carcinogen present in both main-stream and side-stream cigarette smoke, has been found in fetal hemoglobin, indicating that it crosses the placenta.

The importance of smoking as a cause of cancer cannot be overstressed. Lung cancer is now the leading cause of cancer deaths among women in Canada. In 1994, deaths of women from lung cancer approached 5600, while those from breast cancer were about 5400. Between 1982 and 1989, the overall incidence of cancer increased by 0.3% for women and 0.5% for men. In contrast, the lung cancer incidence in women increased by about 43% while in men it increased by about 8%.

Other indoor pollutants include formaldehyde gas (see Chapter 2), other toxic chemicals, particulates such as asbestos fibers and fiberglass wool, and radon-source ionizing radiation (see Chapter 12). Airtight houses and build-ings, constructed during the energy crisis of the 1970s, increase the risk of adverse health effects. Industrial indoor pollution is a special problem. In Ontario, the Ministry of Labor has jurisdiction over levels of air pollutants in the workplace and defines acceptable limits under various conditions (see Chapter 2).

Atmospheric distribution of pollutants

Air pollution generally begins as a local problem, but it can become global if the pollutants enter the atmospheric circulating system. Pollutants can enter the atmosphere in the form of gases, vapors (from volatile liquids), aerosol droplets, or fine dust particles (see Chapter 3 for a discussion of the distribution of pollutants in the biosphere).

Movement in the troposphere

The troposphere is the air mass up to an altitude of about 10 miles (mi). In the upper troposphere the winds are predominantly westerly and average 35 meters/s (mps) to disperse pollutants worldwide in about 12 days. Vertical movement circulates air north and south from the equator in systems called Hadley cells. In a band from 30°N latitude to 30°S latitude, other cells called Ferrel cells circulate air toward the poles. Speeds can reach 30 mps. Micro-scopic particles are retained for 1 or 2 months in the upper and mid-tropo-sphere and about 1 week in the lower troposphere (<1 mi).

Airborne dioxins and similar compounds, chiefly from municipal and industrial incinerators, can be distributed over a distance of 1500 kilometers. Half of the dioxins reaching the Great Lakes came from as far away as Texas. One-twentieth of all sources of dioxins account for 85% of the dioxins deposited in the Great Lakes region. The fetus and the breast-fed infant experience the highest body burden of dioxins.

Movement in the stratosphere

The stratosphere extends from 10 to 30 mi above the Earth. Movement occurs very slowly, at the rate of a few centimeters per second, but particles may stay for 2 or 3 years at an altitude of 20 mi and about 1 year at 11 mi. Certain gaseous pollutants such as freon, chlorofluorocarbons (CFCs), and some rare radioactive isotopes (e.g., krypton-85 from nuclear reactors has $t_{1/2}$ of 10.5 yr) are not readily removed by physicochemical means and may persist in the atmosphere for very long periods. Recent studies suggest that fluorinated gases will persist in the atmosphere for 300 to 2000 years or more, depending on the chemical.

Water and soil transport of air pollutants

The subject of the exchange of pollutants among various components of the biosphere was introduced in Chapter 3. Gaseous atmospheric pollutants can be dissolved in rainwater and solid particles carried in it mechanically. Precipitation thus carries them into the soil and groundwater, and they can reach oceans, lakes, and rivers by runoff and soil erosion and deep aquifers by seepage. The oceans are the ultimate repository for pollutants, and surface evaporation may conduct them back into the atmosphere. Several studies have confirmed this biospheric circulation of toxicants. In the 1950s, atmospheric tests of nuclear bombs resulted in widespread dissemination of radioactive fallout. Of particular concern was the presence of strontium-90, which exhibits chemical characteristics similar to calcium, including deposition in bone. Strontium-90 reached significant levels in cow's milk, in other dairy products, and in fruit and vegetables, and concern about its accumulation in the bones of children was a major factor in the discontinuation of atmospheric nuclear testing. The estimated North American exposure from all anthropogenic radionuclides is estimated now to be <1 mrem/yr. In 1969, contamination of Antarctic snow with DDT was identified. The only way it could have reached there was through precipitation. Presently, the most compelling concern is the problem of acid rain, the pH of which may be less than 4. Acid rain may be deposited far from its source.

Types of pollutants

Gaseous pollutants

These include:

1. Sulfur dioxide (SO_2), which forms acid rain as sulfurous acid;
2. Sulfur trioxide (SO3), which forms acid rain as sulfuric acid;

3. Nitrogen monoxide (nitric oxide, NO), oxidized to nitrogen dioxide (NO_2), a part of photochemical smog and acid rain;
4. Carbon monoxide (CO), a product of incomplete combustion, which forms carboxyhemoglobin which is incapable of transporting oxygen to the tissues;
5. Ozone (O_3), which contributes to photochemical smog;
6. Hydrogen sulfide (H_3S), which is very toxic;
7. Various hydrocarbons (C_xH_y), from automobile emissions;
8. CFCs, freon, vinylchloride, and radioactive isotopes; and
9. Methane from several sources

In 1999, the Sierra Legal Defense Fund (SLDF) drew attention to the fact that gasoline marketed in Canada had one of the highest sulfur contents in the world. This tends to defeat the pollution control systems in automobiles and even destroy catalytic converters. The result is increased emissions of sulfur dioxide, sulfate particles, carbon monoxide, nitrogen oxides, and hydrocarbons. SLDF lawyers have intervened on behalf of Friends of the Earth in a court case. FOE requested data regarding the sulfur content of various makes of gasoline from Environment Canada to give consumers a choice, but five major oil companies took Canada to court to prevent the release of this information, claiming that it would cause them financial harm and jeopardize their competitiveness. The federal government has since announced that it will reduce the sulfur content of gasolines to 30 ppm by the year 2005 as part of a plan to reduce all automotive emissions.

Particulate pollutants

1. Dusts. Fine particle solids may arise from sawdust, cement, grains, metals, rock (in quarrying operations), incomplete combustion of fossil fuels (producing particles of <1.0 μ, i.e., smoke), and any other substance including chemicals (pesticides, etc.) existing in powder form. Particulate emmissions from internal combustion engines are thought to be a major contributing factor to poor air quality in urban centers.
2. Liquids. Any liquid that forms droplets 1.0 to 2.0 μ in diameter will remain in suspension in air as a "mist" (e.g., sulfuric acid). Droplets <1.0 μ are defined as an aerosol. The term is also applied to solid particles of this size. It is important to note that water vapor is by far the most significant greenhouse gas, accounting for about 85% of infrared trapping, but its level fluctuates widely.

Health effects of air pollution

Acute effects

Short-term exposure to hazardous levels of air pollutants may result in irritation to the eyes and the respiratory tract. Populations at high risk include the very young and the elderly, whose respiratory and cardiovascular systems are not fully functional; people with asthma, emphysema, and heart disease; and heavy smokers. These groups had the highest mortality rates during the killer smog in London, England. The accidental release of toxic chemicals from industrial plants has caused serious health problems and death, the most tragic being the release of 40 tons of methyl isocyanate from the American Cyanamid plant in Bhopal, India, in 1984. Nearly 3000 people died.

Chronic effects

Long-term exposure to lower levels of pollution may result in, or aggravate, chronic bronchitis, pulmonary emphysema, bronchial asthma, and lung cancer. Cigarette smoke will cause all of these problems. Excessive secretion of bronchial mucus and a chronic cough are the hallmarks of chronic air pollution effects. Dust and other allergens, including pollen, 1 to 90 μ in diameter, can induce or trigger allergic reactions in susceptible people.

Air pollution in the workplace

Systemic poisoning has occurred in workers inhaling toxic levels of metals such as lead, arsenic, mercury, manganese, zinc, and cadmium, as well as pesticides and drugs. Oxides of all of these metals, those of copper, tin, and nickel, and brass dust can cause a febrile reaction (fever, joint and muscle aches) called metal-fume fever. Cutting with an acetylene torch generates temperatures high enough to vaporize metals, including lead. Workers exposed to vinylchloride gas have a high incidence of hepatic angiosarcoma, an otherwise rare tumor. Pneumoconiosis, or coal miner's lung, results from the inhalation of coal dust with the formation of localized lesions with silica crystals, emphysema, fibrosis, loss of vital capacity and, eventually, right heart failure due to increased cardiac output to compensate for inadequate oxygenation of the blood. Organic solvents may be hazardous because of their CNS-depressing action.

Some recent studies have suggested that the offspring of firefighters have a higher incidence of birth defects in locales where firefighters, or their spouses, are responsible for washing their work clothes. This presumably is the result of the absorption of toxic contaminants on the clothing through the skin, although absolute confirmation of this risk source has yet to be confirmed.

Asbestos

Asbestos workers are exposed to a variety of health hazards, including "white lung syndrome" (asbestosis, a form of fibrotic pneumoconiosis), carcinoma of the lung, mesothelioma (cancer of the pleural and peritoneal membranes), and possibly gastrointestinal cancer, although animal studies have not been able to confirm this. Mesothelioma is a rapidly fatal cancer occurring most often 30 to 40 years after the first exposure. The linings of the chest (pleura) and abdomen (peritoneum) thicken, fluid accumulates, and widespread metastases occur. This cancer occurs rarely in people not exposed to asbestos. There are several forms of asbestos fiber, and not all of them cause mesothelioma. There is no doubt that the form known as crocidolite is carcinogenic, but controversy has centered on whether the form known as chrysotile is also carcinogenic. There is an ongoing study of Quebec chrysotile miners born between 1881 and 1920 and employed for at least 1 month. Over 70% of these have now died, and an estimated 30 cases of mesothelioma would be expected. Seven of these men were also exposed to crocidolite in a small factory, and contamination with tremolite (another form) could account for additional cases. It now seems that the risk is at least much lower for chrysotile asbestos. To cause pleural mesothelioma, asbestos fibers must traverse the lung and appear in the pleura. Chrysotile fibers will do this, and they have been shown to cause mesothelioma-like lesions in experimental animals. The risk associated with chrysotile fibers has not been firmly established in humans. According to some studies, a very large number of fibers must be inhaled for this to occur.

Carcinoma of the lung occurs 60 times more often in asbestos workers who smoke than in those who do not. Asbestos becomes a hazard for the general populace when building insulation begins to break down or is disturbed during construction. Wear of brake linings releases asbestos particles in the air. There is increasing concern that glass wool fibers can cause the same type of cancer as asbestos. An excess in cancer incidence has been shown in workers in the glass wool industry, but no direct evidence linking this to the inhalation of fibers has been uncovered. In the Fiberglass Canada plant in Sarnia, an increased incidence was shown in the 2500 workers but it was not statistically significant. In the United States, NIOSH recommended that allowable air levels of glass fibers in plants be reduced.

Silicosis

Silicosis results from the inhalation of silica particles, silicates, or other mineral fibers. Histiocytes are transformed into fibrocytes; alveoli harden, resulting in loss of elasticity and lung function. Emphysema results, as it does from cigarette smoking.

Pyrolysis of plastics

Prior to about 1980, firefighters did not routinely wear a breathing apparatus unless dealing with a fire involving known toxic fumes. There is some evidence (still largely anecdotal) that firefighters who attended fires involving plastics are beginning to show increased cancer rates. It is now known that when polyurethane smolders, fine particles of degraded polymers are produced which may have toxic chemicals adsorbed to them. These release lytic enzymes into the lungs to cause massive tissue damage and edema.

Dust

Even barn dust can be an environmental hazard in the workplace. It may contain dried fecal material, animal dander, protein from feed grains and hay, skin parasites, and microorganisms. A Scandinavian study found a high incidence of respiratory and other health problems in farm workers who spent a lot of time in hog barns. Thirty percent of workers lost work time due to respiratory problems.

CO and NO_2

Chemicals involved in atmospheric pollution can sometimes become a problem indoors. There is increasing concern over indoor events that involve the use of internal combustion engines. These include tractor pulls, monster truck rallies, and mud races. CO levels have been shown to peak as high as 250 ppm during such events. Peak levels should not exceed 30 ppm. NO_2 levels may also be elevated because of incomplete combustion. CO is colorless, odorless, and non-irritating. It can produce headache, nausea, and mental impairment. NO_2 is irritating and may cause pulmonary edema. High concentrations may be fatal.

There is growing evidence that particle pollution at levels encountered in the environments of most large urban centers may be more hazardous than previously believed. There are elevated incidences of premature deaths, hospital admissions, and a variety of health problems. There is a statistically significant association between acute exposures to particles and increased mortality regardless of the source of the particles or the climatic conditions prevailing at the time of exposure. This seems to suggest that the particles are the primary cause, although the mechanisms involved are not yet known.

The Centers for Disease Control in Atlanta issued a report that 23 million Americans were at risk because of exposure to particles <10 µm in diameter and concentrations >155 µg/m^3 of air (the 24-hr average acceptable level is 150 µg/m^3). The EPA is considering setting new levels at a much lower concentration.

Air pollution in the workplace can take some strange forms. A recent (1999) report dealt with two workers in a cattle breeding station who were found unconscious on the floor of the laboratory in which samples of bull semen were frozen for storage. A tank of liquid nitrogen had been leaking and displacing the air in the room. If they had not been discovered in time, they could have asphyxiated.

Multiple chemical sensitivity

Multiple chemical sensitivity (MCS), as the name implies, refers to a condition in which an individual reacts in an adverse manner to a wide variety of chemically diverse agents with exposure occurring by any portal of entry. Because the initial, sensitizing stimulus is frequently an odor often encountered in the workplace, it seems appropriate to consider MCS here. The disorder has been referred to as 20th century disease, environmental disease, chemical AIDS, "total allergy syndrome," and environmental chemical intolerance (CI). It has been, and remains, highly controversial because of the lack of any confirmatory laboratory test, because of the vagueness of the symptoms and their similarity to many found in other conditions, and because of a lack of an animal model that accurately mimics the condition. Typically, the initial "sensitizing" dose involves an aversive reaction to a chemical odor from any source such as mothballs, volatile solvents (e.g., glues, marker pens, carpeting, or correction fluid), newsprint, perfumes, disinfectants, fuels, particulate pollutants — the list is endless. This initial aversion is followed by an extension of the aversive reaction to other chemical agents and possibly to other portals of entry. The initial exposure may or may not be at a fairly high level, but subsequent exposures at much lower levels will elicit the aversive response.

A working definition of MCS has been proposed based on the following criteria:

1. The initial symptoms are associated with an identifiable environmental exposure.
2. The symptoms involve more than one organ system.
3. Symptoms recur and recede in response to the presentation and withdrawal of predictable stimuli.
4. The symptoms are elicited in response to low-level exposure to a wide variety of diverse chemicals. A hallmark of the condition is that the level of exposure that will elicit a response is far below that which will produce a detectable effect in the general population.
5. No standard test of organ system function can explain the symptoms.

Symptoms may be somatic, such as headache, fatigue, dizziness, nausea, musculoskeletal pain; cognitive, such as difficulty in concentrating, poor memory; apparent neurological, such as clumsiness, parathesias (numbness in various areas); and/or affective such as irritability, depression, and anxiety. There may also be hyperreactivity to sound, light, and touch. The condition is more common in women than in men.

Because of obvious overlaps in symptomotology MCS has been linked to chronic fatigue syndrome (CFS), a subset of Gulf War syndrome patients, and fibromyalgia. The connection with Gulf War syndrome was strengthened on September 8, 1999 when the (U.S.) National Academy of Sciences released a long-awaited report on the health effects of the poison gas sarin, its antidote pyrizostigmine bromide, depleted uranium, and vaccines against anthrax and botulism. The study reviewed research into these agents not involving Gulf War veterans. The findings were generally negative (see http://books.nap.edu/books). Some studies reported long-term health effects following exposure to sarin at concentrations high enough to elicit an intense immediate reaction. No connection could be found between exposure to depleted uranium and kidney disease or lung cancer. A Canadian study of 69 veterans did not find evidence of increased uranium levels compared to the general population.

Higher levels of chlorinated hydrocarbons have been reported in a study of patients with CFS when compared to non-CFS controls, strengthening the possibility that chemical exposure plays a role. Moreover, 20 to 37% of patients with CFS report a significant degree of chemical intolerance, as do 23 to 47% of fibromyalgia patients. Patients with acute anxiety syndrome also have an increased incidence of MCS and fibromyalgia.

Numerous theories have been proposed to explain MCS but none has received universal acceptance. These theories can be categorized roughly as immunologic, psychologic, or neurogenic.

Based on evidence that a variety of laboratory measures of immune function have shown changes (such as depressed leukocyte counts, lymphocyte counts T-cell counts, and changes in complement levels), a popular theory holds that the sum of low (often undetectable) levels of several chemicals constitutes a total body burden that compromises the immune system. There does not, however, appear to be a consistent pattern of laboratory findings suggestive of a specific immune defect. Moreover, there have been no animal experiments that show an immune deficit developing as a result of exposure to low levels of multiple chemicals. The allergy theory suffers from a similar lack of evidence of immune dysfunction. While it is entirely possible that the immune system could be affected in MCS, it seems unlikely that this can be the sole basis for the condition.

Neurogenic theories draw heavily on the phenomenon of limbic kindling, which can be demonstrated experimentally. The amygdala is a part of the limbic system that receives input directly from olfactory pathways. Kindling is the phenomenon whereby repeated exposure of rats to low levels of electrical or chemical stimulation will eventually lead to increased sensitivity to seizures from unrelated stimuli such as handling. The effect has not been demonstrated in humans but it is known that the olfactory bulb is capable of concentrating inhaled chemicals and that the amygdala interacts with the autonomic and endocrine systems. Kindling is felt to be an animal model for temporal lobe epilepsy (TLE) and TLE and MCS share some characteristics, notably a higher incidence of cystic ovary.

Psychological theories are based in part on Pavlovian conditioning, in which the association of a stimulus (the bell) with a somatic response (salivation and gastric secretion when presented with food) will lead to the induction of the somatic response by the unrelated stimulus alone. Thus, a traumatic chemical exposure, or a chemical exposure associated with a traumatic event, could lead to the induction of a noxious response by exposure to very low levels of that chemical or other chemicals. A problem with this theory is that individuals most likely to be exposed to high levels of chemicals do not seem to have a higher incidence of MCS. Other studies have reported that MCS patients have increased incidences of acute anxiety disorder and clinical depression, and some authors have suggested a higher frequency of childhood abuse.

For every theory proposed, there are counter-arguments. No one theory is universally accepted, nor is there incontrovertible evidence for any one of them. There is no question that, for the sufferer, the condition is very real and debilitating. Despite a scientific literature now numbering in the hundreds of papers, we seem no closer to an answer. There is now little doubt that the mind can exert a great influence on somatic function with accompanying neurochemical and endocrine changes. Many years ago, members of the author's department conducted a student lab in which medical students were given the impression that they were volunteering to particpate in a trial of a centrally acting drug. Two treatment groups were formed, one to receive the "active" agent and the other a placebo, on a double-blind basis. Both groups actually received a lactose placebo. Both groups were tested on a battery of tests of psychomotor performance and mentation. Both groups reported the same incidence and variety of side effects, mostly relating to the central nervous system. Symptoms included facial flushing, euphoria, anxiety, irritability, restlessness, inability to concentrate, tremors, headache, sedation, and bradycardia. There is an obvious similarity between many of these symptoms and those associated with MCS. There is no question that odors can be powerfully evocative. Even inherently unpleasant odors can evoke pleasant memories, and vice versa. The smell of gasoline on a crisp spring morning still reminds me of getting my motorbike out of winter storage as a young man. A friend and neighbor cannot enter our house because she cannot tolerate the smell of perfume or any scented product. Some workplaces now ban the use of these because of this problem. We should not be surprised that some people can become ill as a result of these phenomena.

Some hospitals have been establishing environmental suites where no synthetic materials are used and the air is filtered with HEPA filters. Victoria Hospital in Halifax, Nova Scotia, started a pilot project a number of years ago, which led to the opening of a full-time clinic devoted to environmental illnesses. They reported in the news media that one diagnostic test, the SPECT brain scan, held promise. SPECT stands for single-photon emission computed tomography. News reports indicated that when a patient with MCS is exposed to a trace chemical, the SPECT scan showed subtle changes in brain chemistry. To date, no reports have appeared in the scientific literature.

Chemical impact of pollutants on the environment

Sulfur dioxide and acid rain

Over 100 million metric tons of sulfur dioxide from fossil fuels are emitted annually into the atmosphere around the world. Sulfur dioxide plus water (in atmospheric water vapor) forms sulfurous acid and sulfuric acid in a complex series of reactions that involve shifts between gaseous and aqueous phases. A simplified summary is shown below. For a more complete discussion of the chemistry of acid rain consult Baird (1995).

$$2SO_2 \text{ (gaseous)} + O_2 \text{ (gaseous)} \longrightarrow 2SO_3 \text{ (gaseous)}$$

$$SO_3 \text{ (gaseous)} + H_2O \text{ (aqueous)} \longrightarrow H_2SO_4 \text{ (aqueous sulfuric acid)}$$

$$SO_2 \text{ (gaseous)} + H_2O \text{ (aqueous)} \longrightarrow H_2SO_3 \text{ (aqueous sulfurous acid)}$$

Nitric acid also contributes to the acid rain problem. Nitric acid can be formed from the nitrogen oxides (NO_x): nitrous oxide (N_2O), nitric oxide (NO), and nitrogen dioxide (NO_2).

$$NO_x + H_2O \longrightarrow HNO_3$$

Nitrous oxide is released from oceans and during biological processes in soil (the nitrogen cycle). It is a greenhouse gas as well as a source of acid rain.

The average retention time for sulfur dioxide in the troposphere is very short (about 2 to 4 days). The sulfuric acid thus formed is carried to the soil in precipitation (rainwater, snow). A pH as low as 1.7 was recorded in West Virginia in 1979 (battery acid and gastric acid are about pH 1). Core samples of snow in the Arctic regions revealed a pH of 6.8 180 years ago vs 3.8 in recent years.

The anions in acid rain are SO_4^- (70%) and NO^- (30%). This acid mist may affect the respiratory tract of people (and animals). Asthma sufferers and people with allergies are prone to loss of lung function and respiratory disease.

The absorption of acid into the soil solubilizes metals such as aluminum, cadmium, and lead and facilitates their movement into vegetation and water, including drinking water. The accumulation of these metals may contribute to human diseases. Aluminum has been implicated in dementias, lead may affect the development of the central nervous system in infants and children, and cadmium can cause kidney disease (see Chapter 6). Acidification of lakes leads to a complete loss of aquatic life. It is estimated that up to 4000 lakes in Ontario have been so affected.

Paradoxically, although the Mount Pinatubo volcanic eruptions were partly responsible (along with the El Niño) for the extremely cool, wet summer of 1992 in North America, the long-term effects are more likely to contribute to acid rain and global warming. There is some debate, however, about the extent to which air pollution and clouds may negate the effect of increased ultraviolet (UV) radiation exposure due to ozone depletion. The 1991 eruption injected 15 to 30 megatons of SO_2 into the stratosphere which, within 1 month, was converted to $H_2SO_4^{2-}$. This formed an aerosol that was expected to remain in the atmosphere for up to 3 years. The total aerosol load is estimated to be 10 to 20 times that produced by anthropogenic and other biological sources in the same year. Some models predict that ozone will be rendered more susceptible to degradation by atmospheric chlorine, and reflection of long wavelength infrared may increase global warming (see below). In fact, a marked decline in atmospheric ozone began in 1991 but recovery was noted in 1993, and by 1994, it had returned to essentially normal levels. It is not known whether this is a long-term trend, or if the effect was attributable to the Mount Pinatubo eruption.

The chemistry of ozone

In the stratosphere, at an altitude of about 20 mi, short-wave UV radiation converts O_2 to O_3, which, by direct absorption, prevents UV radiation from penetrating the Earth's atmosphere. When longer UV wavelengths are absorbed (>242 nm), O_3 is split back into O and O_2 and thus recycled.

Ozone depletion is of considerable concern because it contributes to climatic change by allowing short-wave UV radiation to penetrate to the Earth's surface. Because this band is the ionizing form of UV radiation, ozone depletion is also associated with an increased risk of skin cancer. The layer is thinnest at the equator, so that the incidence of skin cancer in tropical climates is greater than in the temperate zones. Light-skinned people are at greatest risk. The incidence of skin cancer in Texas is 3.8:1000, compared with 1.2:1000, in Iowa. The incidence of skin cancer is also increasing in northern climates as well, and warnings against unprotected sunbathing will be routine as each summer approaches. Sunscreen factors of 20 or more are recommended, as is avoidance of exposure between the hours of 11:00 a.m. and 3:00 p.m.

The ozone layer is normally maintained at about 1 ppm but it can be depleted by the action of certain pollutants. Nitric oxide (NO) is a major offender in this regard. The following reactions can occur:

$$NO + O_3 \longrightarrow NO_2 + O_2$$

$$NO_2 + O^{\cdot} \longrightarrow NO + O_2$$

Thus, NO will recycle to break down thousands of ozone molecules unless it reacts with another free radical; for example:

$$OH + NO \longrightarrow HNO_3 \text{ (nitric acid)}$$

Chlorine

Chlorine, its oxides, and chlorine compounds such as chlorofluorocarbons (CFCs), widely used as aerosol propellants and refrigerants, also contribute to ozone depletion. For example:

$$Cl + O_3 \longrightarrow ClO + O_2$$

$$ClO + O \longrightarrow Cl + O_2$$

The appeal of CFCs was that they are chemically inert under nearly all conditions, and it was not until the impact of supersonic jet transports was studied that the effect of UV light on them was realized. The calculated ozone loss is presently 1% but this would increase to 10% by the year 2015 if the release of 800,000 tons of CFCs annually were to continue.

CFCs are of particular concern because they are heavier than air. They reach the stratosphere by slow percolation vertically, driven by the concentration gradient. This means that even if their use were to stop immediately, their effect would not begin to decline for many years to come and it has not yet peaked. CFCs are themselves greenhouse gases, further contributing to the problem of global warming.

CFCs have been temporarily replaced by hydrofluorocarbons (HFCs) and hydrochlorofluorocarbons (HCFCs) in some jurisdictions. Although these destroy just as much ozone as CFCs, they dissipate much more quickly in the atmosphere. In November 1992, 86 countries attended the United Nations Ozone Layer Conference in Copenhagen. The conference agreed to an accelerated ban on CFCs by 1995, instead of 1999 as originally proposed. France blocked a similar ban on, or reduction of, HFCs and HCFCs. Ironically, two physicians in Dortmund, Germany may have rendered all of the hydro-, chloro-, fluorocarbon technology obsolete. The substitute, labeled Greenfreeze, is a simple, non-patentable mixture of butane and propane that can be used in refrigerator units and is not harmful to the ozone layer. The technology is available to anyone. Greenfreeze refrigerators are being marketed already by DKK Schjarfenstein in Germany.

Global warming

Water

Water vapor is undoubtedly the greatest single contributor to the greenhouse effect, accounting for up to 75% of heat trapping. High-flying jet aircraft annually produce millions of liters of water as vapor.

Carbon dioxide

Another gas of concern regarding global warming is carbon dioxide (CO_2). CO_2 is a product of combustion and decay of organic matter. Green plants normally consume CO_2; these plants take it up during the day, converting it to O_2 and carbohydrates by photosynthesis. It is a basic rule that the oxidation of one carbon atom yields one molecule of carbon dioxide:

Combustion, decay: $C + O_2 \longrightarrow CO_2$

Photosynthesis: $CO_2 \longrightarrow O_2 + Carbohydrates$

Reaction in atmosphere: $CO_2 + H_2O \longrightarrow O_2 + CH_2O \ polymer$

This last reaction reverses in winter and returns the CO_2 to the atmosphere. Depletion of rainforests for timber and increased CO_2 production from industrial sources and internal combustion engines have led to a dramatic increase in atmospheric CO_2. Solar energy is either reflected from the surface of the Earth or absorbed by it, in which case it heats the Earth. Very little heating of the air occurs, which is why air is cooler at high altitude. Most of the reflected energy passes back out into space but some is at a very long wavelength (50,000 nm) because the Earth is much cooler than the sun, and it is reflected back to Earth by water vapor and CO_2, the "greenhouse effect." This results in additional global warming.

There has been considerable controversy regarding anthropogenic contribution to global warming, in part because satellite measurements have failed to detect significant evidence of it. Very recently, however, an international panel of over 200 scientists reached a consensus that the effect is real and significant. Barring strict controls on greenhouse gas emissions, global average temperatures will rise by 1.5 to 3°C by the year 2050. By 2030, sea levels will rise 8 to 29 cm and continental interiors (the breadbaskets of the world) could go dry in summer. Although considerable uncertainties remain, warming over the past century is real, with increases of 0.3 to 0.6°C since the late nineteenth century. Additional evidence can be taken from the fact that salmon are showing up in the Beaufort Sea and appear to be colonizing the Mackenzie River. More seals are also showing up, and the banks of the Mackenzie River are eroding in places because the permafrost is thawing. For the first time in their oral history, the Inuit at Sachs Harbor are experiencing thunderstorms with lightening and seeing birds such as robins and barn swallows. Melting permafrost has caused mudslides, allowing an entire lake to spill into the sea. The ice in some places has become so thin as to be dangerous.

Jeffrey Severinghaus of the Scripps Institute of Oceanography has been analyzing arctic ice for trapped isotopes of argon and nitrogen (the polar ice caps are also a huge sink for CO_2). His data suggest that there was an abrupt warming, by 16°C, at the end of the last ice age 15,000 years ago. This suggests that climate change may occur much more rapidly than previously thought. His work is published in the October 29, 1999 issue of *Science*. There are also recent indications from Antarctica that the south polar ice cap, which contains 90% of the world's ice, may be thinning in some areas. Recent radar images from a Canadian-American satellite reveal faults in the eastern ice sheet that could allow a large section of it to slide into the sea should rapid global warming occur. It was already known that the western ice sheet was capable of moving large sheets of ice toward the sea (about 212 sq. mi/year). The water displaced by huge volumes of ice sliding into the ocean could have serious consequences for many low-lying coastal areas. See also http://www.bergen.com/morenews/antarctic.

The Greenland ice cap also appears to be shrinking. A recent report released by NASA (July 2000) indicates that more than 50 billion tonnes of water are melting from it yearly, adding to a 23-cm rise in sea level. Even within the last 5 years, records indicate an accelerating rate of ice loss (see also http://www.spacer.com/planetary/icecaps.html and http://www.enn.com/enn-news_archives/1999/).

Some lessons can be learned from history. In the late autumn of 1815, a huge volcano called Tambora in Indonesia exploded with a force many times that of Mount St. Helen. The resulting dust spread over the surface of the Earth and 1816 became known as "the year without a summer." Air temperature dropped an average of 1°C. Crops were severely affected in the United States and it snowed in June in Maine. It was so cold and wet in Switzerland that summer that Mary Shelly challenged her husband to a writing contest for something to do. The result was *Frankenstein*. Fortunately, this situation was temporary, but it illustrates how small a change in temperature is required to effect drastic climatic changes.

Methane

Methane is a greenhouse gas, reflecting radiant energy back to the earth. As a fuel, it has several advantages (see below), but uncombusted methane reaches the atmosphere from fermentation in the intestinal tracts of animals (both wild and domestic) and from the fermentation (rotting) of vegetation. Rice paddies are a major source of methane. It has been suggested that the North American penchant for burgers is a contributing factor to the greenhouse effect by encouraging extensive beef production (and rainforest destruction) in South America. Animals and rice paddies each produce about 100 megatons of methane annually; and because rice is one of the world's staple grains, a switch to a vegetarian diet would not likely result in a net methane saving. The biggest source of methane in the world is its termites, which produce about 200 megatons annually. Because they digest dead wood, their population is expanding because of deforestation, and this is probably the greatest threat to the environment from methane (see also Chapter 13). There has been some controversy over the contribution of wetlands to methane production. While some authorities claim that this has been overestimated because of failure to account for uptake of methane by boreal soil, others have pointed out that fens, bogs, and beaver ponds are significant contributers because of their large area. Estimates of net contributions are 1.5 to 3.5 $g/m^2/yr$ (*Delta*, 6, 1, 1999). Like ozone levels, levels of polluting gases in the atmosphere, notably methane and carbon monoxide, have been returning to normal in recent years, and this too has been suggested to be the result of the declining influence of the Mount Pinatubo explosion.

The ocean, and suboceanic oil wells, may be important and underestimated sources of atmospheric methane. At the low temperature and high pressure encountered on the very deep sea floor, methane hydrates will form and cover the sea bottom, with methane gas trapped below the hydrates. Underwater landslides can scour the hydrates, releasing vast amounts of methane into the ocean and atmosphere. Oil well blowouts can also release huge volumes of methane. Because gasified water will not provide buoyancy, any object on the surface, such as a ship or oilrig, will sink instantly. Over 40 rigs have been lost in this manner, with few survivors. There are also huge reserves of methane trapped as hydrates in arctic ice. These will be released as ice melts in response to global warming.

Subtle greenhouse effects

Significant impact can occur long before the catastrophic events discussed above. As temperatures climb, diseases of humans, livestock, and plants, which are normally constrained by subfreezing winter temperatures, will advance northward. Those depending on insect vectors will gain a foothold more rapidly as insect populations soar with increases in temperature. The cold barrier to Africanized honeybees will move north and their spread will be accelerated. Crop failures are bound to occur when both drought and disease assail crops. This will necessitate increasing dependency on pesticides and irrigation, further stressing the environment. There is already evidence that expanding human populations modify the environment, and that diseases adapt to the new conditions, in some cases evolving to become new threats to human health.

Global cooling: new Ice Age?

Studies of ice core samples from the polar ice cap suggest that the last Ice Age, in which the world was covered with ice to a depth of nearly 1000 m as far south as New York, resulted from a mean temperature drop of no more than 1.5°C. Ice Ages result from irregularities in the Earth's orbit around the sun, which take it further away about every 200,000 years. Interglacial periods last about 10,000 years and there is some evidence that the present one is ending soon.

Complicated, is it not? It is even further complicated by the fact that the oceans act as CO_2 sinks, dissolving vast amounts of the gas. As gas is dissolved, the greenhouse effect is lessened, the temperature drops, and more snow and rain fall, contributing to the polar ice caps. As they expand, they reflect more and more UV rays back into the stratosphere, further lowering the Earth's temperature. The reflective property of materials is known as the "albedo factor" and the efficiency of ice is nearly 100%.

Sulfur dioxide

SO_2, a major contributor to acid rain, may also contribute to global cooling. Aqueous-phase oxidation of SO_2 to SO_3 in clouds (rich in water vapor) occurs, and evaporation of the water releases a particulate aerosol of SO that backscatters solar radiation.

Motor vehicle exhaust

As noted above, the internal combustion engine is a major source of CO_2 and other pollutants such as lead and CO.

Polar ice in Greenland has shown a sharp increase in carbon and lead deposits since the 1950s. Lead rose from 0.03 to 0.20 µg/kg snow or ice. Levels corresponding to the era of 800 B.C. were <0.001 µg/kg ice. In urban areas, the lead content of the air is 5 to 50 times higher than in the country (1.1 vs 0.02 µg/m³). In 1968, the gasoline combustion engine accounted for 181 of the 183 kilotons of lead released annually into the atmosphere in the United States. Other sources included coal and oil combustion and manufacturing processes including lead smelting. Vehicles using tetraethyl lead gasoline emit, on average, 1 kg lead per year. Distribution and eventual inhalation of this lead by humans constitutes a health hazard. At levels of about 10 µg/dL (ingested), it is a potent neurotoxin to infants and children, causing impaired hearing, slowed neuronal transmission, and a variety of behavioral problems including hyperactivity, learning disabilities, and reduced mental capacities (see also Chapter 6). Lead from solder joints in plumbing is also implicated as a source.

Carbon monoxide (CO) is a product of all forms of combustion and it has been responsible for many deaths due to the use of heating devices (camp stoves, space heaters, defective furnaces, fireplaces, etc.) in poorly ventilated sleeping areas, including tents. Because the internal combustion engine is the principal source of CO (it has been used as an instrument of suicide many times over), it is convenient to consider CO toxicity here. Its natural concentration in the atmosphere is extremely low (1 to 2 ppm) and about 100 megatones yearly arise from vehicle engines in the United States. Other sources include natural processes (70 megatones) and other human combustion sources (250 megatones). Urban concentrations can range from 20 to 100 ppm, depending on traffic and weather conditions. Firemen are at risk from CO poisoning if a breathing apparatus is not worn. Combination with free radicals in the environment helps to buffer the atmospheric levels of CO:

$$CO + OH^- \longrightarrow CO_2 \, H^+$$

CO combines irreversibly with hemoglobin (Hb) to form carboxyhemo-globin, which cannot combine with O_2 so that asphyxiation results. Levels greater than 6.4×10^3 ppm cause dizziness and headaches within 2 min, and unconsciousness and possibly death in 15 min. Treatment consists of intra-venous nitrates that cause the formation of methemoglobin. This has a much greater affinity for CO and binds to free the Hb. If a hyperbaric facility is available, O_2 at high pressure (2 atm) will provide adequate oxygenation through dissolved O_2 until enough Hb is free to assume oxygen transport.

Natural factors and climate change

Although much has been made of the impact of human activity on climate change, the influence of natural events is often ignored, as is the fact that nearly all computer simulation models are based on a doubling of CO_2 without reference to other factors, including water vapor, discussed above. Methane is another natural greenhouse gas (see above and Chapter 14).

Contrary to statements in many texts, the Earth's orbit is not a fixed ellipse, but will vary considerably. Moreover, its tilt may vary by several degrees. Both effects are due to the gravitational pull of Jupiter and they can have a dramatic impact on climate. Some estimates claim that they are responsible for up to 85% of climate variation. Solar activity such as flares and sunspots also correlates well with climatic changes. The 1988–1989 sea-son was very active, with over 400 sunspots recorded. The activity of El Niño, the Pacific up-welling of warm water from the depths, also correlates well with sunspot activity and it has a marked effect on climate.

From 1675 to about 1725, there was a "Little Ice Age," with record cold temperatures and snowfalls. There was ice skating on the Thames River in London, England and elsewhere in Europe. Since then, the climate has been warming. Should this trend have reversed? Cooling trends tend to occur fairly slowly, whereas warming trends are much more abrupt. Ice cores taken from the Vostock 4 site in Russia, going back 800,000 years, measured CO_2 levels in bubbles in the ice and these indicated a natural cycle of about 100 years. This cycle appears to be shortening dramatically. The observatory at Muana Loa in Hawaii recorded atmospheric CO_2 levels of 315 ppm in 1958, rising to 355 ppm in 1989. Methane levels are rising even faster, from 400 to 1800 ppb in the same period. There is little doubt that human activity has contributed to these changes, but the seas are a vast sink for CO_2 and a natural warming trend would also release CO_2 from this sink. Increased water vapor from the oceans would increase cloud cover, which would probably have a buffering effect on the warming trend. Nevertheless, some models predict a 4-ft rise in sea level over the next 50 years unless the trend is reversed, submerging Bangladesh and the Maldives. In northwestern Ontario is the Experimental Lakes Area, which has been the subject of climatic, hydrologic, and ecological study for over 20 years. Records show that both air and lake-water temperatures have increased by 2°C during this period and that the ice-free season has lengthened by 3 weeks. The thermocline has also deepened. Evaporation has been higher and precipitation lower than normal, with the result that lake levels have dropped and pollution has concentrated, all of which have ecological consequences.

Glaciers and small ice caps respond much more rapidly to climate changes than do the polar ice caps and thus they are useful for computer modeling of short-term effects of warming. In the last 100 years, mountain glaciers have retreated in most parts of the world and have likely contributed to the 10- to 20-cm rise in sea level during this period. A recent study of 12 such areas where data collection has been in progress for several years indicates that for a 1 Kelvin (K) rise in temperature, glacial melting would account for a rise in sea level of 0.58 mm/yr, which is significantly less than earlier estimates.

Remedies

How all these factors will balance out is anybody's guess, but it is clear that we do not know enough to take chances. As one wag put it, Chicken Little only has to be right once. It is abundantly clear, however, that a reduction in fossil fuel consumption is essential. This means more fuel-efficient cars, changes in transportation habits with more public transit usage, and the development of alternative fuels. One such fuel is natural gas (methane) for heating and for automotive fuel. Although methane obeys the one carbon–one CO_2 rule, and although it is a greenhouse gas itself, it produces more energy per carbon with fewer polluting by-products than petroleum oil-based fuels, so that the negative impact on the environment is much less. Further development of nuclear energy may also be essential. Despite the emotional reactions that it can generate, a well-designed, well-regulated, and well-operated nuclear power generator may be the safest source of electrical energy available. It is worth remembering that not one proven death has resulted in North America from a nuclear generator (see also Chapter 2), but hundreds have died from failures of hydroelectric dams and thousands from coal mining accidents and black lung disease. In 1979, a power dam in India collapsed, killing thousands. *Time* magazine carried a few lines, compared to the several pages devoted to Chernobyl. In 1989 in the (then) U.S.S.R., a gas pipeline explosion killed hundreds of people in two passing passenger trains.

The Atomic Energy Commission of Canada performed a study indicating that nuclear energy is even safer than solar energy because the latter requires materials that need to be mined and refined with energy derived from coal. The calculation determined that solar energy was inherently 500 times more dangerous than nuclear energy. Fossil fuels (coal and oil) were 2000 times more dangerous, and natural gas was the safest of all energy sources. Hydrogen may also become a source of pollution-free energy for transportation.

A theory that was popular a number of years ago is undergoing a renaissance. Known as the Gaia hypothesis, it considers the Earth and its atmosphere as a living organism, with all parts interconnected and in balance — like a mammal in a state of homeostasis. Interference with one component of the system may have far-reaching consequences that may not be foreseen with our incomplete understanding of how the system works. Chapter 13 deals with this topic in more detail. One essential change that must occur is a dramatic reduction in the use of fossil fuels by developed countries, especially in North America. Northern developed countries use 7.5 kilowatts (kW) per person per year vs. 1.0 kW for southern, developing countries. Given the populations of 1.2 billion for the former and 4.1 billion for the latter, the north consumes 2.2 times as much energy as the south. Add to this the fact that 6 million hectares of arable land are lost annually to erosion, development, and salination, whereas clearing, which entails the loss of irreplaceable rainforests, creates 4 million hectares, and it becomes evident that major changes are essential. One of these is a check of the population explosion. The doubling time of the Earth's population is 30 years and of its energy consumption, 20 years. It is evident that we cannot continue in this manner if our species — and the planet — are to survive.

Further reading

Abelson, P.H., Global change, Editorial, *Science*, 249, 1085, 1990.

Baird, C., *Environmental Chemistry*, W.H. Freeman, New York, 1995.

Bates, D.V., Asbestos: the turbulent interface between science and policy, *Can. Med. Assoc. J.*, 144, 554–556, 1991.

Bell, I.R., Baldwin, C.M., and Schwartz, G.E., Illness from low levels of environmental chemicals: relevance to chronic fatigue syndrome and fibromyalgia, *Am. J. Med.*, 105, 74S–82S, 1998.

Brasseur, G. and Granier, C., Mount Pinatubo aerosols, chloroflurocarbons, and ozone depletion, *Science*, 257, 1239–1242, 1992.

Coghlin, J., Gann, P.H., Hammond, S.K. et al., 4-aminobiphenyl hemoglobin adducts in fetuses exposed to the tobacco smoke carcinogen *in utero*, *J. Natl. Cancer Inst.*, 83, 244–280, 1991.

Dockery, D.W., Pope, C.A., Xu, X. et al., An association between air pollution and mortality in six U.S. cities, *New Engl. J. Med.*, 329, 1753–1759, 1993.

Epidemiologic Notes and Reports. Nitrogen dioxide and carbon monoxide intoxication in an indoor ice arena — Wisconsin, 1992, *Morbid. Mortal. Wk. Rep.*, 41, 383–384, 1992.

Fisher, D.E., *Fire and Ice: The Greenhouse Effect, Ozone Depletion and Nuclear Winter*, Harper and Row, New York, 1990.

Friedlander, S.K. and Lippman, M., Revising the particulate ambient air quality standard, *Environ. Sci. Technol.*, 28, 148A–150A, 1994.

Gowdey, C.W., Hamilton, J.T., and Philp, R.B., A controlled clinical trial using placebos in normal subjects: a teaching exercise, *Can. Med. Assoc. J.*, 96, 1317–1372, 1967.

Graveling, R.A., Pilkington, A., George, J.P.K., Butler, M.P., and Tannahill, S.N., A review of multiple chemical sensitivity, *Occup. Environ. Med.*, 56, 73–85, 1999.

Kerr, R.A., The greenhouse consensus, *Science*, 249, 481–482, 1990.

Kerr, A., Pinatubo fails to deepen the ozone hole, *Science*, 258, 395, 1992.

Lelieveld, J. and Heintzenberg, J., Sulfate cooling effect on climate through in-cloud oxidation of anthropogenic SO_2, *Science*, 258, 117–120, 1992.

Lloyd, S., The calculus of intricacy: can the complexity of a forest be compared with that of Finnegan's wake? *The Sciences*, Sept./Oct., 38–44, 1990.

McDonald, J.C. and McDonald, A.D., Asbestos and carcinogenicity, *Science*, 249, 844 (letter), 1990.

Mossman, B.T., Bignon, J., Corn, M. et al., Asbestos: scientific developments and implications for public policy, *Science*, 247, 294–301, 1990.

Multiple chemical sensitivity: clinical, experimental and theoretical considerations. Proceedings of a symposium (several authors), *Toxicology*, 111, 69–145, 1996.

Novelli, P.C., Masarie, K.A. et al., Recent changes in atmospheric carbon monoxide, *Science*, 263, 1587–1593, 1994.

Oelermans, J. and Fortuin, J.P.F., Sensitivity of glaciers and small ice caps to greenhouse warming, *Science*, 258, 115–117, 1992.

Planet Earth — Problems and Prospects. *Notes, Sesquicentennial Symposium of Queen's University*, Kingston, June 1991.

Population at risk from particulate air pollution — United States, 1992, *Morbid. Mortal. Wk. Rep.*, 43, 290–293, 1994.

Ravishankara, A.R., Solomon, S., Turnipseed, A.A., and Warren, R.F., Atmospheric lifetimes of long-lived halogenated species, *Science*, 259, 194–199, 1993.

Schindler, D.W., Beaty, K.G., Fee, E.J. et al., Effects of climatic warming on lakes of the central boreal forest, *Science*, 250, 967–970, 1990.

Rodhe, H., A comparison of the contribution of various gases to the greenhouse effect, *Science*, 248, 1217–1219, 1990.

Robert, O., The threat of barn dust, *Veterinarian Mag.*, 3, 29–30, 1991.

Walker, C.H., Hopkin, S.P., Sibly, R.M., and Peakall, D.B., *Principles of Ecotoxicology*, Taylor & Francis Ltd., London, 1996.

Weiss, B., Neurobehavioral properties of chemical sensitivity syndromes, *Neurotoxicology*, 19, 259–268, 1998.

Yanko, D., Are animal disease patterns changing because of global warming? *Vet. Mag.*, 2(3), 18–21, 1990.

Review questions

1. Which of the following sources, in industrialized countries, accounts for the greatest amount of air pollution?
   a. Industry
   b. Generating electric power
   c. Transportation
   d. Heating
   e. Waste incineration

2. One reason for stopping atmospheric nuclear tests was concern over the deposition of a radioactive isotope in the developing bones of children. The isotope of concern was:
   a. Cobalt-60
   b. Strontium-90

    c. Iodine-125

    d. Cesium-133

    e. Carbon-14

3. The diameter of droplets or particles defined as aerosols is:

    a. Larger than 10 μ

    b. From 5 to 10 μ

    c. Smaller than 1 μ

    d. From 1 to 2 μ

    e. From 2 to 5 μ

For Questions 4 to 7, use the following code:

    Answer A if statements a, b, and c are correct.

    Answer B if statements a and c are correct.

    Answer C if statements b and d are correct.

    Answer D if only statement d is correct.

    Answer E if all statements (a, b, c, d) are correct.

4. Which of the following can be sources of indoor, household pollution?

    a. Radon gas

    b. Cigarette smoke

    c. Asbestos fibers

    d. Formaldehyde gas

5. Pollutants that can contribute to acid rain include:

    a. Sulfur dioxide

    b. Sulfur trioxide

    c. Nitric oxide

    d. Chlorofluorocarbons

6. Acid rain:

    a. May solubilize toxic metals

    b. Can never be neutralized naturally

    c. Can kill aquatic life

    d. Is never the result of a natural phenomenon

7. Which of the following statements is/are true?

    a. The troposphere is the air mass in contact with the Earth's surface and it extends vertically for about 10 miles.

    b. Hadley cells are vertical air movements that circulate air northward and southward from the equator.

    c. The stratosphere extends from about 10 to 30 miles above the Earth.

    d. Pollutants are rapidly cleared from the stratosphere.

8. List four groups of individuals who are at greater than average risk of respiratory problems from air pollutants.

9. List four conditions known to be associated with air pollution in the workplace.

10. List five natural factors that likely contribute to global warming.
11. What do you think is the single most important underlying cause of anthropogenic environmental problems?
12. List three environmental consequences of global warming.
13. Complete the following equations:

$$SO_2 + H_2O \rightarrow H_2SO_3$$
$$NO + OH \rightarrow HNO_3$$
$$NO_2 + O^. \rightarrow$$
$$Cl + O_3 \rightarrow$$
$$ClO + O^. \rightarrow$$
$$HSO_3 + O_2\text{--uv} \rightarrow$$
$$SO_3 + H_2O \rightarrow H_2SO_4$$
$$SO_3 + NO + H_2O \rightarrow$$
$$NO + O_2 \rightarrow$$

Answers

1. c
2. b
3. c
4. E
5. A
6. B
7. A

Refer to text for remainder.

Case study 3

In August 1989, a previously healthy 4-year-old boy developed signs and symptoms that included leg cramps, rash, itching, excessive perspiration, rapid heartbeat, intermittent low-grade fever, personality changes, and peripheral neurological disorders. The interior of the house had been painted 1 month earlier using over 60 L of a latex-based paint. The house was sealed and air-conditioned.

Q. Is there likely to be any connection between the boy's illness and the redecorating?

The signs and symptoms strongly resemble a condition known as acrodynia, a rare form of childhood mercury poisoning. A 24-hr urinary mercury determination revealed a level of 65 µg/L. The boy's mother and two siblings had similar urine mercury levels.

Q. Do these results point to an environmental source of the mercury?

Q. What additional analysis or analytical information would be useful?

Case study 4

In the same year (1989), two farm workers in their 30s were working in an indoor manure pit 25 × 25 × 5 ft deep, attempting to clear a blocked pump intake pipe. Several hours later, both men were found face down in several inches of liquified manure.

In an unrelated incident, five men died after consecutively entering a similar pit, each attempting to rescue those who had entered before. The first person to enter was attempting to replace a shear-pin in a piece of equipment.

Q. What is the underlying cause of death in both of these incidents?

Q. What simple safety measure could have prevented these tragedies?

(See also Case Study 2.)

Case study 5

Two machine shop workers were cleaning and degreasing equipment in a confined, poorly ventilated space using a chlorofluorocarbon known as CFC-113. After about 20 minutes, one worker clutched his chest and collapsed. He was cyanotic and not breathing. The second worker called for help and began to administer artificial respiration (he was not trained in CPR) but he himself collapsed soon thereafter. Both victims were evacuated. The first was pronounced dead on arrival at hospital, and the second recovered in the ambulance after oxygen was given.

Q. What was the portal of entry of the toxicant?

Q. What organ systems were involved in the intoxication?

Q. What was the likely offending agent?

Q. What steps should have been taken to prevent this accident?

Case study 6

Soapstone carving is an economic mainstay for the Inuit of the Canadian North. Recently, the Inuit Art Foundation created a comic-book superhero named Sanannguagartiit ("your carving buddy" in the Inuktitut language). With his flying snowmobile and his loyal sleddog Quimmiq, he crisscrosses the Arctic demonstrating masks and respirators and advising carvers to leave their work clothes at the carving hall and not to take them home.

Q. Why would it have been necessary to institute this program?

Case study 7

Workers in a particular industry have been having a very high incidence of respiratory problems, including:

1. A high frequency (15.7% of workers) of chronic bronchitis
2. A sevenfold increase in restrictive lung function and a threefold increase in obstructive lung function compared to the general population
3. A 20% occurrence of shortness of breath and flu-like symptoms such as fatigue, muscle and joint pain, and general malaise
4. A 30% annual occurrence of absenteeism due to respiratory illnesses
5. A 35% occurrence of wheezing and tightness in the chest
6. A 50% frequency of chronic, productive cough and frequent colds

Q. Is this most likely a primary infection or a pollutant?

Q. If the latter, what is the likely portal of entry?

These men are farm workers in the pork industry.

Q. What contaminants could be present in the air in pig barns?

Q. What corrective measures should or could be taken?

Case study 8

Recently in Wisconsin, an incident was reported to public health authorities in which 11 high school students were treated in two different emergency departments for acute respiratory symptoms, including labored breathing (dyspnea), spitting (coughing) blood (hemoptysis), cough, and chest pain. Two of them required hospitalization. All of them had participated in an ice hockey tournament the previous evening in an indoor arena. Interviews with other players and spectators revealed that many of them had suffered respiratory problems that became progressively worse as the evening progressed, and many had central nervous system symptoms including headache, dizziness, sleepiness, nausea, and vomiting. Of 131 students who were interviewed, 48% reported symptoms. The frequency among players was double that among spectators.

Q. What is the likely portal of entry in this toxic reaction?

Q. What offending agent(s) would you suspect?

Q. What laboratory test would be helpful?

chapter five

Halogenated hydrocarbons and halogenated aromatic hydrocarbons

Introduction

Halogens are the related elements chlorine (Cl), bromine (Br), fluorine (F), and iodine (I). They may exist as gases (Cl_2, F_2), a liquid (Br_2), or a solid (I_2). Halogenated hydrocarbons, also known as organohalogens, are a group of organic compounds of diverse structure to which one of these halogens has been attached. The core structure may be either simple, consisting of one or two carbons, or more complex and aromatic (halogenated aromatic hydrocarbons or HAHs). Because they have been implicated almost universally in toxic reactions in mammals and lower species (including carcinogenesis in some), it is appropriate to consider them as a group despite their chemical diversity. Polycyclic aromatic hydrocarbons (PAHs) are multiringed, planar chemicals that share many of the same toxicological properties as HAHs.

Early examples of toxicity from halogenated hydrocarbons

One of the oldest and simplest of these compounds is carbon tetrachloride (CCl_4), which was used extensively as a solvent and a dry-cleaning agent until its hepatotoxic nature was discovered. In fact, it was used originally as a treatment for hookworm in humans and domestic animals and as a component in fire extinguishers. These uses may still exist in some parts of the world. CCl_4 owes its toxicity to the fact that it is converted in the liver to carbon trichloride (CCl_3), which is a free radical capable of inducing peroxidation of lipid double bonds and poisoning protein-synthesizing enzymes. Older, halogenated hydrocarbons include trichloroethylene and the anesthetics halothane and chloroform (now abandoned because of its toxicity). Figure 21 shows the variety of structural formulae included in this class of compounds. The earliest form of poisoning associated with a halogen was probably "bromism," a condition resulting from the use or abuse of

PBBs   X = Br or H

PCBs   X = Cl or H

(210 possible structures for each)

**TCDD or 2,3,7,8-
tetrachlorodibenzodioxin is
one of 75 possible dioxins**

Figure 21 Structural formulae of a number of halogenated hydrocarbons.

sodium or potassium bromide as a sedative and sleeping potion in the early 1900s. Symptoms included severe headache, stupor, delirium, cardiac problems, very bad breath (from the bromine), and an acneform skin rash of the type now called "chloracne." The word *bromide* is used now to indicate a soothing but meaningless statement of the sort frequently uttered by certain politicians.

Physicochemical characteristics and classes of halogenated hydrocarbons

The characteristics of halogenated hydrocarbons that make them useful for a variety of applications are generally the same ones that make them hazardous to the environment and to humans. These include:

1. High lipid solubility,
2. Ability to survive heat >800°C,
3. High resistance to chemical breakdown, and
4. Toxicity to microorganisms

These agents are used for a variety of purposes (see below).

Antibacterial disinfectants

Hexachlorophene has been used for many years as a surgical scrub and, in a 3% solution, as a hospital disinfectant. It is also used as the active ingredient in deodorant soaps. In the late 1960s, a change was proposed in U.S.FDA regulations to permit the use of hexachlorophene as an antifungal wash for fruit and vegetables.

In light of the then-recent thalidomide tragedy, extensive testing was required for approval to be granted. Rats fed high levels of hexachlorophene developed weakness, ataxia, paralysis, and evidence of a type of brain pathology known as status spongiosus, indicative of axonal degeneration. In 1971, a study was done in which infant monkeys were washed daily for 90 days in 3% hexachlorophene. Neurological symptoms were observed and status spongiosus was seen in all specimens at post-mortem. Significant blood levels have been detected in infants washed with 3% hexachlorophene, and those with severe diaper rash, burns, or congenital skin disorders are especially prone to absorb it, as the natural permeability barrier has been disrupted (see Chapter 1). Autopsies of infants dying from a variety of causes and who had received high exposures showed evidence of status spongiosus. In 1972, hexachlorophene was accidentally added in high concentration to baby powder during its manufacture in France. Forty-one deaths of infants and young children were attributed to this error. A few years later, Dr. Hildegard Halling, a Swedish physician, published a report indicating that nurses who washed frequently in hexachlorophene (10 to 60 times daily) had a higher incidence of birth defects in their offspring (25/460 births) than those who did not (nil/233 births). Although this clinical study was criticized for design flaws, other studies with rats have revealed teratogenic effects. This product is no longer used in nurseries in North America and pregnant women are advised to avoid it.

Herbicides

This group includes 2,4-dichlorophenoxyacetic acid (2,4-D), 2,4,5-trichloro-phenoxyacetic acid (2,4,5-T), and dioxins such as 2,3,7,8-tetrachlorodibenzo-*p*-dioxin (TCDD = dioxin). Agent Orange, used as a defoliant in Vietnam, was equal parts of 2,4,-D and 2,4,5,-T and contained TCDD as a contaminant. One of the best-documented human exposures to dioxin was the explosion at Seveso, Italy (see Chapter 2). Hundreds of individuals suffered from chloracne, which is the hallmark of toxicity of halogenated hydrocarbons.

The herbicides 2,4,-D and 2,4,5-T are used to control broad-leafed plants along highways, railways, and utility rights-of-way. They are hormonal growth promoters and force plants to consume energy at a greater rate than it can be replaced. The humans at greatest risk of toxic exposure are the workers who apply the sprays, and poisoning from dermal and respiratory absorption has occurred as well as from accidental ingestion. Signs and symptoms include peripheral neuritis, muscular weakness, and chloracne. Although 2,4,5-T is a weak teratogen in some animals, it is the presence of TCDD (or dioxin) that is the greatest source of public concern.

Dioxin (TCDD) toxicity

Dioxins are a family of compounds of which TCDD (2,3,7,8-tetrachloro-dibenzo-*p*-dioxin) has received the most public attention. There is significant species variation in TCDD toxicity, with the guinea pig being most sensitive (LD_{50} 1 μg/kg) and the rat quite insensitive (LD_{50} 22 μg/kg). There are other toxic manifestations of dioxin toxicity.

Hepatotoxicity

All species show enlarged livers, and microsomal monooxygenase enzyme induction occurs in most. Rats develop fatty livers with triglyceride deposition. Hepatic fibrosis has been reported in humans. People exposed at Seveso had elevated serum enzyme levels (serum glutamic-oxaloacetic transaminase [SGOT] and serum glutamic-pyruvic transaminase [SGPT]), indicating liver damage, for several weeks after the accident.

Porphyria

Porphyrins are pigments widely distributed in nature and are present in the body as by-products of heme synthesis which is required for the formation of hemoglobin, myoglobin, and cytochromes. Heme is ferrous protoporphyrin IX. Hematin, the iron-containing molecule in catalase and peroxidases, is ferric protoporphyrin IX. The rate-limiting step in the synthesis of heme is ALA synthetase (ALA = gamma-aminolevulinic acid). Dioxin significantly increases the levels of ALA synthetase and, hence, ALA levels and the synthesis of porphyrins. This is not enzyme induction, but probably is due to interference with a feedback control system. The excess porphyrins are excreted in the urine, giving it a port wine color, and they are deposited in the skin, producing pigmentation. Because porphyrins are photoreactive, a condition known as porphyria cutanea tarda develops, characterized by photosensitivity, blistering, fragility of the skin, pigmentation, and hirsutism (hairiness).

Congenital defects in porphyrin metabolism cause the same syndrome and this has been suggested as the explanation for the werewolf and vampire myths of Europe (characterized by hirsutism and the avoidance of sunlight). There is even an explanation of why drinking blood might have a therapeutic effect. Heme is the feedback substance that turns off ALA systhetase. Absorption of sufficient heme might thus inhibit porphyrin synthesis. In the late 1950s, an extensive outbreak of porphyria cutanea tarda occurred in Turkey during a famine as the result of consuming seed grain treated with hexachlorobenzene as an antifungal agent. A simplified scheme of the steps involved in porphyrin synthesis is shown in Figure 22.

Chloracne

This skin disorder is typified by rash, cysts, and hyperpigmentation and is the hallmark of poisoning with all halogenated hydrocarbons. It was the predominant toxic manifestation at Seveso.

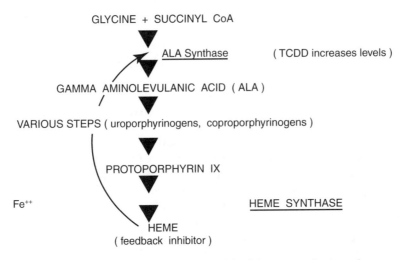

Figure 22 Effect of TCCD on the (simplified) heme synthetic pathway.

Cardiovascular effects

A 10-year mortality study of the population exposed to TCDD after the Seveso explosion of 1976 revealed significantlly increased mortality from cardiovascular events.

Carcinogenicity

Dioxin is a potent hepatocarcinogen in mice and to a lesser extent in rats. There is a latency period before the emergence of liver tumors. Evidence of cancer in several studies of Vietnam veterans, for whom claims of increased incidence of cancer have been made, has been inconclusive.

A retrospective cohort study was conducted by scientists at the (U.S.) National Institute for Occupational Safety and Health (NIOSH) on 5172 workers at 12 U.S. plants in which TCDD was a chemical contaminant of the manufacturing process. Exposure was well documented and serum TCDD levels were obtained from 253 workers. The mortalities from several cancers previously associated with TCDD (stomach, liver, and nasal cancers, Hodgkin's disease and non-Hodgkin's lymphoma) were not significantly different from the overall population, but the incidence of all cancers taken together was slightly but significantly increased. In a subcohort of 1520 workers with more than 1 year of exposure and more than 20 years of latency, mortality from soft tissue sarcoma was significantly higher than for the general population. The authors concluded that the results were not suggestive of the high relative risks of cancer reported for TCDD in previous studies. The slight risk of increased soft tissue sarcoma is weakened by the small numbers involved (only three cases) and confounding factors such as smoking and exposure to other chemicals.

In another epidemiologic study of 754 Monsanto employees exposed to high levels of TCDD in a 1949 accident, of whom 122 developed chloracne, there was no increased incidence of cancer in those who developed chloracne, although they were presumably the group with the highest exposure. Conversely, workers who were also potentially exposed to 4-aminobiphenyl, a potent bladder carcinogen, had increased mortality from bladder cancer, lung cancer, and soft tissue sarcoma. This suggests that TCDD might act as a co-carcinogen or promoter. Again, the effects of confounders such as smoking and exposure to other chemicals could not be ruled out, but recent experimental evidence supports the suspicion that TCDD could act in this way.

Walsh et al. studied the cell toxicity of aflatoxins in cultured human epidermal cells. AFB_1 was markedly toxic at 1 µg/mL. Neither AFB_2 nor AFB_1 dihydrodiol were toxic. TCDD alone was not toxic to the cells but at 5 nM, it dramatically stimulated AFB_1 toxicity at levels as low as 0.1 µg/mL. It also increased the formation of AFB_1 epoxides and a 20-fold increase in DNA adduct formation was observed. AFB_1 is the most carcinogenic of the aflatoxins (see Chapter 10).

The most recent report from Seveso is suggestive of a carcinogenic effect in humans. The exposed population was divided into three groups according to their likely level of exposure. Persons who had left the area were traced with a 99% success rate. They were followed from 1977 to 1986 and cancer incidences were compared to a reference population not exposed to high TCDD levels. The population exposed to the highest levels was small, but nearly 5000 were in the middle exposure area. In these, the relative risk factors compared to the reference population were elevated for several cancers: 2.8 for hepatobiliary cancer, 5.7 for lymphoreticulosarcoma in men, 5.3 for multiple myeloma, and 3.7 for myeloid leukemia in women. In the lowest exposure zone, the incidence of non-Hodgkin's lymphomas and soft tissue tumors was elevated, especially among those who had lived in the area for over 5 years. Paradoxically, the incidences of breast cancer and endometrial cancer were below expected levels.

Weaknesses in this study include the inability to control for confounding factors such as smoking and the lack of hard data regarding real exposure levels. Despite the conflicting results of several epidemiological studies, most authorities now agree that there is a high index of suspicion for TCDD carcinogenicity in humans, especially for non-Hodgkin's lymphoma. The definitive word, however, remains to be heard, and the existing evidence comes from high industrial and occupational exposures that may not be relevant to environmental exposures encountered by the general population if a threshold truly exists because of the TCDD receptor story (see next section and Figure 23).

The mechanism of carcinogenesis in animals appears to be epigenetic. Recent evidence indicates that TCDD binds to a specific receptor, the Ah (for aryl or aromatic hydrocarbon) receptor, and that a minimum number of Ah receptors must be occupied for TCDD to exert its effect (see below). The implication of this is that there is a threshold dose and that the linear multistage carcinogenesis model is inappropriate for TCDD and for any other agent, such as PCDDs and PCDFs, that works by this mechanism (see Chapter 3 and below). The EPA is now reconsidering the use of the linear multistage model, at least for TCDD and perhaps for a few other agents.

Neurotoxicity

Many toxic effects have been observed, including impaired vision, hearing, and smell, depression, sleep disturbances, and others. These were observed in workers exposed to TCDD in the 1949 Monsanto accident.

Reproductive toxicity

Testicular atrophy, necrosis, and decreased spermatogenesis have been seen in laboratory animals.

Metabolic disturbances

Weight loss and depletion of adipose tissue occur in lab animals.

The role of the aryl hydrocarbon receptor (AhR) and enzyme induction

TCDD is one of the most potent inducers of aryl hydrocarbon hydroxylase (AHH) yet discovered. In some species it is effective at 1 μg/kg. It induces synthesis of hepatic microsomal cytochrome P450 (CYP1A1 and possibly 1B1). TCDD uptake into the cell is passive (i.e., concentration gradient dependent). Intracellularly, it binds to the aromatic hydrocarbon cytosolic receptor (AhR), which is the product of a regulatory gene. The unliganded aryl hydrocarbon receptor complex (AhRC) contains the aryl hydrocarbon receptor (AhR). Binding of an aromatic hydrocarbon ligand causes release of the AhR that is translocated to the nucleus by a translocator protein. The unliganded AhRC is heteomeric; but after binding to a halogenated aromatic hydrocarbon (HAH) or a polycyclic aromatic hydrocarbon (PAH), the AhR is released and associates with the AhR nuclear translocator protein to form heterdimeric transformed AhRC. Ligands for the AhR are typically hydrophobic aromatic compounds, including the HAHs dibenzo-*p*-dioxins, dibenzofurans, biphenyls, and PAHs such as benzo[a]pyrene and many other benzo derivatives (products of combustion), naphthalene, naphthacene, and many others (see http://chrom.tutms.tut.ac.jp/JINNO/DATA-BASE/00alphabet.html).

In the nucleus, AhRC activates numerous xenobiotic-responsive structural genes. The information is transcribed to m-RNA and translated to protein synthesis and the production of cytochrome P450 ($1A_1$, $1A_2$, $1B_1$). This is known as a pleiotropic response; that is, it results in more than one phenotypic effect. TCDD is the most potent known inducer of AHH. The consequence of this induction is that several drug (xenobiotic-)metabolizing enzymes are induced, and reactive metabolites may be formed that react with proteins and nucleic acids to cause mutations, teratogenesis, and carcinogenesis, as well as altered drug metabolism. Conversely, the reactive metabolites may be excreted or detoxified, as by conjugation with glucuronide. A schematic representation of this pathway is shown in Figure 23.

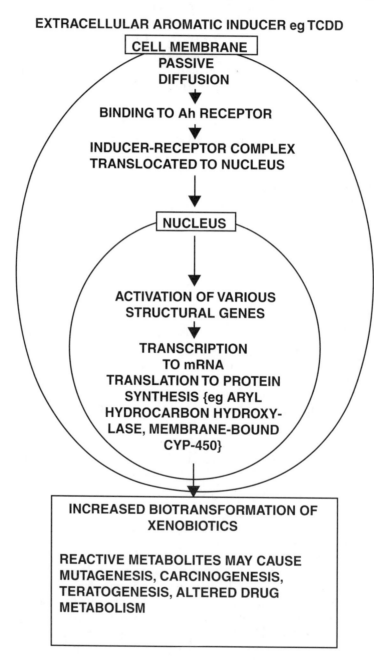

Figure 23 The Ah receptor and enzyme induction.

This system has been most studied in the mouse, where it is inherited as an autosomal dominant pattern. Similar systems have been identified in the rat, the rabbit, and some fish. It is most heavily concentrated in the liver. In humans, there is considerable variation in the Ah locus. To date, no endogenous substrate has been identified for the Ah receptor. AhR "knockout" mice, however, have small, fibrosed livers. In the absence of an exogenous ligand, AhR has been found in the nucleus of cultured Hela cells, behaving as if they were ligand bound. It seems apparent that the AhR has an important physiological role. It is known that primitive species (bacteria, yeasts) utilize polycyclic hydrocarbons as an energy source and possess P450 metabolizing enzymes for them (camphor in *Pseudomonas*, benzo[a]pyrene in yeast), so these may have evolved as a detoxication system. The mechanism of TCDD toxicity is not known, but if there is a natural substrate for Ah receptors, its displacement by TCDD could be involved in the latter's toxicity. It is not clear whether TCDD itself or its metabolite(s) are responsible.

TCDD also has non-receptor-mediated effects, including interference with calcium homeostasis and a variety of membrane-related changes. It is interesting that the chloracne associated with TCDD also occurs with bromides, which could not act as Ah ligands. For more on TCDD toxicity, see Chapter 12.

Paraquat toxicity

This PAH herbicide (1,1'-dimethyl-4,4'-bipyridinium ion) is highly water soluble and therefore poorly absorbed across the skin or gastrointestinal mucosa. It is extremely toxic to humans when inhaled, however, and 5 g may be fatal. Pulmonary congestion, edema, and hemorrhage may result in almost complete functional destruction of the lung. In severe cases, lung transplantation has been tried as a last resort, with disappointing results. Although paraquat is poorly absorbed from the oral route, its highly toxic nature can result in lung toxicity days or weeks later. It is thus one of the "hit-and-run" class of toxicants. Liver and kidney damage and neurological damage also occur (see Chapter 9 on pesticides).

Insecticides

Chemical insecticides are used to increase food crop production, protect livestock and household pets against insect pests (warble fly, bot fly, screwworm, fleas), control disease-carrying insects (anopheles mosquitoes that carry malaria), and control destructive insects such as termites. It was the chlorinated hydrocarbon insecticide DDT (dichlorodiphenyltrichloroethane) that first raised concerns over the impact of pesticides on the environment. In 1961, the author Rachael Carson brought out her book *Silent Spring*, in which she documented the devastating effect that this chemical had on bird life because it weakened eggshells so that the eggs collapsed in the nest or the chicks were abnormal at hatching. The persistence of the chemical (it and its metabolite DDE have a biological $t_{1/2}$ of 50 years) and its high lipid solubility result in its biomagnification up the food chain.

Predatory and fish-eating birds are especially vulnerable to DDT. The product was banned in the United States and Canada in 1972, but it is still used elsewhere in the world, including Mexico, and trace levels may be present in imported products. Levels of from 1 ppb to 1 ppm may be present in fish, oysters, and other seafoods and can contribute to human tissue levels of up to 10 ppm.

Halogenated hydrocarbon insecticides (chlorinated hydrocarbons) are principally neurotoxic, interfering with axonal transmission by altering sodium and potassium transport across the axonal membrane to prevent normal repolarization. Evidence of carcinogenicity also has been obtained in animal experiments (see Chapter 9 on pesticides for more details).

Industrial and commercial chemicals

Biphenyls

Polybrominated biphenyls (PBBs) are used as fire retardants in thermoplastics for TV and office machine casings. The more familiar polychlorinated biphenyls (PCBs) are highly stable and resistant to degradation in acids or bases, by oxidation or heat (to 800°C).

These characteristics make PCBs ideal insulators in electrical transformers, as hydraulic fluid, and in brake linings. They are also used as plasticizers in polymer films. The same characteristics, however, make these agents very persistent in the environment. Exposure to these compounds is largely an occupational hazard, but exposure in the environment can occur as a result of contamination of groundwater from spills, improper storage of waste PCBs from old transformers and capacitors, or fires in storage sites. Although the manufacture of PCBs was banned in the United States in 1977, they remain a problem because of their persistence and resistance to destruction. Forty percent of North Americans have body fat levels of 1 ppm or higher (these agents have high lipid solubility). Prior to 1970, 500,000 tons of PCBs were produced in North America.

Toxicity

1. *Animal.* The LD_{50} in rats may be 1 to 10 g/kg. Chronic toxicity involves skin lesions, hepatotoxicity, immunosuppression, and reproductive dysfunctions. Carcinogenicity has also been reported. Recently, genetic damage in cetaceans (whales, dolphins) and seals in the Baltic Sea has been ascribed to high levels of PCBs.
2. *Human.* Characteristic chloracne, impaired immune response, liver damage, gastrointestinal disturbances (nausea, vomiting, loss of appetite), and CNS disturbances (weakness, ataxia) as well as reproductive problems and cancer have been associated with exposures.

Pharmacokinetics and metabolism

Because of the high lipid solubility, these compounds are well absorbed from the gastrointestinal tract (>90%) and stored in body fat. They are secreted in milk (toxicity has been shown in nursing mice and rats), and they constitute a hazard for the nursing infant (see study of Michigan mothers in Chapter 3). They have been shown to induce cytochrome P448 through the Ah receptor pathway, to form reactive intermediates, and to deplete glutathione. Conjugation with glucuronide and renal excretion are the final detoxification mechanisms.

Biodegradation

Recent studies indicate that solar photolysis (near-UV light) can reduce the $t_{1/2}$ in surface water to 1 to 2 years by dechlorination of PCBs, but this has no effect on bottom sediments that may contain high levels. Fortunately, chemical dechlorination can occur here, and recently, subsequent oxidation by anaerobic bacteria has been discovered in sediments of the Hudson River. These discoveries offer some hope that biodegradation of PCBs may occur at a faster rate than anticipated.

Accidental human exposures

In 1968, nearly 1700 people in the Fukuoka region of Japan developed chloracne as a result of using rice cooking oil contaminated with PCBs. The PCBs in turn were contaminated with tetrachlorodibenzofuran, which is structurally and toxicologically similar to TCDD. These "Yusho" patients (*Yusho* means "oil disease") constitute the largest human population known to be exposed to toxic levels of PCBs, and their health continues to be monitored closely to identify delayed effects. Five years later, 22 deaths were reported in 1200 of these patients, 9 from cancer, 2 involving the liver. Calculated body burdens were 5.9 µg/kg for chloracne and 4.4 µg/kg for nausea and anorexia. These are 200 times higher than average current levels found in North American populations. Similar mortality statistics were reported after an outbreak in Taiwan. In neither case were the cancer deaths considered excessive, and they were not age-adjusted (see also Chapter 3). Several cases of contamination of animal feeds have occurred. In North Carolina, a leaky heat exchanger contaminated 16,000 tons chicken feed, only 10% of which was recovered. No human health problems were directly attributed to the accident, but there was some evidence that children might have been affected *in utero* (see also Chapter 3).

A major human exposure to PBBs occurred in Michigan in 1973 (the year of Seveso). A PBB product intended for use as a fire retardant and marketed as "Firemaster" was accidentally bagged as "Nutrimaster," a magnesium supplement for dairy cattle. Entire herds of dairy cattle were afflicted with loss of appetite, open sores, weight loss, loss of milk production, sterility, and stillbirths. It took several months of detective work to trace the source of the problem; meanwhile, human food supplies were being contaminated by milk containing the PBBs. Meat and eggs were also contaminated. By 1976, nearly 30,000 of Michigan's best dairy cattle had died or been destroyed, along with thousands of sheep and hogs and millions of chickens. Thousands of Michiganders had consumed unknown quantities of contaminated food, and hundreds began to complain of headache, fatigue, joint pain, and numbness in fingers and toes. Farm families were the most severely afflicted. Little was then known about the human toxicity of PBBs and long-term effects have not yet been identified, but could yet emerge (see also Chapter 3).

The problem of disposal

Recent evidence from a disposal site in Great Britain suggested that current incineration temperatures might not be adequate because levels of PCBs were detected in the soil around the site. A complicating factor was the presence of a municipal incinerator within 100 meters, but some authorities feel that incineration temperatures should reach 2700°C. The PCB fire (deliberately set) at St. Basile le Grand in 1988 highlights the dangers of unsecured storage. Public concerns over PCB hazards has led to the NIMBY response (Not In My Back Yard) and resulted in resistance to the establishment of proper disposal sites and to refusal to allow ships containing toxic wastes to unload. Such ships, wandering the seas in search of a berth, create a potential for a marine environmental disaster, with contamination of food fish, that could be far worse than the hazards associated with well-run disposal sites.

Solvents

Carbon tetrachloride, chloroform, and methylene chloride (dichloromethane) are still popular industrial solvents because they are not flammable. Exposure is mainly an industrial problem but these agents may still appear as cleaning fluids in the home. A source of some concern is the presence of trace amounts of chloroform in drinking water as a result of the chlorination process.

Toxicity

These chemicals are hepatotoxic, causing central lobular necrosis with fatty degeneration of adjacent areas. They also can cause renal damage, and chronic exposure has been linked to neoplasms of the lung and liver. Cardiac arrhythmias have been reported, as has nausea and vomiting. The cardiac arrhythmias result from sensitization of the heart to catecholamines such as adrenaline and noradrenaline. Other halogenated anesthetics such as cyclopropane and halothane will also do this, and hepatotoxicity has been reported as a toxic effect of halothane in some patients as well as in anesthetists routinely exposed to the drug. Liver toxicity can be increased in mice exposed to inducers of microsomal enzymes, suggesting that a toxic metabolite might be involved.

Mechanism of toxicity

These substances are metabolized by cytochrome P450 enzymes. Chloroform is converted to chloromethanol, phosgene, and CO_2 + HCl. Phosgene is normally converted to glutathione for excretion; but if glutathione is depleted, covalent binding to proteins may lead to liver and kidney necrosis. The metabolic pathway for chloroform is shown in Figure 24. Methylene chloride is metabolized to CO_2 + CO. Carbon monoxide poisoning through the formation of carboxyhemoglobin can occur. Chloroform is an example of a trihalomethane.

Figure 24 Biotransformation of chloroform.

Trihalomethanes (THMs)

These are halogen-substituted, single-carbon compounds having the general formula CHX_3, where X may be chlorine, fluorine, bromine, iodine, or a combination of these. They are formed during the process of water chlorination from naturally occurring organic compounds. The most common agents found in drinking water are chloroform ($CHCl_3$), bromodichloromethane ($CHBrCl2$), chlorodibromomethane ($CHClBr_2$), and bromoform ($CHBr_3$).

These are liquid at room temperature, rather volatile, and only slightly soluble in water. Their octanol-water partition coefficients range from 1.97 for chloroform to 2.38 for bromoform. All are sensitive to decomposition in air and sunlight. THMs are also released into the environment from industrial sources. Chloroform is the most common THM found in drinking water. Its major toxicity and metabolic transformation are noted above.

As a group, THMs are rapidly and well absorbed from the gastrointestinal tract, metabolized through dihalocarbonyl compounds via the cytochrome P450-dependent, mixed-function oxidases to CO_2 and CO, and eliminated through the lungs. Because of their high lipid solubility, they accumulate in adipose tissue > brain > kidney > blood.

Several epidemiological studies have shown a correlation between chlorination of surface or groundwater and the incidence of many cancers, but correlations with actual measured levels of THMs have been more difficult to demonstrate. Exceptions are pancreatic cancer in white males, rectal cancer in males only, and stomach cancer in both sexes. When population migration patterns were considered, however, the correlation with stomach and rectal cancer could not be demonstrated, and other studies have suggested that other water quality parameters may be involved.

In animal studies, chloroform has been shown to be carcinogenic in rats and mice. Chlordibromomethane has been reported to be hepatotoxic in mice, but no evidence of carcinogenicity was obtained. These agents are probably mutagenic and teratogenic, as indicated in some studies.

In light of these facts, there are efforts directed toward limiting the levels of THMs in drinking water. Standards vary widely throughout the world. In Canada, the current maximum standard is 350 µg/L, not to be exceeded. In the United States, the EPA has set a limit of 100 µg/L; this is an average based on quarterly samples and is therefore more enforceable. The WHO sets a guideline of 30 µg/L, but with a warning that disinfecting efficiency should not be compromised in the pursuit of lower levels. The European Economic Community passed a directive that haloform levels in drinking water should be "as low as possible," which is unenforceable.

Thus, it is evident that the problem of THMs in drinking water is another example of how cost-benefit analysis must be performed to weigh the potential risks of cancer from the chemicals against the known risks of epidemic infections if water supplies are not treated.

Further reading

Axelson, O., Editorial: Seveso: disentangling the dioxin enigma? *Epidemiology,* 4, 389–392, 1993.

Bertazzi, P.A., Pesatori, A.C., Consonni, D. et al., Cancer incidence in a population accidentally exposed to 2,3,7,8,-tetrachlorodibenzo-*para*-dioxin, *Epidemiology,* 4, 398–406, 1993.

Bertazzi, P.A., Zochetti, C., Pesatori, A.C. et al., Ten-year mortality study of the population involved in the Seveso incident in 1976, *Am. J. Epidemiol.,* 129, 1187–1200, 1989.

Collins, J.J., Acquavella, J.F., and Friedlander, B.R., Reconciling old and new findings on dioxin, *Epidemiology,* 3, 65–69, 1992.

Collins, J.J., Strauss, M.E., Levinskas, G.J., and Conner, P.R., The mortality experience of workers exposed to 2,3,7,8,-tetrachloro-*p*-dioxin in a trichlorophenol process accident, *Epidemiology,* 4, 7–13, 1993.

Denison, M.S. and Helferich, W.G., Eds.,*Toxicant-Receptor Interactions,* Taylor & Francis, Philadelphia, 1998.

Environmental Health Directorate, Health and Welfare Canada, Consultation Package on Trihalomethanes, 1989.

Fingerhut, M.A., Halperin, W.E., Marlow, B.S. et al., Cancer mortality in workers exposed to 2,3,7,8-tetrachlorodibenzo-p-dioxin, *New Engl. J. Med.,* 324, 212–218, 1991.

Fingerhut, M.A., Steenland, K., Sweeney, M.H. et al., Old and new reflections on dioxin, *Epidemiology*, 3, 69–72, 1992.

Gough, M., Agent Orange studies, *Science* (letter), 245, 1031, 1989.

Hankinson, O., The aryl hardrocarbon receptor complex, *Annu. Rev. Pharmacol. Toxicol.*, 35, 307–340, 1995.

Johnson, E.F., A partnership between the dioxin receptor and a basic helix-loop-helix-protein, *Science*, 252, 924–925, 1991.

Landers, J.P. and Bunce, N.J., The Ah receptor and the mechanism of dioxin toxicity, *Biochem. J.*, 276, 273–278, 1991.

Roberts, L., EPA moves to reassess the risk of dioxin, *Science*, 252, 911, 1991.

Roberts, L., Dioxin risks revisited, *Science*, 251, 624–626, 1991.

Ryan, J.J., Gasiewicz, T.A., and Brown, J.F., Human body burden of polychlorinated dibenzofurans associated with toxicity based on the Yusho and Yucheng incidents, *Fund. Appl. Toxicol.*, 14, 722–731, 1990.

Walker, C.H., Hopkin, S.P., Silby, R.M., and Peakall, D.B., *Principles of Ecotoxicology*, Taylor & Francis Ltd., London, 1996.

Walsh, A.A., Hsieh, P.H., and Rice, R.H., Aflatoxin toxicity in cultured human epidermal cells: stimulation by 2,3,7,8,-tetrachlorodibenzo-*p*-dioxin, *Carcinogenesis*, 13, 2029–2033, 1992.

Review questions

For Questions 1 to 6, use the following code:

    Answer A if statements a, b, and c are correct.
    Answer B if statements a and c are correct.
    Answer C if statements b and d are correct.
    Answer D if statement d only is correct.
    Answer E if all statements (1,2,3,4,) are correct.

1. Halogenated hydrocarbons are characterized by:
   a. High lipid solubility
   b. Susceptibility to chemical breakdown
   c. Toxicity for microorganisms
   d. Decomposition at temperatures greater than 200°C
2. Dioxin (TCDD) toxicity is characterized by:
   a. Chloracne
   b. Hepatotoxicity
   c. Porphyria
   d. None of the above (a, b, c)
3. Dioxin (TCDD):
   a. Induces the enzyme gamma-aminolevulinic acid (ALA) synthetase
   b. Interferes with feedback inhibition of ALA synthetase
   c. Inhibits porphyrin synthesis
   d. Induces arylhydrocarbon hydroxylase

4. With regard to chloroform:
   a. Phosgene is a major metabolite.
   b. Phosgene is detoxified by conjugation with glucuronide.
   c. Liver necrosis can occur if phosgene escapes the detoxification process.
   d. Phosgene is the only toxic metabolite of chloroform.
5. With regard to the detoxification of PCBs while in the environment (biodegradation):
   a. Sunlight may break down PCBs in surface water.
   b. No breakdown of PCBs occurs in bottom sediments.
   c. Bacteria may detoxify PBCs by oxidation.
   d. Chemical dechlorination does not reduce the toxicity of PCBs.
6. With regard to the aromatic hydrocarbon (Ah) receptor:
   a. TCDD attaches to it.
   b. Occupation of a certain minimum number of receptors is necessary for TCDD to be carcinogenic.
   c. The linear multistage assessment model for carcinogens may not be appropriate for TCDD.
   d. No other chemical is known to attach to the Ah receptor.

For questions 7 to 11, match the chemical listed below to the appropriate use.

   a. Hexachlorophene
   b. Polybrominated biphenyls (PBBs)
   c. Polychlorinated biphenyls (PCBs)
   d. 2,4-dichlorophenoxyacetic acid (2,4-D)
   e. Dichlorodiphenyltrichloroethane (DDT)

7. Insecticide
8. Disinfectant
9. Transformer insulator
10. Fire retardant
11. Herbicide

For questions 12 to 15, answer True or False.

12. Dioxin (TCDD) can cause behavioral abnormalities.
13. Porphyrins are byproducts of heme synthesis.
14. Victims of TCDD poisoning at Seveso showed no evidence of hepatotoxicity.
15. Pentachlorophenol is used as a wood preservative. From its name, one would predict that it would cause chloracne if accidentally consumed.

Answers

1. B
2. A
3. E
4. A
5. B
6. A
7. e
8. a
9. c
10. b
11. d
12. True
13. True
14. False
15. True

Case study 9

Part 1. Three members of a family became dizzy and nauseated within an hour of eating snacks (taquitos) consisting of tortillas wrapped around a meat filling. Two of them subsequently had grand mal epileptic seizures. The snacks were commercially prepared and sold in sealed bags of 48. They had been purchased a few days earlier. In an unrelated case a few weeks later, a 17-year-old male had four closely spaced seizures 30 minutes after consuming taquitos from the same manufacturer and purchased from the same store. The boy was on long-term antiepileptic therapy because he had been diagnosed as an epileptic the previous year. Following the initial episode, the manufacturer had voluntarily removed from shop shelves and destroyed all existing packages of the product.

Q. What organ system seems to be the primary site of toxicity?

Q. Was the most likely cause of the reaction bacterial or chemical?

Q. Was the likely site of contamination the factory or the retail store?

Part 2. Analysis of some remaining taquitos from the first case revealed traces of endrin. No source or trace of endrin was found at the factory.

Q. To what class of compound does endrin belong?

Review the toxicity of this class of chemicals.

A statewide press release turned up several other cases of seizures, including five people who had experienced seizures within 12 hours of consuming taquitos purchased from the same store.

Q. What preventive or remedial measures might you recommend?

Case study 10

A maintenance employee in a factory died after acute exposure to solvent fumes. He had been using a mixture of chlorinated solvents to remove grease from machinery. The principal component was trichloroethane (methyl chloroform).

Q. What was the immediate cause of death?

Q. What steps might you suggest to prevent this type accident?

Q. This substance is similar to carbon tetrachloride. What would have been the nature of the toxic response if the exposure had been chronic rather than acute?

chapter six

Toxicity of metals

"Mad as a Hatter"

Introduction

The process of felting, employed in making hats many years ago, required the use of mercurial compounds and many hatters suffered from the CNS disturbances (including behavioral disorders) associated with mercury toxicity. Metal intoxication as an occupational disease may be 4000 years old. Lead was produced as a by-product of silver mining as long ago as 2000 B.C. Hippocrates described abdominal colic in a man who worked as a metal smelter in 370 B.C., and arsenic and mercury were known to the ancients even if their toxicity was not. In 1810, a remarkable case of mass poisoning with mercury occurred. The 74-gun man-o'-war HMS Triumph salvaged 130 tons of mercury from a Spanish vessel wrecked while returning from South America, where the mercury had been mined. The mercury was contained in leather pouches, which became damp and rotten, allowing it to escape and vaporize. Within 3 weeks, 200 men were affected with signs of mercury poisoning, including profuse salivation, weakness, tremor, partial paralysis, ulcerations of the mouth, and diarrhea. Almost all animals onboard died, including mice, cats, a dog, and a canary. Five men died. When the vessel put in at Gibraltar for cleaning, all those working in the hold salivated profusely.

The common nineteenth-century practice of adulterating foods and beverages (wine, beer, etc.) to increase profit led Accum to publish a treatise on the subject in 1820. Lead, copper, and mercury were frequently detected. Methods were not yet in place to detect arsenic, which was found to be a widespread adulterant later in the century. In 1875, the British Parliament passed the first Food and Drugs Act as a result of these investigations.

In the past it was common to refer to heavy metal toxicity, as it was those metals that first emerged as industrial hazards. Heavy metals are arbitrarily defined as those having double-digit specific gravities and they include platinum (21.45), plutonium (19.84), tungsten (19.3), gold (18.88), mercury (13.55), lead (11.35), and molybdenum (10.22). These are in contrast to iron (7.87), manganese (7.21), chromium (7.18), zinc (7.13), selenium (4.78), and aluminum (2.70). Intermediate are copper (8.96) and cadmium (8.65).

In general, it can be seen that metals with specific gravities less than 8 are mostly essential trace nutritional elements (copper also is one and therefore the exception, as is aluminum, which is not a nutritional element), whereas those having specific gravities greater than 8 are the more toxic ones. It must be stressed once again that dose is all-important. Aluminum, with a specific gravity of 2.70, has toxic properties. Arsenic exists in two solid forms: yellow arsenic (1.97) and grey or metallic arsenic (5.73). Both are highly toxic.

Lead (Pb)

The Latin word for lead is *plumbum*, hence the chemical designation Pb. This word also gave origin to such English ones as plumb bob (a mason's line with a metal ball attached for establishing vertical trueness), plummet (to fall as if leaden), and aplomb (to be as calm and undeviating as a plumb line). Lead was obviously well known to the ancients. In fact, they spent a lot of time trying to turn it into gold (alchemy). Lead toxicity was also familiar to them. Diascorides described its CNS toxicity as delirium.

Despite early knowledge of lead's toxic effects, the low melting point of the metal, coupled with its density, made it popular and useful. Well into the 1940s and early 1950s, it was possible to buy lead toys, and kits were available to cast lead soldiers and lead fishing weights. An 1885 description of chronic lead poisoning is as good as any to be found in a modern text:

> The chief signs of chronic poisoning are those of general ill health; the digestion is disturbed, the appetite lessened, the bowels obstinately confined, the skin assumes a peculiar yellowish hue, and sometimes the sufferer is jaundiced. The gums show a black line from two to three lines in breadth, which microscopical examination and chemical tests alike show to be composed of sulfide of lead; occasionally the teeth turn black. The pulse is slow and all secretions are diminished. Pregnant women have a tendency to abort. There are also special symptoms, one of the most prominent of which is lead colic. This colic is paroxysmal and excruciating.

Modern-day sources of lead are numerous. In the eighteenth century, the industrial West discovered what the Chinese had known for centuries, namely that lead glazes produce crockery with a richer, smoother look. From this source and from lead solder in cans and kettles and water pipes leached by soft (but not hard) water, we consume about 150 µg/day. In some areas, the figure may reach 1 to 2 mg. Children are more vulnerable because all dirt and dust contain lead, especially in cities where lead from auto exhaust (tetraethyl lead) settles out on the ground. This will persist long after the conversion to lead-free auto fuel. Children may also consume old lead-based paint, common in older buildings and which may also be on cheap wooden toys. In children, CNS toxicity is the dominant feature. This starts with vertigo and irritability, progressing to delirium, vomiting, and convulsions. The mortality rate is about 25% if treated and about 65% if untreated. In infants, exposure produces progressive mental deterioration after 18 months, with loss of motor skills, retarded speech development, and hyperkinesis in some cases. In the United States, the Lead Paint Poison Prevention Program was introduced in 1970. Since that time, the mean blood lead level of U.S. children has fallen from over 1 µmol/L (20.7 µg/dL) to less than 0.25 µmol/L (5.2 µg/dL). Only two deaths in children from acute lead encephalopathy have been reported in the past 20 years.

Children are not the only victims of lead poisoning from lead paint. Sandblasting of old, lead-painted buildings may, over time, cause chronic lead poisoning in workers who inhale the dust. Proper respirators and protective clothing are required for sandblasters. Heating of lead paint to a sufficiently high temperature can release lead fumes that can be inhaled. Cutting torches can produce sufficient heat to do this.

In all exposed individuals, subchronic toxicity can involve interference with mitochondrial heme synthesis at several levels, with resultant hypochromic (pale) microcytic (small) anemia. The pathway involved in this is illustrated in Figure 25.

Toxicokinetics of lead

Elemental lead is not absorbed by the skin or through the alveoli of the lungs. Inhaled particulate lead is returned to the pharynx by the bronchial cilia and swallowed. Tetraethyl lead, however, may be absorbed across the skin and alveoli and readily penetrates CNS. Most of it is destroyed in exhaust emissions but sniffing leaded gasoline can result in severe CNS damage.

Gastrointestinal absorption of lead probably occurs via calcium channels as lead is a divalent cation (Pb^{2+}). It first appears in red blood cells, then hepatocytes, and then the epithelial cells of the renal tubules. It is gradually redistributed to hair, teeth, and bones where 95% of it is stored harmlessly. The $t_{1/2}$ in blood is about 30 days; in bone, 25 years. Little reaches the adult brain but much more enters the infant brain. Renal excretion is the main route of elimination.

Cellular toxicity of lead

Lead affects oxidative phosphorylation and ATP synthesis in the mitochondrion. It also increases red cell fragility and inhibits sodium/potassium ATPase. Kidney tubular cells become necrotic and chronic exposure may lead to interstitial nephritis. Nuclear inclusion bodies, consisting of lead bound to a protein, may be formed in renal cells. This may be considered

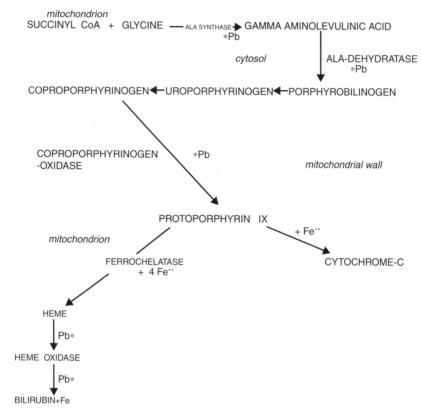

Figure 25 A simplified scheme showing points of interference of lead in heme synthesis. See also Figure 22 for ALA synthase and heme inhibition.

as a protective mechanism. Carcinogenesis has been demonstrated in experimental animals and chromosomal abnormalities have been observed, but evidence of tumor production in humans is scarce. Most of the toxic effects of lead and other heavy metals can be explained by their affinity for thiol groups. This is also the basis of chelation therapy.

Fetal toxicity

A characteristic of all metals is their ability to penetrate the placental barrier, so that fetal toxicity can occur as a result of maternal exposure. Lead (Pb) is considered to be a human carcinogen and pregnant women are generally removed from jobs where exposure may occur.

Prolonged exposure to low levels of Pb leads to impairment of the learning process. Current experimental evidence suggests that Pb is inhibitory to the NMDA receptor complex. Reduced availability of dopamine also could be involved and hypocholinergic function has been described.

Treatment

Pb chelators are the treatment of choice. These bind Pb (and other divalent cations) so that it can be excreted. Calcium/sodium ethylenediaminetetra-acetate (CaNa$_2$EDTA) and dimercaprol (British antilewisite, BAL) are given intramuscularly followed by oral penicillamine for several weeks. BAL was developed during World War II as a treatment for lewisite, a vesicant arsenical poison gas. A newer chelator is *meso*-2,3,-dimercaptosuccinic acid (DMSA). The chemical structures of these chelators are shown in Figure 26. In the case of EDTA, Pb is exchanged for Ca^{2+}, whereas with the others, the Pb is bound to sulfhydryl groups. The complexes are excreted, primarily in urine. A disadvantage of chelation therapy is that it does not remove Pb from the brain very efficiently.

   Despite 50 years of use, objective evidence for the benefit of chelation therapy for Pb poisoning is scanty. It is widely agreed that it has drastically reduced the mortality from Pb encephalopathy if diagnosis and treatment are started early. It also relieves Pb colic, malaise, basophilic stippling, and it rapidly restores red-cell ALA dehydratase. It does not influence the residual manifestations of chronic Pb poisoning such as peripheral neuropathy.

Mercury (Hg)

Mercury (Hg) exists in three forms: elemental mercury, inorganic compounds, and organic compounds. Elemental mercury causes toxicity when the mercury vapor is inhaled, as exemplified by the episode described at the beginning of this chapter. The major source of elemental mercury in the environment is the natural degassing of the Earth's crust. Estimates of the level of mercury reaching the atmosphere range from 25,000 to 150,000 tons/yr, and

Figure 26 Chemical structures of some metal chelators.

the atmosphere represents a major mechanism for global transport of metallic mercury. Conversely, anthropogenic sources account for only 10,000 tons/yr; but because industrial effluent tends to be concentrated, these are the sources usually associated with toxicity. Metallic mercury and its vapor can be an industrial hazard. Mercury is used in the manufacture of chlorine and sodium hydroxide by the mercury cell process, in paint preservatives, and in the electronics industry. It is a by-product of smelting processes (most mineral ores contain mercury), and it is released during fossil fuel combustion.

Elemental mercury toxicity

In vapor form, elemental mercury is well absorbed across both the alveoli of the lungs and the blood-brain barrier. Acute poisoning usually occurs within several hours. Weakness, chills, metallic taste, salivation, nausea, vomiting, diarrhea, labored breathing, cough, and tightness in the chest may ensue. If the exposure is more prolonged, interstitial pneumonitis may develop. Recovery is usually complete except that residual loss of pulmonary function may persist. Chronic exposure to mercury vapor results in CNS disturbances, including tremor and a variety of behavioral changes that can include depression, irritability, shyness, instability, confusion, and forgetfulness. Mercury vapor from mercury nitrate formerly used in the felting process accounted for the "mad hatter" syndrome. The behavioral abnormalities of the "Mad Hatter" in Lewis Carroll's *The Adventures of Alice in Wonderland* were really quite mild, compared with the other characters, which is in keeping with the topsy-turvy world that Carroll created. Thyroid disturbances may also be present.

Inorganic mercurial salts

Inorganic salts such as mercuric chloride can cause severe, acute toxicity. The proteins of mucous membranes are precipitated, giving them an ash-gray color in the mouth and pharynx. Intense abdominal pain and vomiting are common. Loss of blood and fluid from the gastrointestinal tract results from sloughing of the mucosa in the stool and may lead to hypovolemia and shock. Renal tubular necrosis occurs after acute exposure and glomerular damage is more common after chronic exposure. A phenomenon called "pink disease" or acrodynia commonly follows chronic exposure to mercury ions. It is a flushing of the skin that is believed to have an allergic basis.

Organic mercurials

Methylmercury is the most common cause of organic mercurial poisoning and the most important one environmentally. It is extremely well absorbed from the gastrointestinal tract (90%) and deposited in the brain. Because of its high affinity for SH groups, methylmercury binds to cysteine and this may then substitute for methionine and be incorporated into proteins. This can result in the formation of abnormal microtubles required for cell division and neuronal migration. The main signs and symptoms are neurological and consist of visual disturbances, weakness, incoordination, loss of sensation, loss of hearing, joint pain, mental deterioration, tremor, and in severe cases, paralysis and death. Infants exposed *in utero* may be deformed and retarded. Experimentally, methylmercury has been shown in cell cultures to mobilize Ca^{2+} from intracellular stores that are sensitive to inositol 1,4,5-trisphosphate.

Mercury is a waste product of many industrial processes. It is methylated in sediment by bacteria and cyanocobalamin. Several outbreaks of methylmercury poisoning have occurred. The most widely known began in 1953 in Minimata, Japan, near a plant that manufactured acetaldehyde and discharged mercury-containing compounds into Minimata Bay. People who ate mollusks and large fish from the bay developed the symptoms that came to be known as Minimata disease; 900 cases developed and there were 90 fatalities. Because of the high fetal toxicity of mercury, many deformed infants were born. Another source of mercury toxicity is the consumption of seed grains treated with methylmercuric chloride as a fungicide. Several mass poisonings have occurred around the world. In Iraq in 1972, one such episode resulted in over 6500 cases of poisoning and 500 deaths.

Mechanism of mercury toxicity

Mercury toxicity can be explained entirely by its ability to bind with the hydrogen of sulfhydryl (SH) groups to form mercaptides (i.e., X-Hg-SR and $HgSR_2$, where X = an electronegative radical and R = a protein). Organic mercurials such as methylmercury form mercaptides, R-Hg-SR′. The term *mercapto* means "to capture mercury" and refers to sulfur-containing groups. Because SH groups are important components of many enzymes, mercury acts as an enzyme poison and interferes with cell function at many levels. Mercury can also combine with other physiologically important ligands such as phosphoryl, carboxyl, amide, and amine groups. Metallic Hg vapor may be oxidized by catalase enzyme in red blood cells to the less toxic divalent form. Alcohol competitively inhibits this process. Mercury was an important pharmaceutical agent for centuries, and its pharmacological properties also depend on its affinity for SH groups. It was used as an antibacterial agent (for syphilis), as a laxative, in skin creams, and in diuretics. Mercurial diuretics were still in use in the 1960s. They were eventually replaced by safer agents. Aminomercuric chloride may still appear in freckle-removing creams, and daily application for years may result in increases in 24-hr urine mercury excretions from 10 µg to 1 mg and the development of symptoms such as excessive salivation and insomnia.

Treatment of mercury poisoning

Chelation therapy is recommended for elemental, inorganic mercury poisoning. Dimercaprol and penicillamine are SH-containing chelators. Dimercaprol is given intramuscularly and penicillamine, orally. Hemodialysis can also be used, and vomiting may be induced if there has been recent ingestion of mercury. These treatments are of little use in methylmercury poisoning, however. Dimercaprol actually increases brain levels of methylmercury, and penicillamine and hemodialysis do not relieve symptoms. Some success has been achieved with binding resins taken orally. Because there is a significant enterohepatic recirculation of methylmercury (i.e., it is excreted in the bile and reabsorbed from the intestinal tract), binding it to a polythiol resin prevents its reabsorption because it is excreted in the feces.

The Grassy Narrows story

In 1969, Norvald Fimreite, a Ph.D. candidate in the Department of Zoology at the University of Western Ontario, first made public his findings on the mercury contamination of fish in Canadian and border lakes. The highest levels were recorded from a small lake, Pinchi, in British Columbia (10 ppm) and from Lake St. Clair (7.03 ppm) in the Great Lakes waterway. The (Canadian) federal standard for export and consumption was 0.5 ppm. His report was a bombshell, coming on the heels of reports of Minamata disease from Japan. Fimreite estimated that Canadian industry was releasing 200,000 lb of mercury annually into the environment. Most of it came from chloralkali plants and from pulp and paper mills that used mercurials as antisliming (antialgal) agents and chlorine and alkali as bleaching agents. The question of mercury discharge from the Dow (Canada) Chemical plant had been raised 6 years earlier in the Ontario provincial legislature but nothing had been done.

In 1970, the Ontario Water Resources Commission took steps to reduce Dow's output; but in Dryden, near the Manitoba border, the Dryden Pulp and Paper Co. (owned by the British Reed Group) had been emitting mercury vapor since 1962, and some workers developed bleeding gums and muscle twitches. By 1970, it had pumped an estimated 20,000 lb of mercury into the surrounding environment, including discharges described as a brown froth into the Wabigoon River. Raw sewage was also discharged into the Wabigoon River, providing a rich source of anaerobic bacteria to methylate elemental mercury. The Wabigoon River is part of the English River system, and about 50 km downstream lie the Grassy Narrows and White Dog Indian reserves. The residents gleaned a slim but adequate living as fishing guides and lived largely off the land, eating fish, deer, and moose supplemented with garden vegetables. In March 1970, contamination of fish in Lake Erie was detected and the Lake St. Clair and Lake Erie fisheries were closed. Chloralkali plants and pulp mills were ordered to stop using mercury by the end of May after a concerted attack in the Ontario legislature by opposition parties. Mercury, however, is not biodegradable, and it is only when it is buried by uncontaminated sediment that it ceases to be a threat.

In June 1970, the Lamms, owners of Ball Lake Fishing Lodge, hired Fimreite to conduct a survey of mercury levels in the fish of the English-Wabigoon system. The findings were appalling. Levels ranged from 13 to 30 ppm, as high as those from Minamata Bay. The government lifted Fimreite's license to collect specimens for scientific purposes and ignored his appeals to test the residents of the reserves until his data were made public, when it conceded that it had similar findings. A ban was placed on eating fish from the contaminated area but otherwise the government continued to downplay the problem. Tourist fishing dried up and the Indians went on welfare. Blood levels of mercury were not seriously studied until 1973, and ranged from 45 to 289 ppb (normal is about 20 ppb for a city dweller). Some residents were showing signs of mercury poisoning and the incidence of stillbirths was rising.

The social costs of this tragedy were perhaps even greater than the direct effects of mercury. In the years surrounding the discovery of mercury in the Grassy Narrows area, the death rate rose to 1 in 50, three times the national average. Most were alcohol related. Many of the deaths were newborn or very young infants. Violence became rampant. Dr. Peter Newbury, also a graduate of the University of Western Ontario, conducted a study for the Society of Friends (Quaker) and the National Indian Brotherhood and felt that the CNS effects of mercury were a contributing factor in the violence. Gasoline sniffing became common among young people (it remains a problem on many reserves). The Grand Council of Treaty Three District, which includes Grassy Narrows and Kenora, completed a study in 1973. They found that in the preceding 42-month period, there had been 189 violent deaths of native people. They reported 38 from gunshot, stabbing, or hanging, 30 in fires, 42 drownings, 25 from exposure, and 16 from car accidents. In the same year, members of the Ojibwa Warrior Society occupied Anicinabe Park on the outskirts of Kenora. Barricades were erected and manned by armed warriors. The park was claimed as Indian land. The standoff lasted for several weeks but achieved little.

Cadmium (Cd)

Cadmium is present naturally in the environment in very low levels, being solubilized during the weathering of rock (levels are about 0.03 µg/g of soil, 0.07 µg/mL of fresh water, and 1 ng/m^3 of air). Dissolved cadmium may form a number of soluble and insoluble organic and inorganic compounds. Cadmium is chemically similar to zinc and it is present in zinc ore in a ratio of about 1/250. Most cadmium is produced as a by-product of electrolytic zinc plants. It is used in metal plating, in the manufacture of nickel-cadmium batteries, in the manufacture of pigments, in plastic stabilizers, and, in small amounts, in photographic chemicals, catalysts, and fungicides used on golf courses. Environmentally significant emissions come primarily from smelting operations for copper, lead, and zinc, from auto exhaust, and from the manufacture of pigments and alloys (most nickel-cadmium batteries are imported into Canada). Cadmium is readily taken up by plants and stored in the leaves and seeds. It is present in sewage sludge fertilizers (recommended maximum, 20 ppm). Water pollution with cadmium may result in high levels in fish and especially in mollusks. The main sources in the human diet are organ meats (cadmium accumulates in liver and kidney), cereal grains, shellfish, and crustaceans.

Cadmium toxicokinetics

Cadmium intake in Canada averages 50 to 100 µg/day from inhaled and ingested sources. Inhaled, unpolluted air may contribute up to 0.15 µg/day, whereas breathing air near a smelter can raise the level to 10 µg/day. Cigarettes contain cadmium and smoking increases exposure still further. About 50% of inhaled cadmium is absorbed. Only about 6% of ingested cadmium is absorbed, but it contributes most of the daily load. The FAO/WHO recommends a maximum weekly intake of 500 µg. Absorbed cadmium is bound to plasma albumin and cleared rapidly from the plasma. It is found in red cells only after high exposures. It is rapidly distributed to the liver, pancreas, prostate, and kidney, with slow redistribution to the kidney until, over time, it contains most of the cadmium. Renal levels increase up to age 50 and depend on the cumulative exposure. The $t_{1/2}$ in humans is about 20 years. Cadmium is trapped in the kidney and liver by a cysteine (i.e., SH)-rich protein called metallothionein with a high affinity for cadmium and zinc. Cadmium normally binds to matallothionein, the synthesis of which is induced by the presence of the cadmium. High doses, however, exceed the binding capacity of the protein, and the cadmium is free to bind to other essential cell components such as the basement membrane of the renal glomerulus.

Cadmium toxicity

The kidney is the major organ of toxicity. About 200 µg/g wet weight of kidney appears to be the critical concentration in the renal cortex for damage to occur in the form of proximal tubule dysfunction. Once renal disease develops, cadmium is lost from the kidney. Nutritional deficiencies of zinc, iron, and calcium may predispose cadmium toxicity by increasing absorption from the gastrointestinal tract. Calcium deficiency increases the synthesis of calcium-binding proteins and cadmium absorption. Workers in metal refineries may be exposed to high levels of cadmium fumes and develop respiratory difficulties. Chronic exposure may lead to obstructive pulmonary disease and emphysema. A major exposure occurred in Japan in the late 1940s. Effluent from a lead processing plant washed into adjacent rice paddies over decades, and the rice accumulated high levels of cadmium. Because the people were calcium-deficient due to a poor diet, they developed acute cadmium toxicity with severe muscle pain, malabsorption, anemia, and renal failure. The outbreak was named "Itai-Itai" (ouch-ouch) disease. The fetus appears to be protected from cadmium toxicity by placental synthesis of metallothionein, but heavy exposures can overwhelm this defense.

Animal studies have shown cadmium to be carcinogenic and there is a suggestion that it may increase the incidence of prostate cancer in elderly men. Other metals, notably arsenic, chromium, and lead, have also been implicated as carcinogens.

Treatment

Chelation therapy is not effective. Treatment consists of removing the patient from the source of exposure and supportive measures.

Arsenic (As)

Arsenic is an age-old pharmaceutical preparation. It was believed to be a tonic because it causes facial flushing (rosy cheeks) and fullness (edema). It is still used in the treatment of trypanosomiasis. Ehrlich studied organic arsenicals and developed the first effective treatment for syphilis (Ehrlich's 606). The chemistry of arsenic is exceedingly complex because it can exist in metallic form and as trivalent and pentavalent compounds. Trivalent forms include arsenic trioxide, arsenic trichloride, and sodium arsenite. Pentavalent forms include arsenic pentoxide, arsenic acid, lead arsenate, and calcium arsenate. Organic arsenicals also may be trivalent or pentavalent. Arsenic in the environment arises from weathering of rock; emissions from smelting of gold, silver, copper, zinc, and lead ores; combustion of fossil fuels; and the use of arsenicals in agriculture as herbicides and pesticides. Airborne particles may travel considerable distances and penetrate deeply into the lungs. Arsenic is taken up by plants and the degree of uptake varies with the soil type. Fine soils high in clay and organic material inhibit uptake. Arsenic also enters the water system through runoff and fallout. Wells drilled through rock containing arsenic will yield water high in arsenic. Chronic poisoning may result, and this is a problem in some parts of Nova Scotia. Tobacco contains arsenic. The average daily intake of arsenic in North America is about 25 μg.

Toxicokinetics of arsenicals

Arsenic may be absorbed from the gastrointestinal tract and the lungs, and across the skin and mucous membranes. It penetrates intracellularly by an uptake mechanism used in phosphate transport. Like mercury, arsenic binds to sulfhydryl and disulfide groups to poison numerous cell enzymes and respiration. Chromosomal breakage has been observed experimentally. In general, the order of toxicity is organic arsenicals > inorganic arsenicals > metallic arsenic. The trivalent arsenite has a high affinity for SH groups and interferes with the enzyme pyruvate dehydrogenase. Plasma pyruvate levels will increase. Some biotransformation between trivalent and pentavalent forms may occur. The $t_{1/2}$ of arsenic is about 10 hr and excretion is mainly by the kidneys. Arsenic is an effective uncoupler of oxidative phosphorylation.

Toxicity

As noted, the most poisonous forms are arsenic trioxide (As_2O_3) and sodium arsenite ($NaAsO_2$). Arsenic tends to accumulate in the liver, kidney, heart, and lung. It is also deposited in bone, teeth, hair, and nails, and these become important tissues for diagnostic and forensic analysis. The average human intake is about 300 µg/day, but it may be much higher if fish is a large part of the diet, as they accumulate the poison through biomagnification.

Acute arsenic poisoning causes severe abdominal pain and it is rare. Chronic poisoning results in muscle weakness and pain, skin pigmentation, gross edema, gastrointestinal disturbances, kidney and liver damage, and peripheral neuritis with eventual paralysis. The fingernails develop white lines, called Mee's lines, which can be used to determine when exposure occurred.

Arsine is the gaseous form of arsenic resulting from electrolytic processes and it is extremely toxic, producing rapid and often fatal hemolysis. It has a garlic-like odor.

Treatment

Chelation therapy is used for arsenic poisoning. Both dimercaprol and penicillamine have been used.

Environmental effects of arsenic

Arsenic is toxic to a wide range of plants and animals, including marine species. Of the plants, beans, peas, and rice are especially sensitive. Algae are sensitive, as well as some protozoa such as *Daphnia magna*. Finned fish are quite susceptible.

Chromium (Cr)

Chromium is used in the production of stainless steel, chrome plating, and pigments and in the chemical industry. Chromium has two oxidation states: Cr^{3+} and Cr^{6+}. The latter is much more toxic, causing severe respiratory irritation when inhaled and possibly lung cancer after long chronic exposure. Kidney damage also occurs. The trivalent form readily binds to electron-donating ligands such as macromolecules like RNA, but it does not readily cross the cell membrane. Conversely, the hexavalent form readily crosses cell membranes and is reduced to the trivalent form intracellularly. Toxicity is thus normally related to the presence of the hexavalent form in the environment. As the oxyanion, it is taken up by the cell, probably by a sulfate transport system as shown below. Air levels as low as 10 µg/m^3 of Cr^{6+} can produce respiratory irritation. Chromium is distributed in the biosphere much like arsenic and can have similar effects.

Other metals

Virtually any metal, if taken in excessive amounts or by an unusual route, can manifest toxicity. Thus, the inhalation of metal dusts, including aluminum, can cause pulmonary fibrosis. The kidney does not clear aluminum very efficiently and in renal insufficiency it can accumulate to toxic levels. Minute levels of aluminum are present in food and water, and if aluminum-free water is not used in dialysis machines, patients may accumulate high blood levels over time. This can cause microcytic anemia. This can be treated with the chelator desferoxamine, which removes the aluminum from the blood. Before the problem with aluminum and dialysis was identified, some patients developed very high blood levels and signs and symptoms of dementia similar to Alzheimer's disease.

Public concern has developed over a potential source of inhaled manganese, the fuel additive MMT. Methylcyclopentadienyl manganese tricarbonyl is an octane booster employed by the petroleum industry and manufactured by the Ethyl Corporation. Automobile manufacturers object to it because it defeats antipollution devices. Environmentalists object to it because they believe it constitutes a threat to human health. The petroleum industry claims that it would be too costly to eliminate its use. A WHO group conducted a study of workers in Quebec who had been exposed to manganese fumes and dust during the manufacture of metal alloys. The workers demonstrated problems with motor coordination and mentation. Manganism is a condition in which Parkinson-like signs and symptoms appear — tremor, memory loss, irritability, insomnia, difficulties with speech and, in severe cases, insanity. The workers were believed to be suffering from "micro-manganism." Attempts to ban MMT use in both Canada and the United States have foundered on the absence of definitive evidence that it constitutes a hazard and on concerted legal assaults by its defenders.

Uranium is nephrotoxic. It binds to albumin and to bicarbonate anion that is filtered by the kidney where it dissociates. The free uranyl cation binds to proteins in the proximal tubule and damages them. Toxic metals may also substitute for physiological ones, as when lead and strontium-90 are deposited in bones and teeth similarly to calcium.

Even essential metals such as iron can be very toxic, especially to young children who may ingest iron-containing vitamin preparations, mistaking them for candy. Vomiting occurs and vomitus and stool may contain blood. Acidosis and shock develop. Kidney and liver damage can occur. In adults, iron overload sometimes occurs and hemosiderin is deposited in tissues, including muscle. Desferroxime is a specific iron chelator with a low affinity for calcium that is used to remove systemic iron in both children and adults.

Antimony is an industrial contaminant with distribution and toxicity essentially similar to that of arsenic.

Metallothioneins

Metallothioneins are low-molecular-weight (6000–7000 daltons) proteins rich in sulfhydryl groups found in most mammalian cells. There are four classes of metallothioneins (MTs) based on their amino acid sequences. MT-I and MT-II are the most widely distributed, while MT-III and MT-IV are restricted to neuronal and squamous epithelial cells. MTs have a high affinity for a variey of metals including Ag^+, Cu^+, Cd^{2+}, Hg^{2+}, and Zn^{2+}. They are found primarily in the cytoplasm and their normal function is to serve as storage depots and buffers for copper and zinc. The MT-I gene can be induced by cadmium, copper, mercury, and zinc. Such induction makes them important defenses against heavy-metal poisoning, and experiments with knockout mice have shown that the absence of the MT-I and MT-II genes makes them more vulnerable to cadmium toxicity.

The Agency for Toxic Substances and Disease Registry maintains an excellent Web site at www.atsdr.cdc.gov/tfacts22.html.

Carcinogenicity of metals

Arsenic has long been recognized as being associated with an increased risk of skin and respiratory cancer. In 1930, workers in a factory making an arsenical sheep dip were identified as having an excessive incidence of skin cancer. Arsenic levels in air > 54.6 $\mu g/m^3$ were associated with an increase in lung cancer incidence. High water content of arsenic has also been associated with increased cancer risk.

In 1976, a NIOSH study of 300 workers in a cadmium smelter revealed a significantly higher incidence of cancer. The incidence of lung cancer was twice normal and prostate cancer also was high. Some of these workers, however, were also exposed to arsenic. Inhalation of chromium dust by workers has been associated with an increased incidence of lung cancer. Nickel is also a respiratory carcinogen but it lacks other chronic toxic effects. There is some evidence suggesting that lead may be carcinogenic, or perhaps a co-carcinogen, owing to its persistence in tissues. Case reports and epidemiological studies are difficult to sort out because exposures frequently involve more than one metal. This is true in the steel industry where increased cancer frequencies have been observed.

Many cationic metals will form complexes with thiol groups of cell components and the complex will mimic natural substrates to interfere with cell processes, mostly transport systems. Thus, the methylmercury-cysteine complex mimics methionine, and the complex is taken up into the brain by a transport system for neutral amino acids. Inorganic and organic mercury complexes with glutathione and is transported from liver cells into bile. Arsenic and copper do the same thing. Lead can substitute for Ca^{2+} in a number of transport and receptor-mediated processes. Voltage-activated calcium channels will admit a number of metallic cations, a fact that is exploited in research. Cadmium (Cd) and lanthanum (La) act in this way.

Unusual sources of heavy metal exposure

In 1988, the Texas Department of Health investigated illegal sales of drugs manufactured in Hong Kong. The tablets, sold as "chuifong tokuwan," contained a veritable pharmacy of drugs, including diazapam, indomethacin, hydrochlorothiazide, mefanemic acid, dexamethasone, lead, and cadmium! This potpourri of tranquilizer, diuretic, anti-inflammatory agents, corticosteroids, and heavy metals was repackaged and marketed as "The Miracle Herb — Mother Nature's Finest." Twenty-four percent of 93 persons who took this preparation had elevated urine cadmium levels (1.8 µg/mL compared to 0.5 µg/mL for random controls). The upper limit of normal is considered to be 2.5 µg/mL. No elevated (>25 µg/dL) blood lead levels were detected, but 42% of these individuals had elevated urine levels of retinol-binding protein, indicative of renal tubular damage.

Some of the "health" supplements such as bone meal (for calcium) contain high amounts of lead, and some zinc supplements are contaminated with cadmium. See also Chapter 8 for more on herbal remedies.

In Ohio, several members of a household were hospitalized with a diagnosis of acrodynia, a form of metallic mercury poisoning in which neurological and psychological disorders occur as well as hypertension, rash, sweating, cold intolerance, tremor, irritability, insomnia, anorexia, and diminished performance at school. Twenty-four-hour urine collections revealed mercury levels of 850 to 1500 µg/mL (normal <20 µg/mL). Careful inquiry of neighbors indicated that a previous tenant had spilled a large jar of elemental mercury in the apartment. Treatment was instituted with the oral chelating agent 2,3-dimercaptosuccinic acid (DMSA). Some neurological disorders persisted.

Further reading

Angle, C.R., Childhood lead poisoning and its treatment, *Annu. Rev. Pharmacol. Toxicol.*, 32, 409–434, 1993.

Ariza, M.E., Bijur, G.N., and Williams, M.V., *Environmental Metal Pollutants, Reactive Oxygen Intermediates and Genotoxicity*, Kluwer Academic, Boston, 1999.

Blyth, A.W., *Poisons: Their Effects and Detection*, Wm Wood & Co., New York, 1885.

Cadmium and lead exposure associated with pharmaceuticals imported from Asia, *Morbid. Mortal. Wk. Rep.*, 38, 612–614, 1989.

Cherian, M.G., Metallothionein and its interactions with metals, *Handbook of Exper. Pharmacol.*, 115, 121–137, 1995.

Clarkson, T.W., Molecular and ionic mimicry of toxic metals, *Annu. Rev. Pharmacol. Toxicol.*, 32, 545–571, 1993.

Corey-Slechta, D.A., Relationships between lead-induced learning impairments and changes in dopaminergic, cholinergic and glutaminergic neurotransmitter system functions, *Annu. Rev. Pharmacol. Toxicol.*, 35, 291–415, 1995.

Elemental mercury poisoning in a household, *Morbid. Mortal. Wk. Rep.*, 39, 424–425, 1990.

Fimreite, N., Mercury contamination in Canada and its effects on wildlife. Ph.D. thesis, University of Western Ontario Library, 1970.

Hare, M.F. and Atchison, W.D., Methylmercury mobilizes Ca++ from intracellular stores sensitive to inositol 1,4,5-trisphosphate, *J. Pharmacol. Exper. Ther.*, 272, 1017–1023, 1995.

Hutchison, G. and Wallace, D., *Grassy Narrows*, Van Nostrand Reinhold, Toronto, 1977.

Jaworski, J., Ed., Effects of Chromium, Alkali Halides, Arsenic, Asbestos, Mercury and Cadmium in the Canadian Environment. NRC Executive Reports, Publ. # NRC 17585, 1980.

Marquardt, H., Schafer, S.G., McClellan, R., and Welsch, F., *Toxicology*, Academic Press, New York, 1999.

Niesink, R.J.M., Hollinger, M.A., and de Vries, J., Eds., *Toxicology: Principles and Applications*, CRC Press, Boca Raton, FL, 1996.

Walker, C.H., Hopkin, S.P., Sibly, R.M., and Peakall, D.B., *Principles of Ecotoxicology*, Taylor & Francis Ltd., London, 1996.

Review questions

Answer the following questions True or False.

1. The term *heavy metal* usually refers to those with double-digit specific gravities.
2. Aluminum, which has a specific gravity of only 2.7, has no toxic properties.
3. All forms of arsenic are equally toxic.
4. Lead poisoning may be manifested as both gastrointestinal and central nervous system toxicity.
5. Acute, paroxysmal, colicky pain is common in severe lead poisoning in adults.
6. Mental retardation does not occur in lead poisoning in children.

For Questions 7 to 11, use the following code:

> Answer A if statements a, b, and c are correct.
> Answer B if statements a and c are correct.
> Answer C if statements b and d are correct.
> Answer D if statement d only is correct.
> Answer E if all statements (a, b, c, d) are correct.

7. a. Subchronic lead poisoning involves interference with mitochondrial heme synthesis at several levels.
   b. Anemia may accompany lead toxicity.
   c. Elemental lead is not absorbed through the skin or lungs.
   d. Elemental lead is not absorbed from the intestinal tract.

8. a. All metals can cross the placenta and cause fetal toxicity.
    b. Examining a blood sample is the only way of detecting lead in the body.
    c. Chelation treatment with dimercaprol can be used in lead poisoning.
    d. Chelation therapy is of no use in elemental mercury poisoning.
9. a. Elemental mercury poisoning never occurs.
    b. Methylmercury is the most important environmental source of mercury poisoning.
    c. Chelation therapy is useful in the treatment of methylmercury poisoning.
    d. Methylmercury poisoning may involve a variety of central nervous symptoms as well as signs of arthritis.
10. a. Cadmium accumulates in the liver and kidney.
    b. Cadmium induces the synthesis of the cadmium-binding protein metallothionein.
    c. The kidney is the major organ of toxicity for cadmium.
    d. Chronic inhalation of cadmium fumes may lead to pulmonary disease.
11. a. Like mercury, arsenic binds to sulfhydryl and disulfide groups.
    b. Trivalent forms of arsenic are the most toxic.
    c. Arsenic is readily absorbed from virtually all portals of entry.
    d. Arsenic does not accumulate up the food chain.

For Questions 12 to 16, select, from the following list, the correct mechanism for the stated metal (an answer may be used only once):

a. Chelates with organic ligands containing SH groups
b. Complexes with cysteine and competes with methionine in protein synthesis
c. Mimics calcium and is deposited in bone
d. Carried to the kidney by bicarbonate where it dissociates to cause renal damage
e. The hexavalent form enters cells and is converted to the trivalent form that binds to, and poisons, macromolecules.

12. Lead, arsenic
13. Lead only
14. Uranium (uranyl cation)
15. Methylmercury
16. Chromium

Answers

1. True
2. False
3. False
4. True
5. True
6. False
7. A
8. B
9. C
10. E
11. A
12. a
13. c
14. d
15. b
16. e

Case study 11

In early spring, two of five workers employed in demolishing an old iron bridge visited the company's consulting physician complaining of muscle pain (myalgia), joint pain (arthralgia), headache, and nausea. These workers had been cutting up sections of the bridge using oxyacetylene torches. Large sections were lowered to a barge moored in the river below the bridge to be cut into smaller sections for hauling away. When the remaining three workers were questioned, it was discovered that they too had been suffering from similar symptoms as well as memory loss and irritability. A supervisor and a secretary who worked in the construction shack on shore were not affected, nor were four men involved in loading trucks or operating a small boom crane.

Q. What organ systems are involved in the affected workers?

Q. Could this be a toxicant causing the illness? If so, what are its possible sources?

Q. What is the likely portal of entry?

Q. What diagnostic tests might be useful?

Q. What specific treatment might be appropriate?

Case study 12

A 67-year-old man consulted his physician because of severe abdominal pain, weight loss, and fatigue. The doctor initially suspected gastric carcinoma but the patient was severely anemic, his red cells had basophilic stippling, and he had a blood lead level of 70 µg/dL. Six other household members were also affected, including an 8-year-old child. All had elevated blood lead levels. The home was located in a suburban residential area not near any industrial site.

Q. What is the probable portal of entry of the lead in these people?

Q. A search of the home revealed a ceramic jug as the offending agent. Why was it suspected?

Case study 13

During a routine preemployment medical examination, a 46-year-old male was found to have a blood lead level of 50 µg/dL. He was subsequently investigated by a university hospital toxicology clinic that confirmed the same blood lead level 1 month later. Symptoms included numbness of the fingers and palms, tinnitus (ringing in the ears), an apparent decrease in ability to do mental arithmetic, and mild memory deficits. He had been taking ranitidine for indigestion. This is a histamine H2 receptor blocker that suppresses gastric acid secretion.

A detailed personal and employment history was obtained. He had spent 20 years as an electronics technician in the army and in civilian life but had had little exposure to lead from soldering or welding. He had no hobbies that could serve as a source of lead and no history of bullet or birdshot wounds, and he denied drinking bootleg alcohol or using lead additives in his car.

Q. Is the source of lead likely to be work related or from the home environment? How could this be determined?

Q. What is the probable significance of the gastric distress?

Q. What therapeutic approach would be appropriate?

chapter seven

Organic solvents and related chemicals

> " 'Tis a sordid profit that's accompanied by the destruction of health."
>
> **Treatise on the Diseases of Tradesmen**
> **B. Ramazzini, 1705**

Introduction

Organic solvents are common in the workplace where they may constitute an occupational hazard. They also occur in the home in the form of cleaning solutions, paint strippers, and brush cleaners and they can thus be a source of household poisoning as well as being an environmental hazard. Because they are all fat solvents, they may cause local defatting of tissue and irritation on contact. Many also are systemic toxicants, affecting the CNS like volatile anesthetics or, in some cases, the hematopoietic (blood-forming) system. Commercial solvents are frequently complex mixtures having nitrogen- and sulfur-containing elements (e.g., gasoline and other petroleum-based products).

Industrially, the uses of solvents are many and varied. They are used in extraction processes in the food and pharmaceutical industries (e.g., ethyl alcohol and acetone), for the removal of impurities, for degreasing and vapor cleaning (e.g., trichloroethylene and 1,1,1,-trichloroethane), as vehicles for paints, as carriers for pesticides, for printing inks (e.g., toluene and ethyl acetate), and in adhesives (e.g., hexane, toluene, and methylethyl ketone). In short, volatile solvents are used whenever a fast-drying property is desired in order to leave a coating on a surface. They are also used in the chemical industry for a variety of manufacturing processes.

Classes of solvents

Aliphatic hydrocarbons

Aliphatic hydrocarbons are straight-chain or branched carbon-hydrogen compounds. They are often present in complex mixtures in common commercial products such as gasoline, mineral turpentine, and kerosene. These mixtures can also contain smaller amounts of unsaturated and cyclic carbon-hydrogen substances. While the vapors of these agents are generally less toxic than those of most organic solvents, inhaling the fumes can not only cause disorientation and stupor through effects on the CNS, but also can sensitize the heart to adrenaline. The result may be ventricular fibrillation that can be fatal if not corrected quickly. Fatalities have occurred in workmen cleaning storage tanks and rail tank cars. Chemical pneumonia often occurs from aspiration of the low-viscosity liquid (an occupational hazard for those who siphon gasoline from others' cars!). One member of this solvent group of special note is *n*-hexane, the main ingredient of petroleum ether. With a boiling point of 60 to 80°C, it evaporates readily and produces an insidious form of poisoning. An outbreak of industrial poisoning occurred in Japan in 1973 when workers exposed to fumes from a glue used in making sandals lost sensation and function in fingers and toes from impaired nerve conduction. When the exposure was terminated, they all recovered slowly over several months. Hexane neuropathy has both a central and a peripheral component resulting from demyelination of nerve fibers. The actual toxicant is a metabolite of *n*-hexane, hexane-2,5-dione. An intermediate, 2-hexanone, is also more toxic than the parent substance. If the neurotoxicity of *n*-hexane is rated as 1, then 2-hexanone would be 10 and hexane-2,5-dione would be 40.

$$CH_3(CH_2)_4CH_3 \xrightarrow{\text{oxidation}} CH_3\text{-}\overset{\displaystyle O}{\overset{\|}{C}}\text{-}(CH_2)_3CH_3 \xrightarrow{\text{oxidation}} CH_3\text{-}\overset{\displaystyle O}{\overset{\|}{C}}\text{-}(CH_2)_3\text{-}\overset{\displaystyle O}{\overset{\|}{C}}\text{-}CH_3$$

n-HEXANE   2-HEXANONE   HEXANE-2,5-DIONE

Metabolites are produced in the liver by cytochrome P450 oxidases. They appear to condense with lysine in the myelin, causing disorganization of the membrane.

Halogenated aliphatic hydrocarbons

This group of substances containing chlorine substituents includes methylene dichloride, chloroform, carbon tetrachloride, and chlorinated ethylenes (e.g., trichloroethylene), as well as a number of chemicals that contain other halogens such as bromine and fluorine. They are used as anesthetics and refrigerants. Hepatotoxicity and nephrotoxicity characterize this group. The liver converts carbon tetrachloride (CCl_4) to the free radical $\cdot CCl_3$ that attacks the endoplasmic reticulum, causing protein synthesis to cease. Recovery will usually occur from a single toxic exposure, but repeated exposures lead to cirrhosis. Chloroform is converted to phosgene ($COCl_2$), which has similar toxicity to CCl_4. Phosgene was used as a poison gas in World War I. The chemical reactions involved in this process are illustrated in Chapter 4 (see Figure 24).

Trichloroethylene was used as a light anesthetic in childbirth. It and similar agents are used as industrial solvents. Fatalities have occurred when workers in enclosed tanks were overcome. Death usually results from aspiration of vomit. It is also hepatotoxic. Dichloromethane (methylene chloride) was used for many years as a paint stripper. It has a boiling point of 40°C and thus readily forms vapor. It is converted to carbon monoxide (CO), which forms carboxyhemoglobin in the red blood cells (see Chapter 5). Bis(chloromethyl)ether is discussed below under "Solvent-related cancer in the workplace."

Other halogenated hydrocarbons are discussed in Chapter 4, and the chlorofluorcarbons (CFCs) and their impact on the environment in Chapter 5. Some anesthetics are halogenated hydrocarbons. One such is halothane, chemically a haloalkane, a volatile liquid. Although most halothane is eliminated in the expired air, some undergoes hepatic biotransformation to a reactive metabolite, an alkylating radical, which causes lipid peroxidation and is hepatotoxic. It does not deplete glutathione as do some hepatotoxic agents. The reaction is mediated by cytochrome P450 and toxicity is increased by hypoxia. The incidence of this toxic reaction is low, perhaps 1 in 38,000 patients, and it appears to be genetically determined. Other halogen-containing general anesthetics are methoxyflurane, enflurane, and isoflurane. Methoxyflurane is rarely used in North America now because it releases free fluoride during biotransformation. Free fluoride is toxic to the kidney.

Aliphatic alcohols

Ethanol (ethyl alcohol) is present in alcoholic beverages and also in some lotions, perfumes, mouth washes, cough syrups, etc. Although there is a tendency to trivialize ethanol's toxicity, 3 to 6 mL/kg of pure ethanol can be fatal. Because 70 proof liquor contains 40% ethanol, this fatal adult dose equates with as little as 525 mL of liquor. In severe poisoning, the classical signs of inebriation (difficulty with balance, locomotion, and talking) progress to coma, metabolic acidosis, hypothermia, hypotension, and severe respiratory depression. Hypoglycemia may be present, especially in young children. Treatment is largely supportive, with correction of acidosis and hypoglycemia (5% glucose i.v.) and hemodialysis to remove ethanol and its metabolites. Other simple alcohols and glycols produce early effects resembling ethanol intoxication, which is why they are often abused by alcoholics who are at the bottom of the socioeconomic scale.

Methanol is by far the most toxic of the alcohols. In humans and other primates, it is oxidized to formaldehyde, a very reactive substance which the eye cannot convert to the harmless formate anion. As little as 4 mL has caused blindness through retinal damage. The transformation is illustrated below.

$$CH_3OH \xrightarrow{\text{alcohol dehydrogenase}} HCHO$$

Methanol is present in paint thinners and removers, windshield washer fluids, and fuels for small engines. It is also present in pineapple, where the levels may exceed the recommended limit of 2 ppm. Outbreaks of poisoning have occurred as a result of deliberate adulteration of wine. Home-distilled liquor has also caused poisonings, and "skid road" alcoholics may consume substances containing methanol as a substitute for ethanol. Initial symptoms are those of ethanol intoxication. A fairly long latency period may occur (6 to 30 hr), followed by acidosis; delirium; coma; visual disturbances, which may or may not progress to permanent blindness; cardiac disturbances; and death. Treatment consists of administering activated charcoal by mouth, ethanol (orally or i.v.) as a competitive substrate for alcohol dehydrogenase, and hemodialysis to remove methanol, formaldehyde, and formic acid (which causes acidosis). A fatal dose is 60 to 250 mL. An experimental approach to treatment involves the administration of inhibitors of alcohol dehydrogenase, such as 4-methylpyrazole.

Isopropyl alcohol is found in rubbing alcohols, after-shave lotions, and window-cleaning fluids and is a widely used industrial solvent. It is oxidized to acetone and causes acetonemia but not acidosis. Its toxicity lies between that of ethanol and methanol, being about twice as toxic as ethanol, with coma and respiratory arrest resulting from CNS depression.

Higher alcohols are generally less toxic. *N*-butanol vapors can produce eye irritation (conjunctivitis, keratitis) and inhalation may cause pulmonary edema.

Acetone (C_3H_6O), although a ketone rather than an alcohol, produces similar toxic symptoms to those of ethanol and the treatment is the same. Acetone, like the aliphatic hydrocarbons, dissolves lipids from the skin and can be extremely damaging to the cornea by virtue of defatting the epithelial cells.

Note: The hepatotoxicity of the halogenated hydrocarbons can be potentiated by numerous alcohols, including ethanol, methanol, and isopropyl alcohol, and acetone. The mechanism is unclear.

Glycols and glycol ethers

Ethylene glycol is present in antifreeze. An oral dose of 100 mL can be fatal for an adult. Early signs and symptoms resemble ethanol poisoning. After 24 hr, there may be pulmonary edema and myocardial depression; and after 48 hr, renal tubular necrosis and renal failure. Hypoglycemia and hypocalcemia may occur. Treatment is as for methanol.

The toxicity of ethylene glycol is due to its metabolites. It is metabolized by alcohol dehydrogenase, which is why ethanol is given orally or i.v., as it is for methanol poisoning. When hemodialysis is used in conjunction with i.v. ethanol, ethanol is often added to the dialyzer fluid to maintain blood ethanol levels. Otherwise, the ethanol would be dialyzed because of the concentration gradient. One metabolite of ethylene glycol is oxalate, which chelates calcium (hence the hypocalcemia; see also Chapter 11) and which precipitates as calcium oxalate crystals in the kidney, causing tubular necrosis. Other metabolites are aldehyde, glycolate, and lactate, which cause acidosis. A simplified scheme for the metabolism of ethylene glycol is shown in Figure 27.

Figure 27 The conversion of ethylene glycol to calcium oxalate.

Propylene glycol is fairly nontoxic and it is used in lotions and as a vehicle for injectable drugs. Too rapid injection may cause cardiac depression.

A subgroup of this class of compounds is the glycol ethers. Monomethyl and monoethyl ethylene glycol are used extensively in industry, being present in latex paints and as solvents in the manufacture of lacquers, varnishes, and dyes. There is some evidence that they are reproductive toxins, causing teratogenesis in experimental animals. They are not well-absorbed orally, but transdermal absorption and inhalation are the important portals of entry. Human toxicity has not been well established, but there have been reports of kidney damage and bone marrow depression. Precautions are taken to limit industrial exposure of women.

Aromatic hydrocarbons

Benzene is one of the simplest and most toxic of these cyclic, special hydrocarbons. It is highly volatile and exposure in the workplace is primarily by inhalation. Benzene is unique in this group because of its bone marrow toxicity. Chronic exposure causes a progressive reduction in all formed elements of the blood, including red cells, white cells, and platelets. Aplastic anemia, resulting from almost complete destruction of the marrow, may occur and the mortality rate is high, the only treatment being bone marrow transplantation. Bone marrow depression is dose-time dependent. Leukemia can also develop. Several reports have noted a higher than normal incidence of acute myelogenous leukemia in workers exposed to benzene. Although leukemia has not been demonstrated in animals exposed to benzene, solid tumors have been observed.

Chromosomal abnormalities have been seen in animals exposed to benzene. The toxic agent may be a metabolite of benzene because animal studies have shown that toluene, which competes with benzene in its metabolic pathway, reduces its toxicity. A number of metabolites have been identified, including phenol.

Alkylbenzenes are a group of related compounds consisting of the aromatic ring with one or more aliphatic side chains. As a group, they lack the serious toxic effects of benzene because they are detoxified to metabolites with low toxicity, including hippuric acids, which are readily excreted by the kidney. They are, however, potent CNS depressants, acting like general anesthetics. No mutagenic properties have been demonstrated. The group includes toluene (1-methylbenzene) and the xylenes (1,2-, 1,3-, and 1,4-dimethylbenzene), as well as ethylbenzene and cumene (isopropylbenzene).

Dinitrobenzene is a related substance that is used extensively in the manufacture of plastics, dyes, pigments, explosives, and insecticides, and in many other processes. The opportunity for an industrial exposure to toxic levels is thus fairly high. Meta, para, and ortho forms exist, all are crystals in the pure form, and all volatilize with steam and are soluble in boiling water and organic solvents. They are absorbed across the skin and promote the formation of methemoglobin (MetHb), in which the normal ferrous (Fe^{2+}) iron is converted (oxidized) to ferric (Fe^{3+}) iron. Methemoglobin binds irreversibly to O_2 and reduces the oxygen-carrying capacity of the red blood cells. If MetHb exceeds 1% of total hemoglobin, mild hypoxia results. If it exceeds 15%, a severe condition known as cyanosis-anemia syndrome results. Dinitrobenzene is the second most common industrial cause of methemoglobinemia. The chemical structures of some of the aromatic hydrocarbons are shown in Figure 28.

Solvent-related cancer in the workplace

Benzene

As noted above, benzene is capable of causing leukemia. The first case of "benzene leukemia" was observed in 1928 in a worker who was so heavily exposed that others could not work in the same environment without becoming acutely ill, which probably saved them from a similar fate. By 1980, about 200 cases of benzene-related leukemia were reported. In Italy, which banned it in 1963, workers in the shoemaking and rotogravure industries were estimated to have a risk of leukemia 20 times that of the general population. The latency period may be 15 years or more. In Japan, studies of the survivors of Hiroshima and Nagasaki revealed that the risk of leukemia was increased 2.5 times in those who had worked in jobs involving benzene exposure. In the U.S., benzene was used extensively in the rubber industry until it was

CH₃ figure placeholder

BENZENE   TOLUENE   XYLENE
                                                   (ortho form)

ETHYLBENZENE   CUMENE

Figure 28 Chemical structures of some aromatic hydrocarbon solvents.

banned. Overall, workers in this industry had a three-fold increase, and those with high exposures a five-fold increase, in their risk of leukemia.

Bis(chloromethyl) ether (BCME)

BCME is a potent alkylating agent and carcinogen used in a variety of industrial syntheses. An American study conducted by NIOSH (U.S. National Institute of Occupational Safety and Health) of 136 men exposed to high levels of BCME indicated a 10 times risk of lung cancer. Exposures averaged 5 years or more, and cancer rates were highest in those exposed for 10 years or more. Sputum cytology tests indicated a 34% incidence of abnormal lung cells in exposed worker vs. 11% in those not exposed. A much larger study involving 1800 workers in New York found a 2.5 times increase in risk, and in a Philadelphia chemical plant, heavily exposed workers had an 8 times increased risk. The discovery that fumes of HCl and formaldehyde could react spontaneously to form BCME led to regulations prohibiting the use of these agents in the same area without special ventilation.

Dimethylformamide (DMF) and glycol ethers

These agents are widely used in the tanning industry for finishing hides. There have been three clusters of testicular cancer identified in this population of workers. Three cases were identified in a tannery in Fulton County,

New York between 1982 and 1984. Collection and analysis of air samples by NIOSH identified significant levels of the glycol ethers 2-ethoxyethanol, 2-ethoxyethyl acetate, and 2-butoxyethanol. The workers had previously been exposed to high levels of dimethylformamide (DMF) on the job, although none was detected at time of sampling. Two other clusters of testicular cancer in workers in the tanning industry have been identified. DMF has been shown to be a testicular toxin in animal studies, but not a mutagen, and glycol ethers are known to be reproductive toxins in animals. At the present time, the precise cause of these tumors has not been identified.

Ethylene oxide (CH_2CH_2O)

Although it is not strictly speaking a solvent, it is appropriate to consider this chemical here. It exists as a sweetish, ether-like gas at room temperature and becomes a liquid at 12°C. It is used as an industrial chemical in the manufacture of plastics, as a fumigant in agriculture, and as a sterilizing agent for heat-sensitive materials in health care facilities. All of these uses relate to the fact that this chemical is highly reactive with other organic substances, including proteins, because it contains an epoxide. The structural formula is:

$$
\begin{array}{c}
\text{H} \\
| \\
\text{H–C} \\
| \quad\diagdown \\
\quad\quad\quad \text{O} \\
| \quad\diagup \\
\text{H–C} \\
| \\
\text{H}
\end{array}
$$

It exhibits the same toxicity as other organic solvents including CNS depression, local irritation including respiratory irritation, frostbite due to the rapid evaporation of the liquid, and multi-organ toxicity following chronic exposure. It is completely miscible with both organic solvents and water. When exposures are minimal, as they should be when safe handling practices are observed, ethylene oxide poses little risk. Poor safety procedures or accidental exposure carry the risk of a toxic reaction. Populations at risk include workers in health care institutions, exterminators, fumigators, and chemical plant workers. The question of carcinogenicity of this agent remains unanswered. It is highly suspect because of its structure (see Chapter 1 and Figure 4), and animal studies have shown clear evidence of increased incidences of cancers of the adrenal gland, spleen, kidney, skin, lung, stomach, and brain as well as mononuclear cell leukemias. However, a study of workers exposed to 5 to 10 ppm for an average of 10.7 years did not reveal any increased frequency of cancer.

Factors influencing the risk of a toxic reaction

Virtually all of the factors (discussed in Chapter 1) that can influence the response of experimental animals to a toxic agent can also affect toxicity to humans. The risk of a toxic reaction is a function of the toxicity of the chemical and the duration of the exposure, hence the need to establish exposure limits such as the Short Term Exposure Limit (STEL). As noted above, toxicity may also be affected by the presence of other agents such as ethanol. N-hexane and benzene will enhance the nephrotoxicity of chlorinated hydrocarbons. This may be especially important when exposure to mixed solvents occurs, as it often does. The uptake and distribution of a toxic agent will be influenced by its air/water (blood) partition coefficient and by its water (blood)/oil (fat) partition coefficient. Body depot fat may thus serve as a reservoir for highly lipid-soluble solvents, just as it does for some anesthetics (including sodium thiopental; see Chapter 1), and prolong their CNS-depressing effects. Age, the presence of hepatic disease, and the general state of health all influence individual risk.

Non-occupational exposures to solvents

Solvents are present in many household products, including cleaning agents, waxes and polishes, glues, paints, automotive products (cleaners, polishes, etc.), paint removers, thinners, and brush cleaners, even hair sprays and, of course, gasoline. Exposures may be accidental, usually involving skin contact or inhalation in an enclosed space or ingestion by small children, deliberate, as in the ingestion or inhalation of a substance in a suicide attempt, or as a consequence of substance abuse as in gasoline sniffing. A great tragedy of our time is the high incidence of gasoline sniffing, which has resulted in several deaths among native Canadian and American youths living on Indian reserves. Glue sniffing also is a form of substance abuse which can be practiced by adolescents in large cities and where access to other drugs of abuse is restricted.

A somewhat unexpected source of exposure is products used in arts and crafts. These include adhesives, paints, lacquers, and cleaning solvents. Cigarette smoke also contains volatile solvents, including benzene.

Further reading

Glenn, W., Ethylene oxide: another question mark, *Occup. Health and Safety Can.*, 5, 28–33, 1989.

Klassen, C.D., Amdur, M.O., and Doull, J., Eds., *Casarett and Doull's Toxicology: The Basic Science of Poisons*, 5th ed., McGraw-Hill, New York, 1996.

Marquardt, H., Schafer, S.G., McClellan, R., and Welsch, F., *Toxicology*, Academic Press, New York, 1999.

Methemoglobinemia due to occupational exposure to dinitrobenzene, *Morbid. Mortal. Wk. Rep.*, 37, 353–355, 1988.

Niesink, R.J.M., Hollinger, M.A., and de Vries, J., Eds., *Toxicology: Principles and Applications*, CRC Press, 1996.

Testicular cancer in leather workers — Fulton County, New York, *Morbid. Mortal. Wk. Rep.*, 38, 105–114, 1989.

Review questions

For Questions 1 to 8, answer True of False.

1. Aliphatic solvents are straight or branched-chain carbon-hydrogen compounds.
2. Inhaling gasoline fumes can sensitize the heart to adrenaline.
3. Hexane is oxidized to a hepatotoxic metabolite.
4. Monomethyl ethylene glycol is less toxic acutely than ethylene glycol.
5. Propylene glycol is highly toxic.
6. Chloroform is hepatotoxic without being metabolized.
7. Ethylene oxide is liquid at room temperature.
8. Workers in the tanning industry may show an increased frequency of testicular cancer.

For Questions 9 to 12, use the following code:

Answer A if statements a, b, and c are correct.
Answer B if statements a and c are correct.
Answer C if statements b and d are correct.
Answer D if statement d only is correct.
Answer E if all statements (a, b, c, d) are correct.

9. Halogenated aliphatic solvents:
   a. Cause CNS depression.
   b. Are hepatotoxic.
   c. Are nephrotoxic.
   d. Never require biotransformation to exert their toxicity.
10. Methanol:
   a. Is converted to formaldehyde in the body.
   b. Can cause blindness.
   c. Poisoning is treated with ethanol.
   d. May be present in fairly high concentrations in pineapple.
11. Ethylene glycol:
   a. Is metabolized to a toxic substance or substances.
   b. Is toxic in the parent state.
   c. Is metabolized to oxalate, which may precipitate as calcium oxalate in the renal tubules.
   d. Poisoning is not treated with alcohol.

12. Benzene:
    a. Depresses the CNS.
    b. May cause anemia and reduced platelet count.
    c. Is carcinogenic in animals.
    d. May cause fatal aplastic anemia.
13. Ethylene oxide:
    a. Contains an epoxide bond.
    b. Is a suspected human carcinogen.
    c. Depresses the CNS.
    d. May cause multi-organ toxicity following long-term exposure.

Answers

1. True
2. True
3. True
4. True
5. False
6. False
7. False
8. True
9. A
10. E
11. B
12. E
13. E

Case study 14

Five steam press operators in a rubber plant became ill with signs and symptoms including blue discoloration of the lips and nail beds, headache, nausea, chest pain, dizziness, confusion, and difficulty in concentrating. One worker suffered a seizure. Blood methemoglobin (MetHb) levels ranged from 3.8 to 41.2% (normal < 1%). The product they were using was a solvent-borne adhesive.

Q. What information would you want to elicit concerning their working environment?

Q. What industrial chemicals could be responsible for the high MetHb levels?

Several days later, the steam press was operated by a supervisor for 2 hr so that an industrial hygienist could collect air samples. The supervisor's MetHb level was 12.5% at the end of the 2 hr. The air samples were unrevealing, so technical assistance from the National Institute for Occupational Safety and Health (NIOSH) was solicited.

Q. What analysis might next be called for?

Q. What steps might you recommend in the interim?

Case studies 15 and 16

Cast study 15

A 52-year-old woman was hospitalized for routine hemodialysis for chronic kidney failure. She became somnolent after dialysis and developed metabolic acidosis and shock within 12 hr. She died 24 hr after dialysis. The dialysis fluid had been made up with tap water.

Case study 16

Two children, 4 and 7 years of age, were admitted around 7:00 p.m. to a small rural hospital with an acute onset of marked somnolence, vomiting, and weakness. They developed hematuria (blood in the urine) and were transferred to a nearby city hospital with a pediatric intensive care unit. Urinalysis showed calcium oxalate crystals. They were treated with intravenous fluid therapy and recovered uneventfully.

The children had attended a picnic that afternoon held in an park adjacent to a local firehouse. A survey of the nearly 400 guests revealed similar but less serious symptoms in 28 of them, two of whom required hospitalization. Nineteen of the 28 cases, including the two who required hospitalization, were children under 10 years of age. The most common symptoms were sleepiness, fatigue, dizziness, and unsteadiness when walking. A common denominator in those who became ill appeared to be consumption of a non-carbonated beverage made from crystals with tap water taken from the firehouse. A few individuals who had not consumed the beverage, but had drunk water, also became ill. The severity of symptoms correlated with the amount of beverage or water consumed.

Q. Do you think the offending agent in these cases is:
    a. A pathogenic organism?
    b. A toxic chemical?
    Give reasons for your choice.

Q. Do the calcium oxalate crystals mentioned in Case 16 provide a clue?

Q. What organ system was most affected in both cases?

In both of these cases, there was a direct connection between the general water supply and chilled-water air conditioning systems. Could this have been a factor in these illnesses?

chapter eight

Food additives, drug residues, and other food toxicants

"The food that to him now is as luscious as locusts, shall be to him shortly as bitter as coloquintida."

— W. Shakespeare

Food additives

Food and drug regulations

All developed countries have regulations governing the ingredients and additives that can legally be present in foodstuffs. The nature of the regulations may vary from country to country, but those of Canada are fairly typical.

In Canada, regulations governing foodstuffs and food additives are part of the Food and Drug Act and are enforced by the Health Protection Branch of Health Canada. The following are some important definitions:

- **Food:** Any substance, whether cooked, processed, or raw, that is intended for human consumption, including drinks, chewing gum, and any substance used in the preparation of a food, but not including cosmetics, tobacco, or any substance used as a drug. This definition includes food additives.
- **Food Additives:** Any substance, including any source of radiation, the use of which results in, or may reasonably be expected to result in, it or its by-products becoming a part of, or affecting, the characteristics of a food. This definition does NOT include:
  — Any nutritive material that is commonly recognized or sold as a food
  — Vitamins, minerals, and amino acids

— Spices, seasonings, natural flavorings, essential oils, oleoresins, and natural extracts
— Accidental contaminants such as pesticides, or drugs administered to farm livestock
— Food packaging materials or components thereof

The regulations lay out which food additives are permitted, in what foods they can be used, and what the maximum allowable amounts are. A food additive must do at least one of the following:

1. Improve nutritive value,
2. Extend shelf life,
3. Prevent spoilage during shipment,
4. Enhance appearance or palatability, or
5. Assist in the preparation of the food or in the maintenance of its physical form.

It has been estimated that about 1 billion pounds of food additives are consumed annually in North America, or 3.5 pounds per person. The vast majority of these are harmless, but demonstrated toxicity in some experimental animals for some of these and public concern about harmful effects of man-made chemicals have created pressure on governments to tighten regulations controlling their use and establish stricter limits on allowable levels in foodstuffs. As is the case for any xenobiotic, the use of such agents should only be undertaken on the basis of a cost/benefit analysis. If the advantage is trivial, such as enhancement of texture, then any associated risk would be unacceptable. On the other hand, prevention of spoilage or of the growth of pathogenic microorganisms might justify the acceptance of a slight risk. It is this area that generates the greatest conflict between environmental groups and growers' and manufacturers' lobbies, who may differ markedly on the definition of acceptable risk.

Some types of food additives

Food additives are used for a variety of purposes. The following are some of the major ones.

Acidifiers or **acidulants** provide tartness and act as preservatives by lowering pH. They may also improve viscosity.
Adjuvants for flavor facilitate the action of the principal flavoring agent.
Aerating agents (propellants, whipping agents) are used to produce a foam, as in whipped toppings, etc.
Alkalies control pH, neutralize high acidity foods (tomato products, some wines), and may improve flavor.

Antibiotics are used to prevent bacterial spoilage during storage and transportation.

Anti-browning agents prevent oxidation on the surface of some foods, such as lettuce, which may cause brown spots.

Anti-caking agents are added to powdered or crystalline products (drink mixes, powdered spices, salt, cake mixes, etc.) to prevent caking (formation of lumps).

Anti-mold agents are added to foods (bread, baked goods, dried fruit, cheeses, chocolate syrup) to prevent mold growth. They are also called **antimycotic** or **antirope agents** (liquid or viscous products that become moldy are described as being "ropy").

Antioxidants prevent the oxidation of fatty acids that causes rancidity, and of vitamins, which lose potency.

Antistaling agents prevent bread, etc. from going stale.

Binders are substances used to maintain "body" and hold a product together (e.g., in processed meat, snack foods).

Bleaching agents are used to whiten flour and some cheeses.

Buffers are used in many processed foods.

Chelators or **sequestrants** are used to bind metallic ions that can hasten oxidation of fats and shorten shelf life.

Coating agents (**glazing** or **polishing agents**) are used to coat the skins of fruits and vegetables to prevent bruising, drying, or spoilage, and to coat candies and tablets.

Defoaming agents (**antifoaming agents, surfactants**) are used to prevent excessive foaming in beverages when bottle filling.

Emulsifiers disperse fat droplets in an aqueous medium (e.g., salad dressings, milk shakes, whipped cream, and toppings in pressurized cans).

Extenders (**fillers**) are natural substances (casein, starch, soybean meal) used to add bulk to a food product.

Fixatives maintain the color of meat and processed meat.

Flavor enhancers intensify the natural flavor in soft drinks, fruit drinks, jams, and gelatine desserts.

Flavors (artificial): Any flavoring that does not occur in nature, even if the ingredients are all natural, is defined as an artificial flavor. When something is described as "chocolatey" rather than as chocolate, it indicates that the flavor is artificial, not natural, chocolate. This advertising ploy gets around the regulations prohibiting false advertising and is widely used in North America.

Food colors are added to many products, including some fruit (oranges), to restore color lost in processing or transportation. Most (90%) are synthetic. Food colors that are bound to aluminum hydroxide are known as lakes. All synthetic food colors are highly water soluble. Only vegetable dyes are lipid soluble. This, of course, affects their absorption from the gastrointestinal tract.

Fumigants are toxic gases used to kill pests in harvested dried grains and nuts.

Fungicides prevent fungal growth on the surface of some fruits.

Humectants (**hydroscopic agents**) retain moisture and prevent drying in some candies and ice cream.

Maturing agents (**dough conditioners**): Flour is better for baking if it is aged. Bleaches and other agents speed up the process.

Plasticizers (**softeners**) are used in chewing gum, candies, and edible cheese coatings to maintain pliability.

Stabilizers (**suspending agents**) prevent cocoa, orange pulp, and solids in ice cream from settling out.

Sweeteners (non-nutritional, artificial) are used in low-calorie and diabetic diets.

The above is a partial list of the uses of food additives. It is probably unrealistic to expect that the use of such agents, many of which are synthetic chemicals, can be completely eliminated. Some of them at least are essential to allow the shipment of fresh fruit and vegetables over long distances, as is necessary if these foods are to be available in areas with a short growing season. As is so often the case, public perceptions of risk cloud the issue of artificial food additives. Consumer advocacy groups continue to campaign for tighter controls on such agents. But when saccharin was banned because animal tests had shown the development of bladder tumors in rodents fed high doses, public outcry, originating from a perceived need for this product (vanity is a powerful motivator), resulted in a partial removal of the ban (see below). It is now generally accepted that saccharin is not a carcinogen for humans. The remainder of this chapter concentrates on the more common, or more controversial, food additives.

Artificial food colors

The common, public perception is that synthetic food dyes are inherently more toxic than natural ones. In fact, they are highly purified chemicals, most of which have received extensive toxicity testing. Moreover, they are highly water soluble, so that absorption from the gastrointestinal tract is minimal. In contrast, natural dyes are complex mixtures of compounds that are generally more lipid soluble and therefore better absorbed. Because most natural food additives have been in use for decades, they are on the U.S. "Generally Regarded As Safe" (GRAS) list and have not been extensively tested. Most are lipid-soluble carotenoids (reds, oranges, and yellows); are present in carrots, squash, yams, etc.; and can be regarded as harmless. The possibility of an allergy to any food component cannot be discounted, however. Natural colors tend to be more subdued and to fade more quickly than synthetic ones. Thus, synthetic colors are often preferred. (In Highland lore, "Ancient" tartans are those dyed with the original, more muted, vegetable dyes.)

Some synthetic dyes have been banned. Orange No. 1 and Red No. 3 caused diarrhea in children who consumed large amounts of candy, carbonated beverages, and other confections where these colors were used extensively. More recently, Red No. 2 (amaranth) was banned because embryotoxicity was demonstrated in rats. Considerable controversy surrounds the question of whether synthetic food dyes contribute to hyperactivity in children. Dr. Benjamin Feingold postulated that artificial colors and flavors, together with "salicylate-like" natural substances (present in apples, oranges, peaches, raisins and many berries, and in cucumbers and tomatoes) contribute to behavioral problems such as shortened attention span, easy distractability, compulsiveness, and hyperactivity. Several studies have been conducted to examine this problem. In one type, groups of children were fed the "Feingold" diet or a normal diet and crossed over to the alternate diet after several weeks. This is called a double-blind crossover study. Neither the observers nor the subjects were aware of the treatment group to which they were assigned. Behavior was rated subjectively by parents and teachers. In studies at the Universities of Pittsburg and Wisconsin, 25 to 33% of hyperactive children showed improvement when shifted to the Feingold diet, according to the parents' ratings. Teachers' ratings showed much fewer differences. Moreover, shifts from the Feingold diet to the normal one were not accompanied by behavioral changes. In other studies, children received the Feingold diet throughout the experiment, but doses of a blend of eight certified colors were added periodically. Doses were 26 to 150 mg. The latter corresponds to the intake of the 90th percentile of American children. Both subjective estimates of behavior and objective tests of behavior and learning performance were used. At levels above 100 mg in a Canadian study, there was a slight deterioration in learning in 17 of 22 hyperactive children but no change in behavior. Two of 22 were affected at 35 mg. An Australian study employing a parent questionnaire reported substantial improvement on the special diet, whereas a similar U.S. study did not.

Citrus Red No. 2 is restricted to surface use on oranges not to be processed. It has been shown to be carcinogenic in animals. Orange B is restricted to use in the casings of sausages and hot-dogs. It is related to amaranth (Red No. 2). The manufacturer has discontinued production because of evidence of a carcinogenic contaminant. Red No. 40 was imputed to cause cancer but the evidence was ruled inconclusive. Yellow No. 5 (tartrazine) has been associated with allergic reactions, sometimes severe, and must be on the U.S. ingredients list (candies, desserts, cereals, dairy products). Cross allergenicity with aspirin is common. Canada does not have a compulsory ingredients list. This is a matter of great concern to individuals, and their parents in the case of children, who have life-threatening allergies to foodstuffs and additives.

Blue Nos. 1 and 2, Green No. 3, and Yellow No. 6 are considered safe but the WHO has raised questions about the adequacy of testing.

Emulsifiers

Carrageenin (Irish Moss) is extracted from several species of red marine algae. It contains a variety of calcium, sodium, potassium, and ammonium salts plus a sulfated polysaccharide. It is widely used as an emulsifier and thickener in ice cream, milkshakes, and chocolate drinks. It keeps milk proteins in suspension. Estimated daily intake is 15 mg per person. Only the undegraded form is permitted. Rats fed 2000 mg/kg showed fetal deaths and young with underdeveloped bones. Increased vascular permeability and interference with complement have been shown experimentally. Although potentially serious, especially in ill people (complement is essential to the immune system), carrageenin is not well-absorbed and the FAO/WHO Committee on Food Additives has established an acceptable level of 500 mg/day. Long-term testing is probably indicated.

Furcelleran is a similar substance derived from a red seaweed and is similarly used. Brominated vegetable oil is also used as an emulsifier to keep flavoring oils in suspension in soft drinks. A study in 1976 indicated that the daily U.S. intake averaged less than 0.2 mg/person. These products have been in use for over 50 years, but toxicology studies in the early 1970s showed that doses of 2500 mg/kg of cottonseed BVO caused, in rats, heart enlargement and fatty deposits in heart, liver, and kidney after a few days. Doses as low as 250 mg/kg caused fat deposition in the heart. Maximum daily intake for a child probably does not exceed 0.05 mg/kg but it may occur over a prolonged period. Corn oil BVO fed to rats and pigs at 20 mg/kg for weeks caused deposits of brominated fat in liver and other tissues. The UN Joint FAO/WHO Committee on Food Additives recommended in 1971 that BVO not be used as an additive. The U.S. FDA removed it from the GRAS list pending further safety studies by the manufacturer. These were judged to be faulty by the FDA, and its use has been discontinued.

Preservatives and anti-oxidants

Butylated hydroxyanisole (BHA) and butylated hydroxytoluene (BHT) are synthetics used to prevent premature rancidity in oil- and fat-containing foods. Total daily intake for a child could approach 0.5 mg/kg, which is the maximum recommended by FAO/WHO. Animal studies have consistently shown that high doses (over 500 times the average human consumption) caused liver enlargement and induction of microsomal enzymes. Less than this had no effect. Recent studies have revealed evidence of carcinogenicity in offspring of mice fed BHA. These agents may act as promoters through the enzyme induction mechanism. More work is required before a decision is made. As free radical scavengers, these agents actually may have anticarcinogenic properties.

Sodium nitrate, sodium nitrite, and potassium nitrate are used as curing agents in meats such as bacon and smoked meats. They are always used in combination with salt. Nitrite also inhibits the growth of *Clostridium botulinum,* the organism responsible for botulism. Nitrate is converted to nitrite by bacterial and enzymatic action in the intestine. The average U.S. daily intake from food additives is approximately 11 mg. The major concern is that nitrites can combine with amines to form nitrosamines that are carcinogenic. This process is accelerated by cooking. However, it is important to note that nitrites from food additives account for less than 20% of daily intake, the rest coming from nitrates in drinking water and in vegetables such as celery, spinach, and other leafy vegetables. This is partly due to the use of nitrogen fertilizers and partly from natural sources. Saliva contains nitrate, perhaps providing over 100 mg/day to be converted to nitrites in the lower gastrointestinal tract through bacterial action. Thus, the total nitrate load is about 90 mg from natural dietary sources, 100 mg from saliva, and only 11 mg from food additives. The use of nitrates and nitrites is restricted to the minimum levels required to inhibit the growth of *C. botulinum*.

A particular concern has been the poisoning of infants by nitrates in well water. There are now thousands of such cases, and many deaths, reported, because the formation of methemoglobin from nitrites from nitrates impairs the oxygen-carrying capacity of the blood. A high percentage of rural dwellers in North America and Europe draw their water supply from shallow wells supplied by groundwater. These are vulnerable to contamination by surface runoff and hence by nitrates from fertilizer. Typically, newborn infants in these rural areas would develop, after days or weeks, a syndrome that included cyanosis (blue baby), hypotension (nitrates and nitrites are potent vasodilators) and, eventually, coma and death. Most recovered when they were removed from the home and hospitalized. Invariably, these infants were being fed formula made with well water that contained 20 to 1000 ppm of nitrates. The latency period varied according to the degree of contamination. The conversion of nitrates to nitrites occurred as a result of bacterial action either in the well or in the gastrointestinal tract of the infant, which is virtually neutral (pH ≈ 7) and hence favorable to bacterial growth. With the decline in popularity of formula feeding and its replacement by breast-feeding, nitrite poisoning in infants has almost disappeared in North America.

Artificial sweeteners

Sodium saccharin is several hundred times sweeter than sucrose but leaves a bitter aftertaste. For many years it was the only artificial sweetener in common use. The average daily intake is about 6 mg/kg but some individuals habituated to soft drinks may consume much more. Theodore Roosevelt first required a review of its safety in 1912. In the early 1970s, two studies, one by the FDA, reported that high doses (2500 mg/kg) caused an increase in the incidence of bladder cancer in rats. It was not known whether this was due to an impurity. In 1977, the Canadian Health Protection Branch confirmed that the saccharin was the causative agent, but only when rats were exposed *in utero* to high doses. It was concluded that saccharin is a weak carcinogen and probably a co-carcinogen, but public outcry blocked its recall in the United States. In Canada, it is available in tablet form in pharmacies but it cannot be used as a sweetener in prepared beverages or as a sweetener in restaurants. It is also available in Great Britain. The International Agency for Research on Cancer, an agency of the United Nations, has not deemed it necessary to place saccharin on its list of proven carcinogens. It is doubtful if saccharin poses a real risk. In May of 2000, the U.S. National Institutes of Health released its Ninth National Toxicology Report on Carcinogens in which it dropped saccharin from its list of suspected carcinogens in light of recent studies that failed to show a clear association between the sweetener and cancer.

Xylitol is a natural ingredient of many fruits and berries. It has the same caloric value as glucose but it does not affect blood glucose levels and so it can be used by diabetics. It does not cause caries because it is resistant to fermentation by plaque microorganisms. It has been used in "sugar-free" gum. Some evidence of carcinogenicity has been obtained in rats fed very high doses. Xylitol has been given i.v. to humans as a source of energy. Kidney, liver, and brain disturbances have occurred, along with some fatalities. Use in the United States has been voluntarily stopped pending a review. It has been replaced by aspartame. Sorbitol is another sugar substitute that is equal in calories to sucrose but which will not raise blood glucose levels. Its uses are the same as for xylitol. It is also used as a humectant in jellies and baked goods, and in canned bread to prevent browning. Nausea, cramps, and diarrhea have occurred in some individuals. Acesulfame potassium (an oxathiazinondioxide) was introduced in 1988 as a noncaloric sweetener. It is chemically similar to saccharin. No adverse reactions have yet been reported, but its long-term safety is unknown.

Cyclamate was introduced after saccharin as a sweetener. It has the advantage of not leaving a bitter aftertaste. Both were included in the 1959 "Generally Regarded as Safe" (GRAS) list of the U.S. Food and Drug Directorate. Because cyclamate was fed along with saccharin in one study showing increased bladder tumors, it was removed from the GRAS list and is not available in the United States or Great Britain, but it is in Canada because a Canadian study had shown saccharin to be the culprit.

Aspartame is a dipeptide consisting of the amino acids aspartic acid and phenylalanine. Extensive testing has not revealed any carcinogenic potential and it has no aftertaste. It has largely replaced other sweeteners in soft drinks, gum, and other dietetic foods. It should not be used by people with phenylketonuria (PKU), a hereditary defect of the enzyme phenylalanine hydroxylase, which converts phenylalanine to tyrosine (see Figure 29). These people use an alternate pathway that causes the accumulation of phenylpyruvic acid, which deposits in the brain and causes mental retardation in infants. All newborns are tested for this and, if positive, are placed on a diet free of phenylalanine.

Because tyrosine is a precursor in the synthesis of catecholamine neurotransmitters, there was some speculation that high doses could cause

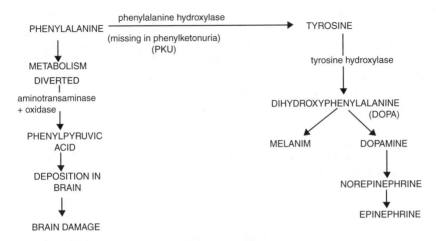

Figure 29 Phenylalanine metabolism in infants with phenylketonuria.

behavioral changes but no evidence of this has been forthcoming in either children or adults. A recent addition to this group is sucralose (tradename Splenda). It is a chemical modification of sucrose in which a number of H and OH groups are replaced by Cl. Sucralose cannot be broken down by digestive enzymes and it is not absorbed. Thus, it cannot enter metabolic pathways for glucose, will not elevate blood glucose, and provides no calories. It is rated as 600 times as sweet as sucrose. It has been pronounced safe by WHO and the Joint Expert Committee on Food Additives. It has been approved in Canada and the United States. A wide variety of products is presently in the laboratory testing stage of development, many of "natural" origin, and some of these will undoubtedly be making their appearance in the near future.

Flavor enhancers

Monosodium glutamate (MSG) has been identified as the offending agent in the "Chinese restaurant syndrome." Subjective symptoms of numbness and tingling of the mouth and tongue have been reported, but double-blind studies with doses up to 3 g failed to confirm an effect. It appears that certain individuals are highly sensitive. Because glutamate is an excitatory amino acid, MSG has been given in doses up to 45 g daily to mentally retarded patients with no behavioral changes or ill effects. Animal studies have shown hypothalamic lesions and infants under 6 months of age may be especially susceptible to MSG toxicity.

Drug residues

The high-risk nature and narrow profit margin that are typical in the live-stock industry have led to the extensive use of antibiotics, sulfa drugs, hormones, and other pharmaceuticals to improve productivity by increasing the rate or extent of weight gain per unit of food consumed. They are also used to prevent or treat disease. Many of these are medications that are also used in human medicine; others are unique to the agricultural field but they may have pharmacological or toxicological consequences for people as well. There are three main concerns about the possibility that traces of these substances might enter our food supply:

1. They may serve as a source of allergic sensitization.
2. Anti-infectives may contribute to the development of resistant strains of pathogenic bacteria.
3. They may exert direct toxic manifestations such as teratogenesis and carcinogenesis.

While there is ample evidence that 1 and 2 occur, the occurrence of 3 is mostly speculative.

Antibiotics and drug resistance

In the late 1950s, Thomas Jukes of Berkeley University reported that the antibiotic tetracycline, at 50 ppm in animal feed, significantly improved the rate of weight gain and the gain-to-food consumption ratio in livestock. Subsequent studies showed that this effect was not related to the prevention of disease and occurred even under optimal conditions of husbandry and hygiene. The effect was confirmed later for other antibiotics and the mechanism remains elusive. There is by now a long list of antibiotics and other anti-infective agents that has been employed as growth promotants. Some of the more common ones are shown in the Table 8.

Antibiotics are also used to prevent the outbreak of disease. Prophylactic use involves higher levels than those used for growth promotion. A typical mixture for preventing dysentery in swine in the 1960s contained chlortetracycline 100 g/T of feed, sulfamethazine 100 g/T, and penicillin G 50 g/T. Concern over the potential dangers of drug residues in food mounted over the next 2 decades. It began with a report in 1959 from Japan of an outbreak

Table 8 Drugs Used as Animal Growth Promotants

Antibiotics
  Several tetracyclines
  Erythromycin
  The aminoglycosides neomycin and streptomycin
  Lincomycin
Antifungals
  Nystatin
Synthetic chemotherapeutics
  Sulfa drugs
  The nitrofurans furazolidone and nitrofurazone

of *Shigella* dysentery in a hospital nursery. The outbreak was unique in that the infecting strain of the bacteria was resistant to several antibiotics and sulfa drugs, some of which had never been used in that hospital. The term *multiple drug resistance* was coined for this phenomenon and in the next few years many reports emerged of MDR in livestock. In 1965 an outbreak of salmonellosis in Great Britain resulted in six deaths. It was traced to the consumption of veal from calves that had been treated with several antibiotics for an organism that demonstrated MDR. In 1971, the Swann Commission in the United Kingdom recommended greater controls over the use of antibiotics in livestock. Government legislation was passed and many other countries, including Canada, followed suit.

During this period it was discovered that the pattern of resistance typical of one type of bacteria could be passed to other, unrelated bacteria and species, and the term *infectious drug resistance* came into use.

Infectious drug resistance (IDR)

For many years it was thought that IDR involved Gram-negative enteric organisms exclusively. *Gram-negative* refers to the histochemical staining characteristics of the bacteria (one of the major means of classifying them, it is related to the composition of their outer cell wall), and *enteric* refers to the fact that they are common inhabitants of the intestinal or enteric tract of both animals and humans. This group includes strains of *Escherichia coli*, *Salmonella* spp., *Shigella* spp., and *Klebsiellia* spp. It is now known, however, that most — and possibly all — bacteria are capable of developing multiple drug resistance through transference of genetic information from one cell to another.

The mechanism of IDR hinges on the fact that bacteria possess extra-chromosomal units of genetic information called *plasmids*. These are rings of DNA that are capable of replication independent of the chromosomes and that can be passed intact from one bacterial cell to another. Genes can be inserted into these plasmids from other sources, including bacterial chromosomes, and this is dependent upon the existence of discrete sequences of 800 to 1800 base pairs of amino acids called *insertion sequences*. When a gene for drug resistance is included between two insertion sequences, the unit is called a *transposon* or, more often, an *R* (for resistance) *factor*. The existence of plasmids has provided the means for genetic engineering and the bacterial synthesis of human insulin and other substances. There are several methods by which plasmids can be transferred from one bacterial cell to another. These include:

1. *Transformation.* The lysis of a cell may release plasmids into the environment that may subsequently be absorbed by other cells. This is a highly species-specific phenomenon.
2. *Transduction.* This involves the participation of phage viruses that incorporate bacterial genetic information and transfer it to other cells. This also is very species specific.

3. *Conjugation*. The plasmids of many bacteria possess a gene, called a *fertility* or *F factor*, which regulates a form of sexual reproduction. A fine tubule or "pilus" is formed between cells and intact plasmids may then be passed, along with their complement of genetic information, from one to the other. This process is not species specific and it is the basis for IDR because R factors will be passed along as well. In this way, resistance genes can be shared among all enteric organisms that contact each other and a multiple resistance pattern acquired. This process is illustrated in Figure 30.

The process of conjugation is clinically important because:

1. Nonpathogenic organisms may serve as a reservoir of drug resistance to be passed on to more virulent ones. There are few species barriers to transfer.
2. The process of natural selection can, when an antibiotic is used therapeutically, result in the emergence of a strain of bacteria with MDR as susceptible strains are killed or inhibited.
3. Resistance patterns, or resistant organisms, may be passed from animals to humans. Many organisms are infectious for both.
4. The process is favored by exposure to low (nontoxic) levels of antibiotics.

The question is: how serious is the problem? In the 1970s there were reports that the enteric organisms of farm workers had identical resistance patterns to those of the livestock they tended. Some consumer advocates have asked, Why not just eliminate the use of antibiotics entirely? A study in 1983 claimed that for an investment of $271 million in animal feed additives, the American consumer saved $3.5 billion in food costs. The result has been a lobbying war between the pharmaceutical and agricultural industries on the one hand and various consumer and environmental groups on the other. In the last few years the debate has heated up as outbreaks of salmonellosis have become more common, especially among patients of nursing homes and psychiatric hospitals, where disease resistance may be lower and personal hygiene less rigorous.

In 1984, the U.S. Centers for Disease Control issued a report on 52 outbreaks of salmonellosis between 1972 and 1983. The source of the infection was identifiable in 38 cases, and in 17 of these (44%) it could be traced to food animals (poultry, veal, hamburger). These accounted for 69% of all outbreaks involving strains with MDR. The mortality from these strains was 4.2% vs. 0.2% from nonresistant strains. In March and April of 1985, the most massive outbreak of salmonellosis in U.S. history occurred in Chicago, with 14,000 people eventually infected. There were only two deaths, yielding a mortality of 0.014%. This was markedly lower than the previous rates reported for either resistant or susceptible strains (the 4.2% mortality rate would have caused 588 deaths), and a follow-up revealed that over 16,000 people were eventually

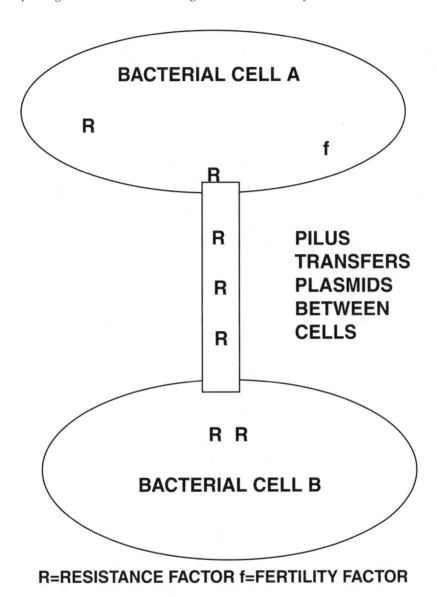

Figure 30 Bacterial conjugation and multiple (infectious) drug resistance.

infected for a final mortality of 0.012%. Both fatalities, however, involved a tetracycline-resistant strain. The controversy continues. Thomas Jukes has criticized efforts to ban antibiotics as additives on the grounds that evidence is inconclusive, but Canada has banned the use of chloramphenicol for any purpose in food animals. A study conducted in Ontario in the 1970s indicated that MDR was very common in cattle, that the most common resistance pattern was against not less than six drugs, and that chloramphenicol was frequently one of these.

Avoidance of meat is no guarantee that multiple resistance patterns can also be avoided. In one study, vegetarians had a higher frequency of MDR bacteria than did meateaters. The use of manure as fertilizer for vegetables might serve as an efficient means of transfer when the vegetables are eaten raw. Meatpackers also tend to acquire MDR patterns.

It must be remembered that the use of antibiotics in agriculture is not the only source of MDR strains. In one study, resistance patterns of coliform organisms in the sewage effluent from a general hospital, a psychiatric hospital, and a residential area were compared. The results are shown in Table 9. It is evident that the incidence of MDR closely parallels the use of antibiotics in a particular setting. Thus, it is higher in the general hospital than in either the psychiatric hospital or the residential area. Moreover, resistance to chloramphenicol, which is reserved for life-threatening situations where less-toxic antibiotics will not work, was very low in all situations, but the correlation still held. In 1985, chloramphenicol was withdrawn as an approved drug for use in food animals.

Table 9 Percent Coliforms Resistant to:

Effluent Source	Chloramphenicol	Streptomycin	Tetracycline
General Hospital	48.8	0.4	24.3
Psychiatric Hospital	9.5	0.03	0.04
Residential area	0.6	0.0007	0.1

Resistance to antibiotics is not the only type of genetic information that can be passed in this manner. Resistance to metals and bacterial virulence is also regulated by plasmid genes. Both vertical (within a species) and horizontal (among species) transfer is now thought to be commonplace in the microbial kingdom. Even transfer between Gram-negative and Gram-positive organisms has been demonstrated.

One of the concerns regarding MDR is that nonpathogenic organisms can pass their transposable resistance to pathogenic ones. This has already been demonstrated in livestock. In 1996, the use of the antibiotic avoparcin was banned because of a link with vancomycin-resistant enterococci (VRE), an organism that has caused disasterous hospital infections.

Sweden banned the use of all antibiotics as growth promotants for livestock in 1986, and the European Union banned tylosin, zinc spiramycin, and virginiamycin in 1998. In 1996, the U.S. Government created the National Antimicrobial Resistance Monitoring System (NARMS) to monitor changes in susceptibility of pathogens from human and animal specimens, healthy animals, and carcasses of food animals. These data will be used to assist in future policy decisions. The question of antibiotic use for growth promotion has already created international trade disputes over the import/export of agricultural products. Additional information on antibiotic resistance can be found at http://www.usda.gov; http://www.fda.gov/; http://www.health.gov.au; and http://www.cfia-acia.agr.ca

Infectious diseases

No discussion of environmental hazards would be complete without passing reference to the hazard of infectious diseases. Not only has the emergence of antibiotic resistance led to clinical problems in treating old diseases such as tuberculosis, which has developed multiple drug resistance and is on the rise, but many organisms seem to be developing more virulent strains and new ones are emerging. The AIDS virus, the *Hantavirus*, the bacteria of Lyme disease, *Legionella*, and bovine spongioform encephalopathy (BSE) and its possible connection to Cruetzfeldt-Jakob disease are but a few examples. Perhaps this should not surprise us, given the rapidity of cell division of bacteria and the adaptability that has allowed them to survive for eons. Next to malnutrition, infectious diseases are the leading cause of death worldwide. There is an obvious relationship between the two.

Allergy

Legislation in most Western countries prescribes withholding periods during which livestock may not be shipped for food following the administration of an antibiotic or other drug by any route. This is designed to prevent the entry of drug residues into the human food chain. One reason for this is the possibility that such residues could result in either allergic sensitization or an allergic reaction in individuals who are already sensitive. Significant levels of drugs could remain at the injection site, which is usually a large muscle mass that also constitutes a preferred cut of meat. Certainly, prior to the introduction of withholding times, such reactions did occur. Of the drugs used in agriculture, those with the highest allergic potential are the penicillins. Penicillin G is still available without prescription in many countries for use in livestock, and it is commonly employed for the treatment of mastitis (udder infection) in dairy cattle. It may thus be present in milk, although testing procedures are usually mandated by law. The test is not specific for a particular drug but employs a strain of bacteria that is extremely sensitive to inhibition by almost all antibiotics. More recently, Enzyme-Linked Immunosorbant Assay (ELISA) has been used.

It may be significant that 30% of patients suffering severe allergic reactions when treated with penicillin have no history of previous exposure, suggesting that sensitization occurred through environmental exposure. Although allergic reactions from contaminated food are rare and difficult to trace, they may still occur when drug residues escape detection. Sulfa drugs may also be offenders in this regard.

Diethylstilbestrol

Diethylstilbestrol (DES) is one of a number of synthetic estrogens first synthesized by Charles Dodds in 1938. Because the work was conducted under the auspices of the British Medical Research Council, the chemicals were not subjected to patent restrictions and they rapidly became widely available. As a group, they revolutionized endocrinology, made possible the development of oral contraceptives, and had a major impact on the practice of gynecology. The possibility of sex without pregnancy and the consequent socio-sexual revolution of the 1960s can be attributed directly to the development of these compounds and their chemical manipulation by the pharmaceutical industry.

DES mimics natural estrogens but is much more potent and more toxic, and it is effective orally, which natural ones are not. It is very cheap to produce and by 1939 it was already popular in the United Kingdom, France, Germany, Sweden, and North America. It was used to treat menopausal complaints, amenorrhea, and genital underdevelopment and to suppress lactation in mothers who did not wish to nurse their infants. By 1942, agricultural use had begun on a limited scale. Early studies showed that rats fed DES gained weight more rapidly with no increase in food consumption and subsequent tests in cattle yielded similar results. Moreover, the "marbling" (diffuse fat distribution) was greater and this was considered to be a desirable flavor enhancer at the time. Athletes such as Ben Johnson were not the first to take steroids to increase weight gain!

By 1945, DES was being widely used in beef and poultry production. There were two methods of administration. One was a feed additive and the other a slow-release pellet that could be injected under the skin of a steer's ear or the neck of a broiler chicken. The former method had the disadvantage of loss through spillage and the accidental ingestion by breeding stock caused some serious fertility problems. The necks and ears containing the remains of the pellet were supposed to be discarded at slaughter, but chickens are frequently sold with the necks attached; and in 1959, a report emerged of a 40-year-old male chef in New York who developed signs of feminization, including loss of facial hair, mammary development, fat deposition of the female pattern, and a high-pitched voice. The problem was traced to the consumption of chicken necks taken from the restaurant and eaten as an economy measure. The use of pellet implants in poultry was banned that same year.

The carcinogenic nature of estrogens has been known for some time. Experimentally, natural estrogens were shown to produce carcinomas of the vagina, cervix, and uterus in mice, and carcinogenic potency was directly related to estrogenic potency. There were sporadic reports of DES inducing cancer in mice, but in 1981 Herbst et al. reported a case of clear-cell adenocarcinoma in the vagina of a young woman who had been exposed to DES *in utero*. A DES registry was established in the United States and by 1984, 500 cases of clear-cell adenocarcinoma had been reported. It is now estimated that females exposed to DES *in utero* have somewhat less than a 1/1000 chance of developing it.

The use of DES in pregnancy was based on data indicating that it improved the vascularity of the uterus and promoted the synthesis of progesterone, both of which would help to maintain pregnancy in the face of a threatened abortion. Other conditions that have been causally associated with exposure to DES *in utero* include adenosis (a self-limiting condition involving the presence of cervical cells in the wall of the uterus), infertility, carcinoma of the cervix and vagina, and breast cancer. Cervical malformations occur in 18 to 25% of DES daughters, but most appear to correct themselves with age. Infertility is by far the most common problem experienced by DES daughters. Pregnancy may be difficult to achieve and maintain. Since the first wave of DES daughters is just now entering middle age, the possibility that increased frequencies of other forms of cancer might emerge cannot be discounted; however, the association with breast cancer is becoming more tenuous.

The fetal risks of exposure to DES appear to relate to the fact that the mechanisms that detoxify natural estrogens cannot handle DES. In rats, a glycoprotein called alpha-fetoprotein binds natural estrogens but not DES. Its presence has not been confirmed in humans but human fetuses have a very active process for sulfonating natural estrogens that may be less effective for DES. Active metabolites of DES may have a high affinity for estrogen receptors in the reproductive system and oxidative metabolites may damage DNA. Reactive metabolites may form after attachment to the receptor by the action of peroxidase enzymes present in estrogen-dependent tissues.

Attempts by the FDA to ban the use of DES in beef cattle were blocked by court actions launched by the beef producers until 1979 when liver and kidneys from beef cattle showed significant levels of DES. Illicit use continued thereafter; and in 1980, the USDA conducted 115 prosecutions against violators. It must be emphasized that all of the health problems associated with DES in North America (barring the chef noted above) are of iatrogenic origin, that is, they arose from medical treatment. None has been attributed to the consumption of meat from treated animals. This has not been the case elsewhere.

In the late 1970s and early 1980s two epidemics occurred in which infants and children under 8 years of age displayed signs of abnormal sexual development such as breast development and precocious puberty. In Puerto Rico, over 600 children were involved. DES was available without prescription and meat inspection controls were poor. In Italy, high levels of DES were detected in baby foods, especially those containing veal. Several hundred infants were affected. Occurrences declined after 1979 due to tougher controls and greater public awareness. In 1985, the DES Task Force of the U.S. Dept. of Health and Human Resources released its findings.

1. The risk of uterine carcinoma *in situ* was about twice as high in DES daughters as in nonexposed women (15.7/1000/year vs. 7.9/1000/year), although there were problems with the study.
2. The risk of genital herpes was also about twice as high (11.8% vs. 6.3%).
3. The risk of breast cancer was difficult to assess because of other predisposing factors.

The question of increased risk of breast cancer was recently addressed by an American multicenter epidemiological study. It compared over 3000 women who had DES prescribed during pregnancy with 3000 similar unexposed women. The risk of breast cancer per 100,000 was 172.3 for exposed women and 134.1 for unexposed women. The rate increased markedly in both groups between 20 and 40 years post-entry to the study. The results were statistically significant and it now seems conclusive that there is a real increase in the risk of breast cancer in DES-exposed women, but probably not in their daughters.

The psychological costs to DES daughters are difficult to assess but probably quite high. Increased incidences of psychiatric problems and suicide have been reported due to feelings of helplessness and frustration. Recently, one small report suggested a higher incidence of homosexual preference in DES daughters. It is well documented that experimental exposure to sex hormones before birth influences brain development and sexual behavior. In rats and guinea pigs, alpha-fetoprotein normally protects the brain from estrogens but it does not bind to DES and prenatal exposure resulted in abnormal patterns of sexual behavior.

There is some suggestive evidence that males exposed to DES *in utero* may have a higher incidence of testicular cancer and underdeveloped male genitalia, but the evidence is far from conclusive. There is conclusive experimental evidence, however, that exposure to DES *in utero* has these effects in male experimental animals.

With regard to the question of DES used as a growth promoter, the U.S. Delaney clause, which prohibits the use of anything as a food additive that has been shown to be carcinogenic in any animal, specifically exempts substances used as animal food additives provided their residues are not detected in the final foodstuff by currently acceptable analytical techniques. The increase in sensitivity of these over the past decade may render the agricultural use of DES impossible.

Concern over drug residues continues. In 1989, Canada's Health Protection Branch ordered the withdrawal of ethylenediamine dihydroiodide, a product used to treat certain infections in cattle. There was concern that the product might contribute to excessive iodine levels in people with consequences for thyroid function. It is estimated that some Canadians might consume as much as six times the recommended daily intake.

Bovine growth hormone

Currently, controversy centers on the use of growth hormones in food animals. Bovine somatotropin, or bovine growth hormone (BGH), is used both to promote growth in beef cattle and to increase milk production in dairy cattle. It is produced by recombinant gene technology (rBGH). Studies indicate that it can increase lean content by at least 5% and feed efficiency by 8%. In dairy cattle, it can prolong the lactation period by several weeks, leading to greater profit. The consumer, however, remains understandably leery of their use. As long as profit margins in agriculture remain narrow, the pressure to use growth promotants will remain strong. The use of growth hormone in the United States has resulted in attempts to ban American beef in the European Economic Community. Canada has not approved its use and issued a notice of noncompliance in January 1999. This decision was based largely on evidence of adverse effects in cattle, including increased incidences of mastitis, lameness, and reproductive problems.

Concern relates primarily to the fact that dairy cows injected with rBGH produce more of the natural hormone peptide insulin-like growth factor-1 (IGF-1), and IGF-1 is secreted in milk. Increased blood levels of IGF-1 have been associated with an increased risk of breast cancer in women. The debate focuses on whether IGF-1 in milk can be absorbed from the gastrointestinal tract. As a polypeptide, the expectation is that it would be broken down by proteolytic enzymes. Some authorities claim that this is prevented by the presence of casein to which IGF-1 is bound. While high-molecular weight polypeptides and proteins are not generally absorbed from the gastrointestinal tract, there is some evidence that this can occur in some people. Experiments in rats showed that they can absorb IGF-1 from their intestinal tract, which is of course quite different from the human one (see Chapter 1).

A British study, published in 1998, of 397 women with breast cancer and 620 controls found that in premenopausal women under 50 with the highest levels of IGF-1, cancer risk was increased sevenfold over controls. The relative risk for other premenopausal women was 2.88. There was no significant increase in risk for any other subgroup. There is also evidence that elevated serum IGF-1 levels are associated with an increased risk of prostate cancer. A case control study within the (U.S.) Physician's Health Study involved 152 patients and 152 controls. There was a strong positive association between serum IGF-1 levels and risk for prostate cancer. Men in the highest quartile had a relative risk of 4.3. This association was independent of baseline prostate specific antigen (PSA) levels.

While it seems clear that increased IGF-1 levels constitute a risk factor for both breast cancer and prostate cancer, the extent to which the hormone can be absorbed from rBGH milk remains open, at least for humans. Meanwhile, it would seem prudent to invoke the precautionary principle, especially as milk production in developed countries has a capacity already in excess of the demand.

Natural toxicants and carcinogens in human foods

Through the acquisition of folk knowledge, human beings have learned to avoid eating rhubarb leaves, daffodil bulbs, and other plants that contain toxic chemicals. Under some conditions, however, foods that are normally safe may become toxic, at least for some individuals, and carcinogens may be more prevalent in human foods than previously realized.

Some natural toxicants

Favism

The broad bean (*Vicia fava*) may induce acute hemolytic anemia in individuals with the hereditary defect glucose-6-phosphate dehydrogenase (G6PD) deficiency. This is especially true for those individuals with the Mediterranean phenotype. It is a problem primarily for males under 5 years of age. The exact nature of the mechanism remains obscure, but there is evidence that these individuals may be deficient in hepatic glucuronide conjugase and hence unable to detoxify the offending ingredient. Susceptibility in the same individual varies from time to time, and all those with the same defect are not necessarily affected.

Toxic oil syndrome

In 1981 in Spain, there was a remarkable epidemic that eventually affected over 20,000 people. In the first 12 months of the epidemic, there were 12,000 hospitalizations and over 300 deaths. It began in a small town on May 1, when a boy was admitted to hospital with acute respiratory failure that rapidly led to death. The epidemic peaked in June, at which time 2000 new cases were being reported every week. The patients usually presented initially with respiratory distress, cough, low-grade fever, oxygen deficiency, pulmonary infiltrates and pleural effusions, and a variety of skin rashes. Nausea and vomiting were sometimes present, as were enlarged liver, spleen, and lymph nodes. Virtually all patients had elevated eosinophil counts (>500 to >2000/mm^3; normal value, 0 to 500/mm^3). The condition often progressed to severe muscle pain, muscle and nerve degeneration, and even paraplegia.

An astute medical clinician traced the problem to the consumption of a cheap, unlabeled cooking oil sold as pure olive oil in the open markets in small towns and villages. The oil consisted of low-grade olive oil mixed with various seed oils, including a rapeseed oil that was imported in a denatured form by mixing it with aniline to render it unfit for human consumption. Two significant clues were the facts that the condition affected low-income families almost exclusively, and nursing infants were never affected. Geographically, the epidemic was limited largely to central and northwestern Spain.

Chemical analyses of suspect oil samples, and comparison with non-suspect samples, failed to reveal the presence of known toxicants such as heavy metals, but the suspect oil contained significant levels of aniline and fatty acid anilides, reaction products of the oil with the aniline. A dose-response relationship between the degree of contamination and the severity of signs and symptoms was also noted.

The toxicity of aniline is well known because of its heavy industrial use and although there are some similarities with the toxic oil syndrome (e.g., skin lesions), aniline toxicity involves CNS symptoms (vertigo, headache, mental confusion) and blood disorders including methemoglobinemia and anemia. It was felt, therefore, that the offending agent was likely a reaction product. To date, however, the syndrome has not been reproduced in an animal model.

The toxic oil syndrome might be regarded as a historical curiosity were it not for the fact that in 1989, a similar problem emerged in the United States. Called the eosinophilia-myalgia syndrome, it affected over 1500 people and it appeared to be caused by consuming contaminated L-tryptophan as a food supplement. Again, the eosinophil count was usually elevated above $2000/mm^3$. Like the toxic oil syndrome, there was also inflammation of muscles and their nerve supply. There was a notable absence of acute respiratory symptoms, unlike the toxic oil syndrome.

The Centers for Disease Control determined that the product came from one supplier and that it had a contaminant, the di-tryptophan aminal of acetaldehyde. This was either the toxic agent or a marker thereof because there was a strong association between the level of this substance and the incidence of the disease. Efforts at identifying the toxic agent and developing an animal model for the condition are continuing.

Herbal remedies

In recent years there has been a tremendous upsurge of interest in, and use of, herbal remedies. One in five Canadians and one in three Americans indicated in a recent survey that they used herbals, often in conjunction with prescription drugs and often without informing any healthcare professional. This has created a $5 billion industry with poor controls for efficacy and potency. The phenomenon appears to be another manifestation of the simplistic "natural is good, synthetic is bad" philosophy that arose following public concern over the presence of toxic chemicals in the environment. This concern was fueled by episodes such as the Love Canal situation and perhaps it is primarily based on the assumption that, because they are natural, they must be safer. This assumption ignores the fact that, if they possess enough of the active ingredient to have a pharmacological effect, they must also carry the potential to cause adverse reactions and to interact with other pharmacological agents, including other herbals. Of course, if they do not possess such potency, they are unlikely to be effective. Because these preparations are presently defined as food supplements, they are not subjected to the rigorous regulations governing prescription and patent medicines.

Leaving aside the question of potency, which is often an unknown quantity, there is mounting evidence of the potential for some herbals to do harm. A wide variety of herbals possesses diuretic activity with associated potassium loss. They may thus interact with prescription diuretics used to treat hypertension and cause excessive potassium loss and low blood pressure. St. John's Wort, cherry stems, and parsley are but a few examples. Others may interfere with antihypertensive therapy and cause a dangerous elevation in blood pressure. Ephedra, the active ingredient of which is ephedrine, a known hypertensive agent, is present in natural weight loss products including fen-phen, and these have been associated with over a dozen deaths and hundreds of cases of illness reported to the U.S. Food and Drug Administration. Licorice can cause pseudoaldosteronism in which sodium is retained, with a resulting increase in blood pressure. Numerous other products may depress blood pressure to dangerous levels in patients receiving antihypertensive therapy. Cat's claw, devil's claw, and garlic fall into this category. *Rawolfia serpentina* is present in many natural products. Its active ingredient is reserpine, a potent depressor of blood pressure. There have been reports of tissue rejection in heart transplant patients taking St. John's Wort. This herb can induce CYP450 to accelerate the metabolism of cyclophosphamide, hence negating its immunosuppressive action. In one study, over 80 commercial herbal extracts and tinctures were examined for effects on CYP450 isoforms. Approximately 95% of these, and 50% of pure compounds examined, inhibited CYP3A4 metabolite formation.

The agents listed above are but a small sample of the host of herbal agents with the potential to cause untoward side effects and drug interactions. Public demand for these products will force regulatory agencies to divert billions of dollars to analyze and test them, with the predictable result that many will be banned and others reclassified as drugs. This will no doubt lead to an even greater public outcry. See also Chapter 6 for more on natural remedies.

Natural carcinogens in foods

Bracken fern, "fiddleheads"

Considerable evidence exists that bracken fern produces bladder cancer in cattle that eat excessive amounts when better fodder is unavailable, and in rats fed large amounts of it. Because the young shoots, called *fidddleheads* because of their curled shape, are eaten as a delicacy in many parts of the world, including Canada and Japan, there has been concern over the potential for carcinogenic effects in humans. At one point it was suggested that the relatively high incidence of bladder cancer in Japan might be related to the consumption of bracken fern. Epidemiological studies, however, have failed to demonstrate such an association, and it is now felt that eating fiddleheads does not constitute a risk factor for cancer.

The economic significance of bracken fern as an agricultural toxicant has been clearly demonstrated. It is a radiomimetic substance, causing bone marrow depression in cattle and other species and thiamine deficiency in horses.

Others

Natural carcinogens and precursors have been detected in many other foods. In their 1987 review, Ames et al. list the carcinogenic TD_{50} for nitrosamines present in many foodstuffs as 0.2 mg/kg for rats and mice. In contrast, PCBs, with a similar daily dietary intake of 0.2 μg, had TD_{50} values of 1.7 to 9.6 mg/kg in these species. Other substances shown to be carcinogens in animal tests, but for which evidence of a risk to humans is weak, include allyl isothiocyanate (in kale, cabbage, broccoli, cauliflower, horse-radish, and mustard oil), safrol (nutmeg, cinnamon, and black pepper), and benzo[a]pyrene (produced during cooking, especially charcoal broiling). By extremely conservative methods, the risk for benzo[a]pyrene has been estimated at 1.5/100,000 at high levels of consumption.

It has been stated that exposure to carcinogens is an unavoidable fact of life, but that the levels in foods are so low that further reductions would not have a significant effect on cancer incidence. This "bad news" is offset by the "good news" that there are probably many more anti-carcinogens in natural foods than there are carcinogens. In one study of human dietary habits, individuals who ate meat but not vegetables on a daily basis had a colon cancer risk (per 100,000 population) of 18.43; those who ate vegetables but not meat on a daily basis had a risk of 13.67; whereas those who ate both had a risk of only 3.87. Vitamins A, C, and E and carotenoids have been shown to be protective against cancer, probably because of their anti-oxidant properties. Dietary fiber is protective, and even meat has been shown to have anti-cancer properties. Indoles and isothiocyanates present in cruciferous vegetables such as cabbage, cauliflower, broccoli, and Brussels sprouts have been shown to be anti-carcinogenic.

Carcinogenic mycotoxins abound in nature and these will be dealt with in Chapter 10, while irradiated foods are discussed in Chapter 12.

Further reading

Ames, B.N., Magaw, R., and Gold, L.S., Ranking possible carcinogenic hazards, *Science*, 236, 271–280, 1987.

Asulfame — a new artificial sweetener, *Med. Lett. Drug Ther.*, 30, 116, 1988.

Baily-Klepser, T., Doucette, W.R., Horton, M.R. et al., Assessment of patients' perceptions and beliefs regarding herbal therapies, *Pharmacotherapy*, 20, 83–87, 2000.

Baily-Klepser, T. and Klepser, M.E., Unsafe and potentially safe herbal therapies, *Am. J. Health-Syst. Pharm.*, 56, 125–138, 1999.

Boullata, J.L. and Nace, A.M., Safety issues with herbal medicine, *Pharmacotherapy*, 20, 257–269, 2000.

Berkelman, R.L., Bryan, R.T. et al., Infectious disease surveillance: a crumbling foundation, *Science*, 264, 368–370, 1994.

Centers for Disease Control. Analysis of L-tryptophan for the etiology of eosinophilia-myalgia syndrome, *Morbid. Mortal. Wk. Rep.*, 39, 581–591, 1990.

Chan, J.M., Stampfer, M.J. et al., Plasma insulin-like growth factor-1 and prostate cancer risk, *Science*, 279, 563–566, 1998.

Colton, T., Greenberg, E. et al., Breast cancer in mothers prescribed dithylstilbestrol in pregnancy: further followup, *JAMA*, 269, 2096–2100, 1993.

Davies, J., Inactivation of antibiotics and the dissemination of resistance genes, *Science*, 264, 375–382, 1994.

Ehrhardt, A.A. et al., Sexual orientation after prenatal exposure to exogenous estrogens, *Arch. Sexual Behav.*, 14, 57, 1985.

Everett, M., Growth hormones will certainly affect the meat industry, *Vet. Mag.*, April, 14–15, 1989.

Flamm, W.G., Pros and cons of quantitative risk analysis, in *Food Toxicology: A Perspective on the Relative Risks*, Taylor, S.L. and Scanlon, R.A., Eds., Marcel Dekker, New York, 1989, 429–446.

Foster, B.C. et al., Effects of Chinese herbal products on cytochrome P-450 drug metabolism, *Can. J. Infect. Dis.*, 11 (Suppl. B), 144P-145P, 2000.

Freydberg, N. and Gortner, W.A., *The Food Additives Book*, Bantam Books, Toronto, 1983.

Gans, D.A., Behavioral disorders associated with food components, in *Food Toxicology: A Perspective on the Relative Risks*, Taylor, S.L. and Scanlon, R.A., Eds., Marcel Dekker, New York, 1989, 225–254.

Grasso, P., Carcinogens in food, in *Toxic Hazards in Food*, Conning, D.M. and Lansdown, A.B.G., Eds., Croom Helm, London, 1983, 122–144.

Hall, R.L., Dull, B.J. et al., Comparison of the carcinogenic risks of naturally occurring and adventitious substances in food, in *Food Toxicology: A Perspective on the Relative Risks*, Taylor, S.L. and Scanlon, R.A., Eds., Marcel Dekker, New York, 1988, 205–224.

Hankinson, S., Willet, W.C. et al., Circulating concentrations of insulin-like growth factor I and the risk of breast cancer, *Lancet*, 351, 1373–1375, 1998.

Hayes, A.W., Ed., *Principles and Methods of Toxicology*, 2nd ed., Raven Press, New York, 1989.

Herbst, A.L., Hubby, M.M. et al., Reproductive and gynecologic surgical experience in diethylstilbestrol-exposed daughters, *Am. J. Obstet. Gynecol.*, 141, 1019–1028, 1981.

Holmberg, S.D., Osterholm, M.T. et al., Drug-resistant Salmonella from animals fed antimicrobials, *New Engl. J. Med.*, 311, 617–622, 1984.

Kilbourne, E.M., Posada de la Paz, M. et al., Toxic oil syndrome: a current clinical and epidemiologic summary, including comparisons with the eosinophilia-myalgia syndrome, *J. Am. Coll. Cardiol.*, 18, 711–717, 1991.

Kimura, T., Murakawa, Y. et al., Gastrointestinal absorption of recombinant human insulin-like growth factor-I in rats, *J. Pharmacol. Exper. Ther.*, 283, 611–618, 1997.

Klassen, C.D., Amdur, M.O., and Doull, J., Eds., *Casarett and Doull's Toxicology: The Basic Science of Poisons*, 5th ed., Macmillan, New York, 1996.

Loizzo, A. et al., Italian baby food containing diethylstilbestrol three years later, *Lancet*, ii, 284, 1984.

Mashour, N.H., Lin, G.L., and Frishman, W.H., Herbal medicine for the treatment of cardiaovascular disease, *Arch. Int. Med.*, 158, 2225–2234, 1998.

Moats, W.A., Ed., *Agricultural Use of Antibiotics*, American Chemical Society Symposium Series, 320, Washington, D.C., 1986.

Munro, I.C., A case study: the safety evaluation of sweeteners, in *Food Toxicology: A Perspective on the Relative Risks*, Taylor, S.L. and Scanlan, R.A., Eds., Marcel Dekker, New York, 1989, 151–167.

Oberrieder, H.K. and Fryer, E.B., College students' knowledge and consumption of sorbitol, *J. Am. Dietetic Assoc.*, 91, 715–717, 1991.

Pariza, M.W., A perspective on diet and cancer, in *Food Toxicology: a Perspective on the Relative Risks*, Taylor, S.L. and Scanlan, R.A., Eds., Marcel Dekker, New York, 1989, 1–10.

Philp, R.B., Real and Potential Problems Associated with the Use of Anti-Infective Agents in Domestic Livestock, Brief to the Ontario Ministry of Agriculture and Food, prepared for the Ontario Veterinary Association, 1979.

Poirier, L., Lefebvre, J., and Lacouriere, Y., Herbal remedies and their effects on blood pressure: careful monitoring required, *Hypertension Canada*, 65, 1–8, 2000.

Report of the DES task force, *J. Am. Med. Assoc.*, 255, 1849–1886, 1986.

Ruschitzka, F. et al., Acute heart transplant rejection due to St. John's Wort, *Lancet*, 355, 548–549, 2000.

Schacham, P., Philp, R.B., and Gowdey, C.W., Antihematopoietic and carcinogenic effects of bracken fern *(Pteridium aquilinum)* in rats, *Am. J. Vet. Res.*, 31, 191–197, 1970..

Spika, J.S., Waterman, S.H. et al., Chloramphenicol-resistant *Salmonella newport* traced through hamburger to dairy farms, *New Engl. J. Med.*, 316, 565–570, 1987.

Sun, M., Use of antibiotics in feed challenged, *Science*, 226, 144–146, 1984.

Thomas, A.P., Precocious development in Puerto Rico, *Lancet*, i, 1299–1300, 1982.

Use of EDDI in Cattle, (Cdn.) Health Protection Branch Information letter No. 757, Feb. 10, 1989.

Williams, D.E., Dashwood, R.H. et al., Anticarcinogens and tumor promoters in foods, in *Food Toxicology: A Perspective on the Relative Risks*, Taylor, S.L. and Scanlan, R.A., Eds., Marcel Dekker, New York, 1989, 101–150.

Review questions

For Questions 1 to 13, use the following code:

Answer A if statements a, b, and c are correct.

Answer B if statements a and c are correct.

Answer C if statements b and d are correct.

Answer D if statement d only is correct.

Answer E if all statements (a, b, c, d) are correct.

1. With regard to dethylstilbestrol (DES):
   a. It was used mainly to synchronize the estrus cycle of dairy cows.
   b. It was given to livestock as a growth promotant in feed or as subcutaneous pellets.
   c. In North America and Great Britain, its use as a feed additive for livestock resulted in significant human health problems.
   d. It was given parenterally to pregnant women to prevent impending abortion.

2. Human health problems associated with diethylstilbestrol (DES):
   a. Have been reported in some countries in infants who were fed formula containing high levels of DES.
   b. Affect women who were exposed *in utero* to high levels from their mother's blood.
   c. Generally involve abnormalities of the genitourinary system.
   d. Have caused widespread problems in people who eat meat containing pellet residues.

3. With regard to salmonella infections:
   a. It is commonly transferred to humans by eating or handling undercooked meat.
   b. It is most often a problem for the general public.
   c. It sometimes involves multiple drug-resistant strains.
   d. It has a high mortality rate in normal individuals.
4. Which of the following statements is/are true?
   a. Estrogens, including DES, have been shown to be carcinogenic.
   b. Women exposed to DES *in utero* have an increased incidence of cervico-uterine deformities.
   c. DES is not metabolized significantly by the human placenta or fetus.
   d. The human fetus cannot deactivate natural estrogens.
5. Regarding the use of antibiotics in agriculture:
   a. They are used exclusively for *treating* infections in animals.
   b. They may be used prophylactically to *prevent* infections in animals.
   c. There are no regulations governing their use in agriculture.
   d. Very low levels added to feed might have a growth-promoting effect.
6. Multiple Drug Resistance (MDR) is:
   a. Characteristic of Gram-negative enteric bacteria.
   b. Possible even when the organism has not contacted all of the anti-infective agents to which it is resistant.
   c. Determined by extrachromosomal DNA.
   d. Seen only in farm livestock exposed to antibiotics.
7. The process of bacterial conjugation:
   a. Refers to the release of DNA from lysed cells.
   b. Requires the participation of plasmids.
   c. Refers to the transfer of DNA by phage viruses.
   d. Requires the participation of an F (fertility) factor.
8. Which of the following statements is/are true?
   a. Salmonella infection is rarely a problem in nursing homes.
   b. The rank order of frequency for resistant forms of intestinal infections is: general hospital > psychiatric hospital > extrahospital environment.
   c. Multiple Drug Resistance (MDR) occurs only in enteric organisms from animals.
   d. The frequency of MDR increases with antibiotic exposure.
9. The legal definition of food additives in most countries includes:
   a. Vitamins
   b. Food colors
   c. Spices
   d. Nitrates
10. Which of the following substances used in or on foods have been associated with a high degree of allergic reactions?
    a. Tartrazine

    b. Carageenen
    c. Sodium metabisulfite
    d. Saccharin
11. Which of the following statements regarding nitrosamines, nitrates, and nitrites is/are true?
     a. The major source of nitrosamines for people is the nitrates and nitrites used as meat preservatives.
     b. Nitrosamines have been shown to be carcinogens in animals.
     c. There are no natural food sources of nitrates and nitrites.
     d. Nitrates and nitrites inhibit the growth of *Clostridium botulinum*, the cause of ptomaine poisoning.
12. Regarding artificial sweeteners:
     a. Individuals with phenylketonuria should avoid aspartame.
     b. There is no convincing evidence that aspartame in normal amounts causes behavioral problems.
     c. Saccharin has been shown to cause bladder cancer in rats exposed *in utero* to very high levels.
     d. Cyclamate or a metabolite is a proven carcinogen for humans.
13. Toxic oil syndrome:
     a. Involves respiratory distress, eosinophilia, and myalgia.
     b. Produced a high mortality rate in an outbreak in Spain.
     c. Resembles a disease traced to tryptophan consumption in the United States.
     d. May be due to a reaction product of aniline and a component of a cheap, low-grade mixture of cooking oils.

Answers

    1. C
    2. A
    3. B
    4. A
    5. C
    6. A
    7. C
    8. C
    9. C
    10. B
    11. C
    12. A
    13. E

Case study 17

Three men were admitted to the emergency department of a large hospital, all suffering from similar symptoms. Inquiry revealed that they had just finished dining in a nearby Chinese restaurant. All had a severe headache, cardiac palpitations, facial flushing, vertigo (dizziness), and perfuse sweating. The symptoms commenced shortly after finishing their meal. Their faces and chests were reddened, blood pressures slightly below normal, and pulses were quickened.

Q. What is the likely portal of entry of this apparent toxicant?

Q. Is this likely due to:

    a.   Food poisoning of bacterial origin?
    b.   Contamination with a pesticide?
    c.   A food additive?

Q. In a call to the restaurant in question, the attending physician inquired as to whether the trio had consumed mushrooms or fish. Why was this question asked?

The response to the above question was no. It was revealed that many other patrons had eaten similar food without trouble, the exception being pork chow yuk. These three were the only ones to consume this dish.

Q. How can this information be helpful?

chapter nine

Pesticides

So naturalists observe, a flea has smaller fleas that on
him prey; And these have smaller still to bite 'em; And
so proceed *ad infinitum*.

— Jonathan Swift

Introduction

The term *pesticides* refers to a large body of diverse chemicals that includes
insecticides, herbicides, fungicides, rodenticides, and fumigants employed
to control one or more species deemed to be undesirable from the human
viewpoint. Pesticides are of environmental concern for two main reasons.
Although considerable progress has been made with respect to their selective
toxicity, many still possess significant toxicity for humans, and many are
persistent poisons, so that their long biological $t_{1/2}$ allows bioaccumulation
and biomagnification up the food chain (see Chapter 3). Thus, there is the
possibility that, besides constituting an ecological hazard, they may enter
human food supplies. By their very nature, pesticides must have an impact
on any ecosystem because they are designed to modify it by their selective
elimination of certain species. As is always the case in considering chemicals
used in the service of humankind, there is a complex risk–benefit equation
that must be taken into account when making decisions regarding the use
of pesticides. There is no question that they have increased agricultural
production when used properly, and they have, in the past, been highly
effective in controlling the insect vectors of human diseases such as malaria
and yellow fever spread by mosquitoes and African sleeping sickness which
affects both humans and animals and is spread by the tse-tse fly. As will be
seen, however, these gains have not been without their problems.

Efforts to control agricultural pests probably evolved in parallel with
cultivation techniques. Early methods included manual removal of weeds
and insects, rigorous hoeing to prevent weed growth, and the use of traps
for animal and insect pests. The first chemical controls to be used against
agricultural pests were the arsenical compounds. In 1910, Ehrlich discovered
that arsphenamine was an effective treatment for syphilis. This was the first

chemotherapeutic agent for a bacterial infection and the first example of a structure–activity relationship. It opened the door to the entire field of chemical control of both infections and of pests. Paracelsus had introduced the use of inorganic arsenicals, notably arsenic trioxide (As_2O_3, white arsenic), into medicine in the sixteenth century but its use was limited by its extreme toxicity. Ehrlich's discovery revived interest in these compounds. In 1824 the Colorado potato beetle was discovered east of the Rockies and its eastward spread accelerated the search for an effective control. As_3O_3 was found to be effective and came into widespread use. Other arsenicals were developed, including Paris green (copper arsenite) which is still used as slug bait. Being a heavy metal, arsenic is persistent in the environment, the significance of which was not appreciated when it was being widely used.

Natural-source insecticides also evolved fairly early on. Certain plants have been employed as fish poisons in Southeast Asia and in South America for centuries and in 1848 a decoction of derris root was used to control an insect infestation in a nutmeg plantation in Singapore. By 1920, large amounts of derris root were being imported into North America. The active ingredient is rotenone, and it has the advantages of low mammalian toxicity and a short $t_{1/2}$ in nature. Pyrethrum flowers (chrysanthemums) have been known for their insecticidal properties for centuries. Commercial manufacture began in 1828. In 1945, the United States imported 13.5 million pounds. By 1954, this had fallen to 6.5 million because of the widespread use of DDT, the banning of which has led to a resurgence of use of pyrethrin compounds. Nicotine sulfate (Blackleaf 40 is a 40% solution) from tobacco is used to control aphids and other insects. It has a short biological $t_{1/2}$ but significant mammalian toxicity.

The mechanization of farming led to a second agricultural revolution by making possible the planting and cultivation of vast tracts of land. Pest control techniques also changed from the small-scale operations of the past to include mechanized spraying from the ground and the air. This involved a marked increase in the use of pesticides and it coincided with the introduction of the first modern synthetic insecticide, DDT.

Dichlorodiphenyltrichloroethane, or DDT, was first synthesized in 1874 but its insecticidal properties were not recognized until 1939. Its structural formula is shown in Figure 31. Its first major use occurred in Sicily in 1943, where it was used to halt an epidemic of tick-borne typhus.

Figure 31 DDT: chemically 1,1,1-trichloro-2,2-bis- (p-chlorophenyl) ethane.

Sometimes called the grandfather of all chlorinated aromatic hydrocarbons, DDT was the first of such agents to arouse environmental concern. Rachael Carson's *Silent Spring* called attention to the ecological damage caused by DDT and led to its banning in the United States and Canada in 1972. Prior to that, however, its use had led to the eradication of malaria in 37 countries and dramatically reduced its incidence in a further 80, providing relief to 1.5 billion people. Its effectiveness in controlling agricultural pests, coupled with its low mammalian toxicity (oral LD_{50} = 113 mg/kg, dermal LD_{50} = 2.5 gm/kg), resulted in extensive use in North America. U.S. production reached 50,000 metric tonnes annually. The availability of cheap surplus aircraft after World War II resulted in the spraying of huge areas to control pests of not only agricultural but human as well. Organochlorines, including the cyclodienes, dominated the insecticide field until the early 1960s, when organophosphates and carbamates were developed. These, plus the development of more disease-resistant hybrid crops, led to the Green Revolution of the 1960s, with dramatic increases in food production.

Classes of insecticides

Organochlorines (chlorinated hydrocarbons)

As already discussed, the parent compound of this group is DDT. Its human toxicity is extremely low. In one rather heroic experiment, volunteers were fed 35 mg/day for up to 25 months without obvious ill effects. Another study of 35 male workers who had DDT levels in fat and liver 80 times the American average, and who had worked in a manufacturing plant for up to 19 years, showed no ill effects. DDT is, however, a potent inducer of cytochrome P450 hepatic microsomal enzymes and may thus affect the rate of biotransformation of other chemicals and drugs. Extremely high doses cause neurological signs and symptoms, including numbness of the tongue, lips, and face; dizziness; hyperexcitability; tremor; and convulsions.

DDT has very high lipid solubility and it is sequestered in body fat. Virtually everyone who was alive after 1940 has DDT in body fat. In the 1960s, significant amounts were found in people all over the world from Sri Lanka to North America. In 1970, the mean concentration in human fat was 7.88 ppm. After the ban, it fell to 4.99 in 1975. There is no evidence that chronic exposure to DDT has resulted in any health problems. In insects, DDT opens up ion channels to prevent normal axonal repolarization. Disorganized neuronal function leads to death.

Other life forms are not as resistant as humans. Fish are extremely vulnerable, and dieoffs have occurred after heavy rains washed DDT into streams and lakes. Deformities also occur. Predatory birds at the top of the food chain are very vulnerable as well. Reproduction is disturbed in a number of ways. DDT induces cytochrome P450 to increase estrogen metabolism and DDT itself has estrogenic activity that affects fertility. Ca^{2+}-ATPase is inhibited, as is calcium deposition in eggshells. This effect is largely due to stable metabolites, notably DDE (dichlorodiphenyldichloroethane). Some bird species are only now recovering. The limited use of DDT against the tussock moth was re-approved in the United States in 1974 and its use in malarial areas has continued without interruption, so that DDT exposure on a worldwide basis still occurs.

The cyclodienes are a subgroup of the organochlorines. This group includes aldrin, dieldrin, heptachlor, and chlordane. Their mechanism of insecticidal action is the same as for DDT, but their toxicity for humans is much greater because of more efficient transdermal absorption. Signs of excessive CNS excitation and convulsions occur before less serious signs appear. Several deaths, mostly in those who handled the pesticide, have occurred. These agents are also persistent in the environment. There is concern about their potential for carcinogenicity because this has been shown in some animals. However, Ribbens reported on a study of 232 male workers who had been exposed to high levels of cyclodienes in a manufacturing plant in Holland for up to 24 years (mean = 11 years). Mortality and cancer incidence were compared to the means for the Dutch male population of the same age group. The observed mortality in the group was 25, which was significantly lower than the expected mortality of 38. Nine of the deaths were from cancer, as opposed to an expected incidence of 12. These workers had been exposed to very high levels of cyclodienes in the early days of manufacture, with recorded dieldrin blood levels of up to 69 µg/L at some time in their history.

Other organochlorines include methoxychlor, lindane, toxaphene, mirex, and chlordecone (kepone). Mirex and kepone are extremely persistent, toxic to mammals (CNS toxicity), and carcinogenic in animals. They also induce cytochrome P450. They are no longer used in North America. Lindane shares the same toxicity but is much less persistent and is used to treat head lice. Lindane (chemically 1,2,3,4,5,6-hexachlorocyclohexane) is the active isomer of benzene hexachloride. Toxaphene induces liver tumors in mice and is fairly toxic; its use is declining. Methoxychlor is similar to DDT but it is much less persistent and less toxic to mammals, which can metabolize it. It also is stored in fat to a much lesser degree. Its formula, along with that of lindane, is shown in Figure 32. The carcinogenic and reproductive toxicity of organochlorines is dealt with in Chapter 12.

Figure 32 Chemcial structures of methoxychlor and lindane.

Organophosphorus insecticides

These insecticides, often referred to as organophosphates, are the most frequent cause of human poisonings. The group includes parathion, dichlorvos (present in Vapona strips), and diazinon. They all act as irreversible inhibitors of acetylcholinesterase, so that the neurotransmitter acetylcholine is not inactivated following its release from the nerve terminal. Signs and symptoms are those of a massive cholinergic discharge and include dizziness and disorientation, profuse sweating, profuse diarrhea, constricted pupils, and bradycardia (slowing of the heart), possibly with arrhythmias. Parathion has a dermal LD_{50} of 21 mg/kg and an oral LD_{50} of 13 mg/kg in male rats, but the NOEL in both rats and humans is only 0.05 mg/kg. Parathion itself is not toxic but it is transformed in the liver to para-oxone, its oxygen analog (see Chapter 1, Figure 3).

The following is a typical case history of organophosphorus poisoning. A 52-year-old white, male farmer was admitted to a hospital emergency department following a highway accident in which his tractor collided with the rear of a motor vehicle about to make a turn. He incurred numerous lacerations and contusions and a fractured right humerus. He was restless and incoherent and required physical restraint. His pupils were bilaterally constricted, his heart rate was 55 beats/min, and he was sweating profusely. His clothing had a strong, chemical odor. His wife volunteered that he had had several episodes of visual difficulty over the preceding 2 weeks. Further questioning revealed that he had been spraying organophosphorus insecticides during this period (organophosphorus poisoning is frequently delayed). Atropine was given intravenously in repeated small doses until the signs of cholinergic discharge abated. Another drug that can be used is pralidoxime, which complexes with the phosphate component of the organophosphorus and releases the cholinesterase. The principal advantage of the organophosphates is their short life in the environment. The sites of action of organophosphates, atropine, and pralidoxime are shown in Figure 33.

Delayed neuropathy is another form of organophosphorus poisoning. In the 1930s, during prohibition, thousands of people in the southern United States became paralyzed after drinking an alcohol extract of Jamaica ginger. The preparation was found to be contaminated with triorthocresylphosphate. The first clinical cases associated with the insecticides occurred in a pilot plant making an organophosphorus insecticide (mipafox) in 1953. Administration of the chemical to chickens caused similar neurological disturbances, including paralysis, and pathology of the axons. In humans, the clinical picture is one of severe polyneuritis commencing a few days after a sufficient single exposure or cumulative exposures. Initially, mild sensory disturbances, weakness, and fatigue, especially of the legs, are seen. The condition can progress to flaccid paralysis and eventually to spastic paralysis. The medium and large peripheral nerves are damaged, with axonopathy being the principal lesion and demyelination also occurring. Treatment involves massive doses of atropine. Inhibition of acetylcholinesterase is not

Figure 33 Sites of action of organophosphate insecticides, atropine, and pralidoxime. Although the neurotransmitter site is labeled a synapse, atropine is primarily a muscarinic receptor-blocking agent, acting at parasympathetic effector junctions. Acetylcholinesterase is present there, as well as in all ganglia, at the neuromuscular junction, the brain, and the adrenal gland.

involved in this toxic reaction. Current theory is that some, but not all, organophosphorus agents inhibit an esterase essential for axonal function. Only certain triarylphosphates and fluorine-containing alkylphosphates appear able to induce the lesions, both clinically and experimentally. The esterase has been named neuropathy target esterase (NTE). Evidence for this enzyme is indirect because it has not been isolated in active form and no physiological role for it has been established. It appears to be tightly bound to the nerve membrane.

Carbamate insecticides

Carbamates (e.g., Sevin) are also inhibitors of acetylcholinesterase, but they do not require metabolic activation and they are reversible. They are not persistent in the environment. Because they lack the phosphate group, pralidoxime cannot be used for treatment of poisoning. In fact, it is contraindicated because it may tie up more reactive sites on the enzyme and increase the degree of inhibition. This group includes aldicarb (Temik®), carbaryl, and Baygon®. The dermal LD_{50} for aldicarb in male rats is 3.0 mg/kg. It is also fairly toxic to humans. Although these agents are generally not persistent in the environment, aldicarb may be an exception. Under certain conditions (sandy soil over aquifers), it may reach water supplies and persist for a considerable time. In Long Island, New York, it has been estimated that the levels of 6 ppb may persist for up to 20 years.

Botanical insecticides

The more common botanical insecticides were discussed briefly above. While it is commonly felt that natural-source insecticides are safer than synthetic ones (another example of the "nature knows best" syndrome), this is not necessarily so. Pyrethrins and rotenone have oral LD_{50} values of about 600 to 900 mg/kg and 100 to 300 mg/kg, respectively. Nicotine is quite toxic, with an oral LD_{50} of 10 to 60 mg/kg. The main problem with pyrethrins has been the rapidity with which they are destroyed in the environment. Newer ones have been isolated with longer half-lives to permit more effective kills.

Herbicides

Chlorphenoxy compounds

These agents, characterized by 2,4-D and 2,4,5-T, act as growth hormones, forcing plant growth to outstrip the ability to provide nutrients. They are employed as a variety of salts and esters. The acute toxicity of these agents is relatively low, with LD_{50} values of 300 to 1000 mg/kg reported for several species of mammals. The dog may be more sensitive (LD_{50} = 100 mg/kg). Ventricular fibrillation appears to be the immediate cause of death. Acute toxicity in humans is manifested largely as chloracne.

The main concern about 2,4,5-T is the likelihood of its contamination with dioxin (TCDD). This subject is dealt with in Chapters 2 and 4. The chemical structures of these compounds are shown in Figure 34.

Dinitrophenols

Several substituted dinitrophenols are used as herbicides, the most common probably being Dinoseb (see Figure 35). It has been reported to have an LD_{50} of 20 to 50 mg/kg in rats. Dinoseb, first registered in 1947, is out of favor because handlers may be at considerable risk for teratogenic effects,

2, 4-D
(2 , 4-dichlorphenoxyacetic acid)

2 , 4, 5-T
(2 , 4, 5-trichlorphenoxyacetic acid)

Figure 34 Chemical structures of 2,4-D and 2,4,5-T.

Figure 35 Chemical structures of Dinoseb and DNOC.

cataracts, and male reproductive disturbances, even when protective clothing is worn. The U.S. EPA suspended all use in 1987.

4,6-Dinitro-*o*-cresol (DNOC, see Figure 35) has caused acute poisoning in humans with signs and symptoms including nausea, vomiting, restlessness, and flushing of the skin, progressing to collapse and coma. Hyperthermia may occur. Death may ensue in 24 to 48 hr. Uncoupling of oxidative phosphorylation is probably the mechanism of toxicity. Atropine is contraindicated because there is no anticholinesterase activity and the CNS effects of atropine may complicate the outcome. Treatment is symptomatic and includes ice baths to reduce fever, intravenous fluids, and the administration of O_2.

Bipyridyls

Paraquat and diquat are the most familiar members of this group (see Figure 36). Both are toxic, but their toxicity differs. The principal organ of toxicity for paraquat is the lungs, although the liver and kidney may also be damaged. Respiratory failure may be delayed for several days after the ingestion of paraquat. It appears to be selectively concentrated in the lungs by an energy-dependent system. Paraquat is believed to undergo conversion to superoxide radical (O_2^-), which causes the formation of unstable lipid hydroperoxides in cell membranes. Widespread fibroblast formation occurs and O_2 transfer to capillary blood is impaired.

Figure 36 Chemical structures of paraquat and diquat.

Treatment consists of attempts to remove or neutralize any paraquat remaining in the gastrointestinal tract by gastric lavage, cathartics, and Fuller's earth as an adsorbant. In complete lung failure, double lung organ transplant offers the only hope for recovery.

In contrast, diquat toxicity is centered on the liver, kidney, and gastrointestinal tract. Superoxide anion formation is also believed to play a role in these organs. Poisoning with paraquat is far more common and it has been used as an instrument of suicide on numerous occasions.

Carbamate herbicides

Unlike the insecticide carbamates, the herbicides do not possess anticholinesterase activity. They have low acute toxicity. Dithiocarbamates are used as fungicides and have similar low acute toxicity; LD_{50} values for these agents are in the g/kg range for rodents.

Triazines

This group, typified by atrazine, is also characterized by low acute toxicity. Amitrole is a herbicide somewhat related to the triazines. It has similar low acute toxicity, but has peroxidase-inhibiting activity and has been associated with tumor formation in the thyroids of rats fed the chemical for 2 years.

Fungicides

A wide variety of agents has been used for their fungicidal properties, some of them quite toxic. Seed grains treated with mercurials have sometimes entered the human food supply with disastrous results (see Chapter 6). Pentachlorophenol and hexachlorobenzene are halogenated hydrocarbons with the toxicity typical of that group (see Chapter 4). Thiabendazole is a fungicide of low toxicity as evidenced by the fact that it is also used as an anthelmintic in domestic animals and humans for the eradication of roundworms.

Dicarboximides

Captan® and Folpet® are agents of some concern. Structurally similar to thalidomide, they have been shown to possess similar teratogenic properties in the chick embryo. Captan has been shown to be mutagenic, carcinogenic, and immunotoxic in animals. The EPA has judged Folpet to be a probable human carcinogen with a lifetime risk of cancer of 2 per million for lettuce and small fruits, and a total of 5.5 per million when all food sources are combined.

Newer biological control methods

The earliest form of biological control no doubt was the development of strains of plants and animals with a high degree of resistance to disease, through selective breeding. Observant farmers probably began this process soon after the domestication process began, and it continues today. Over 40 years ago, as a high-school student, this author worked with Professor Waddell who developed, at the Ontario Agricultural College, the first strains of wheat to be resistant to wheat rust, a fungal infestation. Recently, a strain of American elms with a high degree of resistance to Dutch elm disease has been developed. Ladybugs have been bred and released to control the cottony cushion scale on oranges in California, and *Bacillus thuringiensis* is used to control forest pests.

One of the earliest, high-tech biological controls was developed in the 1950s and involves sterilization by radiation of millions of male insects that are then released to mate with the females. In species in which the female only mates once, this results in a high frequency of infertile unions with a resulting decline in the insect population. This method was first used successfully to control the screwworm fly in the southern United States. This fly lays its eggs in wounds in the skin of cattle and other livestock. The larvae then live on the flesh of the unwilling host. By 1966, the screwworm had been successfully eradicated in the United States and northern Mexico. It recently resurfaced in Libya, creating a political dilemma for the United States. Withholding technological assistance could result in massive infestations throughout Africa (the fly will also lay its eggs in wounds on humans), but the alternative was to offer help to Quaddafi. Humanitarian considerations prevailed. This form of biological control has also been used more recently to control the Mediterranean fruit fly in California.

Analogs of insect hormones have been developed that are highly specific to a given species. These hormones trigger the molting metamorphosis in the larval stage so that the larva cannot develop normally and dies.

Pathogenic bacteria exist that can be cultured in commercial quantities and released to control specific pests. Some agents have been genetically modified for this purpose, but public concerns about "superbugs" have blocked approval of all but a few of these. Given that there are no known bacteria that are infectious for both insects and mammals (as opposed to insects being vectors for infection), this fear seems unjustified. A more legitimate concern is that beneficial or harmless species may also be attacked by the organisms (see also Chapter 14).

Government regulation of pesticides

Most governments have regulations governing the use of pesticides. The Canadian regulations are fairly typical of those in place in industrialized countries. The Pest Control Products Act, administered by Agriculture Canada, regulates the introduction of new pesticides. The risk-benefit principle is applied to decisions, that is, the degree of risk must be acceptable in light of the potential benefit to be derived from pest control with the new agent. Its relative safety and effectiveness compared to existing pesticides will influence the decision.

Problems associated with pesticides

Development of resistance

Insects, like microorganisms, possess the most important characteristics for the evolution of resistant strains: an extremely short reproductive cycle and the production of vast numbers of progeny. Most species of insects can go through many generations in one season, producing millions of offspring. There is thus the capacity for multiple, sequential mutations to occur, and a good chance that some of these will be resistant to one or more insecticide. The development of resistance requires the presence of the appropriate insecticide to select out the resistant strain (by killing off the susceptible ones) with the means of detoxifying the chemical or excluding it from absorption. Only 3 years after the introduction of DDT, houseflies and mosquitoes were showing signs of resistance; and by 1951, DDT, methoxychlor, chlordane, heptachlor, and benzene hexachloride (of which lindane is the active isomer) no longer had any effect on houseflies which proliferated abundantly. By the end of 1980, 428 species of insects and acarines (mites, ticks) were classified as resistant.

In the pre-DDT era, relatively few species developed resistance. This has been attributed to the multi-site mechanisms of action of earlier pesticides (making single mutation resistance unlikely) and to their ionic nature, making detoxification by metabolism nearly impossible.

Closely related insecticides can be detoxified by the same mechanism and generally act at the same target site, so that if resistance evolves to one, either by the evolution of a detoxification process or by modification of the target molecules, cross-resistance to the others will occur. This type of resistance tends to be under the control of a single gene allele or closely linked genes. Cross-resistance to DDT and methoxychlor, aldrin, and heptachlor has developed in this manner.

Multiple resistance

In common with bacteria, protozoa, and cancer cells, insects can develop resistance to several insecticides. This is referred to as multiple resistance. It is the result of the existence of several, independent gene alleles producing resistance to unrelated agents (e.g., organochlorines and synthetic pyrethrins, called pyrethroids) with different modes of action and different detoxification pathways. It can be a very serious problem of insect control. The mechanism of multiple resistance is obviously quite different from that of bacteria, which is discussed in Chapter 8.

Nonspecificity

Broad-spectrum pesticides, as most are, make no distinction between true pests and species that are harmless or even beneficial. They, therefore, may disrupt the natural competition between species to permit the proliferation of one previously held in check or, as has been observed, they may kill off predator species and permit the expansion of a prey species, which then becomes a pest. This has happened with spider mites.

Environmental contamination

A greater danger than direct toxicity to humankind is probably the contamination of the environment and the subsequent bioaccumulation and biomagnification that occur with persistent pesticides. These chemicals may end up in soil, water, air, or all three, depending on their characteristics (see Chapters 3 and 5). The Great Lakes are accumulating hazardous chemicals as a result of agricultural runoff and industrial discharges, frequently accidental ones. It must be stressed, however, that the greatest source of chemical contamination is residential sewage. Even after treatment, phosphates and other household chemicals may enter the water system. The water table in many areas has been contaminated with the herbicide atrazine, commonly used in cornfields.

Callous disregard for the environment can be the result of greedy individuals attempting to increase their profit margin. There have been numerous cases in Toronto of trucks dumping toxic wastes into Lake Ontario after dark, in violation of provincial and municipal laws. In the United States, careless dumping of chlordecone in the James River by a chemical company resulted in a ban on fishing. This cyclodiene is used in ant and roach baits.

Balancing the risks and the benefits

The widespread use of pesticides means that there are trace quantities present in or on almost all foodstuffs. Major advances in analytical techniques over the past 20 years mean that chemicals can now be detected at levels never before possible. Headlines proclaiming that dioxins (or PCBs, etc.) have been detected in Lake Erie fish seldom go on to say that the quantities were at the parts-per-billion level. The detection of dioxins in milk, leached from the carton paper, is an example of this. Actual levels were comparable to a drop in an Olympic-sized swimming pool.

Nonetheless, there is a growing feeling among the public that the use of pesticides should be greatly curtailed because the risks are unacceptable or, what is almost as bad, unconfirmed. In recognition of this, the Ontario Ministry of Agriculture and Food has established the Foodsystems 2002 program, which will attempt to reduce the use of pesticides in agriculture in Ontario by 50% by the year 2002. However, the impact of pests on food production is so great that temporary approval is sometimes granted to new agents before all of the required tests have been completed. In 1977, it was discovered that an American testing company, Industrial Bio-Test Laboratories, had misrepresented toxicological data on numerous agents. Of the 405 pesticides registered in Canada, 106 had been approved partly on the basis of Bio-Test data. In 1983, Health and Welfare Canada announced that five of these were to be withdrawn, one being the fungicide Captan, which was found to be teratogenic and carcinogenic.

Other pesticides that have been banned or that are under investigation include:

Chlordane. This cyclodiene was withdrawn in 1986 because it was considered to be an epigenetic tumor promoter. It is still registered for use against termites.

Alachlor. This herbicide was withdrawn in 1985 because of evidence of carcinogenicity in animals. It was introduced in 1969 and widely used on corn and soybean crops.

Cyhexatin. Dow chemical voluntarily withdrew this insecticide because of evidence of teratogenicity (hydrocephaly) in rabbits.

Toxicity of pesticides for humans

Pesticide applicators are at risk primarily from inhalation and dermal contact with pesticides. Protective clothing and equipment are important means of reducing risk. Nonoccupational poisonings occur largely from oral ingestion of contaminated food although dermal exposure has resulted in poisonings in infants (the pentachlorophenol treatment of hospital linens) and inhalation exposure from spray drift can occur. Household pesticides can cause poisoning by all three routes.

There have been a few isolated fatalities in North America from acute pesticide poisoning, but elsewhere in the world, many cases of mass poisonings have occurred. Consumption of seed grains treated with hexachlorobenzene and organic mercury has resulted in mass outbreaks of poisoning in Turkey, West Pakistan, Iraq, and Guatemala. Accidental contamination of foodstuffs such as flour, sugar, and grain with parathion and others has occurred in several places around the world.

Effects of long-term exposure to very low levels of pesticides on human health remain conjectural but evidence is accumulating that they exist. In one study in Great Britain, a battery of neuropsychological performance tests was administered to two groups of males, 16 to 65 years of age. One was a group of 146 sheep farmers exposed to organophosphates in sheep dip over several years. The other was a group of matched quarry workers not exposed to organophosphates. The farmers performed significantly worse on tests of short-term memory and cognitive function. The tests included simple reaction time, symbol digit substitution, digit span, serial word learning, and others. There is also the possibility that contaminants may emerge as a greater risk than the pesticide itself, as was the case with TCDD.

There is no doubt that persistent poisons can have a catastrophic effect on the environment and this is probably the most compelling argument for limiting their use. In 1996, over 4000 Swainson's hawks were found dead in a 50 km² area in Argentina. The deaths were attributed to the spraying of crops to kill grasshoppers. The hawks follow tractors that stir up the grasshoppers and the hawks then feast on them. Ecologists feared that losses may reach 20,000 birds out of a world population of about 400,000.

There is some hope that nature may be developing her own protective processes against pesticides. There is some new evidence that fields sprayed with the same chemicals year after year may develop a population of bacteria that break down the pesticides and may even adapt to the point that they use them as a food source. Tests on prairie soils indicated that the organism *Rhodococcus* breaks down thiocarbamate insecticides in the test tube within 2 hours. It may be possible through gene splicing to develop plants that protect themselves against pesticide residues.

Further reading

Aldridge, W.N., *Mechanisms and Concepts in Toxicology*, Taylor & Francis Ltd., London, 1996.

Carson, R., *Silent Spring*, Fawcett Crest Books, Greenwich, 1962.

Claus, E.P. and Tyler, V.E., Jr., *Pharmacognosy*, Lea and Febiger, Philadelphia, 1967.

Flanders, R.V., Potential for biological control in urban environments, in *Advances in Urban Pest Management*, Bennett, G.W. and Owens, J.M., Eds., Van Nostrand Reinhold, New York, 1986, 95–129.

Garner, R.J., *Veterinary Toxicology*, Bailliere, Tindall, London, 1957.

Georghiou, G.P. and Saito, T., Eds., *Pest Resistance to Pesticides*, Plenum Press, New York, 1983.

Hardman, J.G. and Limbird, L., Eds. in Chief, *Goodman and Gilman's The Pharmacological Basis of Therapeutics*, 9th ed., McGraw-Hill, New York, 1996.

Hall, S.H. and Dull, B.J., Comparison of the carcinogenic risks of naturally occurring and adventitious substances in food, in *Food Toxicology: A Perspective on the Relative Risks*, Taylor, S.L. and Scanlon, R.A., Eds., Marcel Dekker, New York, 1988, 205–224.

Klassen, C.D., Amdur, M.O., and Doull, J., Eds., *Casarett and Doull's Toxicology: The Basic Science of Poisons*, 5th ed., McGraw-Hill, New York, 1996.

Murphy, S.D., Toxic effects of pesticides, in *Casarett and Doull's Toxicology*, 3rd ed., Klassen, C.D., Amdur, M.O., and Doull, J., Eds., Macmillan, New York, 1986, 519–581.

McEwen, F.L. and Stephenson, G.R., *The Use and Significance of Pesticides in the Environment*, Wiley Interscience, New York, 1979.

Palca, J., Libya gets unwelcome visitor from the west, *Science*, 249, 117–118, 1990.

Ribbens, P.H., Mortality study of industrial workers exposed to aldrin, dieldrin and endrin, *Int. Arch. Occup. Environ. Health*, 56, 1985, 75–79.

Schneider, M-J., *Persistent Poisons*, New York Academy of Science, New York, 1979.

Stephens, R., Spurgeon, A., Calvert, I.A., Beach, J., Levy, L.S., Berry, H., and Harrington, J.M., Neuropsychological effects of long-term exposure to organophosphates in sheep dip, *Lancet*, 345, 1135–39, 1995.

Williams, D.E., Dashwood, R.H. et al., Anticarcinogens and tumor promoters in foods, in *Food Toxicology: A Perspective on the Relative Risks*, Taylor, S.L. and Scanlan, R.A., Eds., Marcel Dekker, New York, 1989, 101–150.

Review questions

For Questions 1 to 8, use the following code:

Answer A if statements a, b, and c are correct.
Answer B if statements a and c are correct.
Answer C if statements b and d are correct.
Answer D if only statement d is correct.
Answer C if all statements (a, b, c, d) are correct.

1. Which of the following statements is/are true?
   a. Lindane is commonly used for the control of head lice.
   b. Toxicological testing is usually performed only on the active ingredient of an insecticide.
   c. Lindane is an organochlorine.
   d. A cause-and-effect relationship for progressive chronic disease resulting from prolonged pesticide use is well-established.
2. Which of the following is/are true?
   a. The herbicide "atrazine" has contaminated the water table in many areas.
   b. Eggshell strength is adversely affected by contact with pyrethroids by the female bird.
   c. The fungicide "Captan" is a teratogen and carcinogen in experimental animals.
   d. Natural pesticides are always safer than synthetic ones.
3. The main reason(s) for the carcinogenicity in animals of the herbicides 2,4,-D and 2,4,5,-T is/are:
   a. Their action as growth hormones.
   b. Their mutagenicity.
   c. Their photosensitizing properties.
   d. The presence of dioxin (TCDD) as a contaminant.
4. The mechanism of TCDD carcinogenicity in humans may involve:
   a. Its pleiotropic response to the Ah locus.
   b. Its conjugation with glutathione.
   c. Its lack of gene restriction.
   d. Its ability to cause chloracne.
5. The insecticide parathion:
   a. Is an organochlorine.
   b. Is biotransformed by a mixed-function oxidase to its toxic form.
   c. Biomagnifies in the environment.
   d. Causes symptoms that can be treated with atropine.

6. Which of the following statements concerning DDT is/are incorrect?
   a. It has a low acute LD_{50}.
   b. It is very persistent in the environment.
   c. It is very water soluble.
   d. It is still used for mosquito control in many places.
7. Which of the following statements is/are true?
   a. Organophosphates are very persistent in the environment.
   b. Organophosphates are inhibitors of acetylcholinesterase.
   c. Carbamate insecticide poisoning can be treated with pralidoxime.
   d. Carbamate insecticides act as reversible inhibitors of acetylcholinesterase.
8. Which of the following statements is/are true about insect resistance to insecticides?
   a. Over 400 species of insects, mites, etc. showed resistance to pesticides by 1980.
   b. Resistance to pre-organic insecticides is especially prevalent.
   c. Cross-resistance to closely related chemicals may occur.
   d. Resistance is more likely to occur if an insecticide has several sites of action.

Answers

1. A
2. B
3. D
4. A
5. C
6. C
7. C
8. B

Case study 18

A 43-year-old male crop-duster was admitted to the emergency department of a rural hospital following an accident in which the aircraft he was attempting to land on a grass strip hit hard, collapsed the undercarriage, and nosed over. The pilot suffered numerous lacerations and bruises but no serious injuries. He was restless and incoherent and he had to be physically restrained. A rapid breath alcohol test was performed and it was negative. His pupils were constricted, his heart rate was slowed, and he was sweating profusely. His ground assistant volunteered the information that the pilot had complained of visual disturbances on several occasions during the previous few days. His clothing smelled strongly of a chemical.

Q. Some of these symptoms could be due to a head injury or to chemical intoxication. What facts point to the latter?

Q. What information would you want to seek from his assistant?

The ground crew revealed that the pilot had been spraying crops with parathion during the preceding two weeks.

Q. What class of pesticide is this?

Q. What drugs would be indicated for treatment?

Q. How would treatment differ if a carbamate insecticide such as Sevin had been used?

Q. What blood test might assist in confirming the diagnosis?

Case study 19

During the months of June and August of 1993, 26 men, 19 to 72 years of age, were admitted to three different local hospitals with an array of symptoms that included nausea, vomiting, dizziness, visual disturbances, muscle weakness, abdominal pain, headache, sweating, and excessive salivation. The men all worked in apple orchards, 19 different ones in total.

Q. What do these symptoms suggest?

Q. What inquiries would you want to make of these men?

Q. What inquiries would you want to make of the orchard operators?

chapter ten

Mycotoxins and other toxins from unicellular organisms

> Anything green, that grew out of the mould,
> Was a wonderful drug to our fathers of old.

> **— Anonymous**

Introduction

Mycotoxins are a group of chemically diverse and complex substances present in a wide variety of filamentous fungi (molds). They are secondary metabolites of fungal metabolism of uncertain function. Some may have a survival advantage by virtue of their toxicity to competing organisms in the microenvironment. The biological function of others is unclear, but many have significant biological activity and several are toxic to mammals. They therefore have significant public health and economic implications. It is estimated that 25% of the world's annual food crops are contaminated by mycotoxins.

Some mycotoxins are of interest to pharmacologists and toxicologists because they have served as research tools to study cell function and to identify various types of neurotransmitters and blocking agents. Poisonous mushrooms are solid (not filamentous), spore-forming fungi that also constitute a health hazard and which have historic, pharmacological significance, but they are considered separately in Chapter 11.

Some human health problems due to mycotoxins

Reports of toxicity from molds are as old as recorded history.

Ergotism

The oldest recorded source of fungus poisoning is ergotism. Ergot is the common name for the fungus *Claviceps purpurea* that affects cereal grains, especially rye, and that produces a number of very potent pharmacologically active agents that cause toxic reactions when people eat bread made from contaminated flour. Periodic epidemics of ergotism have occurred throughout history, and in medieval times these were often attributed to supernatural causes. The earliest reference to ergot seems to have been on an Assyrian tablet circa 6000 B.C. which refers to "a noxious pustule in the ear of grain." The Parsees, an ancient religious community in India, referred in their writings to noxious grasses that caused women to abort and to die shortly thereafter. The ancient Greeks escaped the scourge of ergotism because they never developed a taste for rye bread, which they referred to by a phrase that translates roughly as "that filthy Macedonian muck." Because rye bread did not reach Europe until after the decline of the Roman Empire, few if any references to ergotism exist in Roman writings.

There was an early association of ergot with St. Anthony of Egypt, who lived sometime between 250 and 350 A.D. St. Anthony is considered to be the founder of Christian monastic life and spent long sojourns in the desert, experiencing visions and hallucinations. These frequently involved attacks by Satan in various guises (wild beasts, soldiers, and women) that either physically attacked him or tempted him. Contemporary witnesses claimed that he behaved like an individual being physically abused. Because the signs and symptoms of ergot poisoning resembled his attacks, sufferers in the Middle Ages attached religious significance to them and they came to be known as "St. Anthony's Fire" (or sometimes "Holy Fire"). In 1100, the Monastery of the Hospitillers of St. Anthony was established at La Motte in France. It became the site of pilgrimages by those afflicted with this malady. The signs and symptoms included intense burning pain in the extremities followed by a blackened, necrotic appearance; hence the association with fire. Epidemics of madness in the Middle Ages may also have been the result of ergotism.

Ergot was employed as an abortifacient long before it was known to be the cause of St. Anthony's Fire, and this use continued well into the nineteenth century when it was finally abandoned because of its highly toxic nature. One of the components of ergot is still employed to stop *postpartum* hemorrhage because it constricts blood vessels and contracts the uterus. To summarize, the symptoms of ergotism include drowsiness, nausea, vomiting, muscle twitch, staggering, gangrene, hallucinations, and abortion. Ergot contains a veritable potpourri of pharmacological agents that were characterized by Sir Henry Dale in the early 1900s and contributed to the development of many new drugs.

The active components of ergot, the ergot alkaloids, are all derivatives of lysergic acid. They bear some molecular resemblance to adrenaline (epinephrine), dopamine, and serotonin — all central neurotransmitters. Methysergide is the precursor of LSD. Two derivatives still have medicinal application. Ergometrine (ergonovine) is the one used to control *postpartum* bleeding. Ergotamine is a powerful vasoconstrictor responsible for the gangrene of ergotism. In very small doses it can be used to treat migrane.

Ergotism also affects farm livestock, causing similar signs and, in addition, loss of milk production in cattle and necrotic combs, feet, and beaks in poultry.

Aleukia

Aleukia (literally "absence of white cells") occurs when millet and other grains that have become moldy are consumed. The toxin is a tricothecene from species of Fusarium and it damages bone marrow. Because of severe food shortages in Russia during World War II, the eating of moldy grain resulted in several epidemics of aleukia, including a very large one in 1944. The mortality rate of those with marrow damage was 60%. Other symptoms included hemorrhages in the skin and mucous membranes. The condition is also referred to as alimentary toxic aleukia.

Aflatoxins

Aflatoxins are a family of heterocyclic, oxygen-containing compounds secreted by the molds *Aspergillus flavus* and *A. parasiticus*. These molds grow abundantly on many kinds of plants in very hot conditions. Peanuts are especially prone to infection. Taste is not affected and the toxins are heat stable. Aflatoxin B_1 (AFB_1) is the most frequently encountered member of the group and it is a potent carcinogen in experimental animals (rodents, birds, and fish). Rats fed a diet containing 15 ppb AFB_1 develop hepatitis, which is often followed by cancer of the liver. Epidemiological studies of populations in Uganda, Kenya, and Thailand have shown a close correlation between the incidence of liver cancer and the consumption of food containing aflatoxins. It is estimated that the risk factor for liver cancer is increased 10 times by AFB_1. Infectious hepatitis also increases the risk by a factor of 10. If present together, the risk is increased 100-fold. Other aflatoxins include B_2, G_1, and G_2. Lactating cattle excrete hydroxylated metabolites M_1 and M_2 in milk. These have lower toxicity than the parent compounds but are of concern because of the large amount of milk consumed by infants and young children.

Thus, this natural carcinogen, at least as potent as TCDD in animal studies, is strongly implicated as a cause of human cancer, unlike the latter synthetic agent, that has received much more media coverage. Experiments employing human liver microsomal preparations indicate that cytochrome P450 (Cyp) 3A4 is largely responsible for the formation from AFB_1 of the 8,9-*endo*-epoxide, a genotoxic metabolite. Cyp 1A2 also forms this epoxide but is much less active. It forms other, nongenotoxic metabolites and thus may play a role in detoxification. An inhibitor of Cyp 1A2, alpha-naphthoflavone, increased the formation of the 8,9-epoxide. The *exo*-epoxide of AFB_1 (i.e., preformed before ingestion) is detoxified by conjugation with glutathione. Experimental evidence suggests that species differences in toxicity are due to differences in the rate and extent of conjugation with glutathione.

Whereas rat liver cytosol conjugated endo-epoxide, mouse liver cytosol conjugated the *exo*-epoxide almost exclusively and both were much more active than the cytosol of human liver cells. The mucus cells of the intestine also make the 8,9-epoxide, but because they are sloughed frequently, this does not constitute a risk factor for cancer.

Other aflatoxins found together are B_2, G_1, and G_2. Acute toxicity also can occur in humans. In 1974, an outbreak in India resulted in about 100 deaths, and in Kenya 12 died in an outbreak in the early 1980s.

The aflatoxin story began in 1960 with a serious and mysterious outbreak of a disease in turkeys in Great Britain. Turkey poults developed loss of appetite, feeble fluttering, lethargy, and frequently died within a few days. Necropsies revealed hemorrhage and necrosis of the liver and kidney. The disease was dubbed "X" disease. Outbreaks also occurred in ducklings in Europe, the United Kingdom, and Africa. The common denominator was groundnut (peanut) meal used as feed and subsequently shown to be contaminated with *A. flavus*. Livestock (mammals) may also be affected. Death has occurred in humans consuming an estimated 6 mg/day of aflatoxin B_1.

In North America, there is concern about the long-term effects of consuming low levels of aflatoxins and monitoring systems are in place. Fortunately, the mold is not adapted to the colder northern climate so that Canadian-grown peanuts are free of the toxins. The majority of peanuts and peanut butter are still imported from subtropical climes, however. Peanuts are not the only source of these toxins. Because of drought conditions in the United States in the summer of 1988, up to 30% of the corn crop may have been contaminated. The Quaker Oats Co. was turning away almost one truckload in five at its Cedar Rapids (Iowa) plant in the fall of 1988. The company tests six samples from each truck. Inspections of the 340,000 tons of corn crossing the Canada–U.S. border annually were stepped up in 1988–1989 by Agriculture Canada. The general structure of aflatoxins is shown in Figure 37.

Fumonisins

Produced by *Fusarium moniliforme* and *F. proliferatum*, these mycotoxins are ubiquitous in many parts of the world, including South Africa where they were first identified and many states bordering the Great Lakes. All of the fumonisins (FB_1, FB_2, FB_3) are potentially carcinogenic. Many experts feel that it will become the most significant mycotoxin for human health. FB_1 has been shown to be carcinogenic for animals (a promoter and initiator of liver cancer in rats), and there is a high degree of correlation between the incidence of esophageal cancer in humans and the presence of FB_1 in corn in specific areas of South Africa. The fungus infects corn, millet, sorghum, and rice around the world. Fumonisins are toxic for many species of animals, especially horses, and outbreaks have caused numerous deaths in horses and

AFB₁

AFB₁ 8,9-Epoxide

Fumonisin B₁: R₁ = 0-PTCA*, R₂ = R₃ = OH
Fumonisin B₂: R₁ = 0-PTCA, R₂ = H, R₃ = OH

***PTCA = Propanetricarboxylic acid**

Figure 37 Chemical structures of aflatoxins and fumonisins.

swine in Texas, Iowa, and Arizona. Severely infected corncobs may contain up to 900 mg/kg. Symptomatology in various species is as folllows:

- *Swine*: vomiting, convulsions, sudden death, abortion, pulmonary edema (porcine pulmonary edema syndrome). Symptoms may occur at levels above 20 to 50 mg/kg of FB_1.
- *Poultry*: ataxia, paralysis, sudden death, stunted growth. Toxicosis occurs at levels of contamination of 10 to 25 mg/kg of FB_1.
- *Cattle*: poor weight gain, liver damage.
- *Horses*: equine leukoencephalomalacia (ELEM), brain degeneration with focal necrosis, blindness, wild behavior, liver damage, staggering, ataxia. Symptoms may occur at levels of 10 mg FB_1/kg over 40 days.

FB_1, FB_2, and hydrolyzed FB_1 have been shown to be specific inhibitors of *de novo* sphingolipid synthesis and sphingolipid turnover. FB_1 has been shown to inhibit sphingosine (sphinganine) N-acetyltransferase, leading to the accumulation of sphyngoid bases. This has been shown to stimulate DNA synthesis, and it is hypothesized that this interference with normal cell function could account for the toxicity of fumonisins.

The fumonisins are themselves analogs of sphingosine, with structural similarities to the phorbol esters, which are known carcinogens (see Chapter 11). The general structure of fumonisins is shown in Figure 37.

Other mycotoxic hazards to human health

Several mycotoxins may be potential health hazards by virtue of direct toxic effects or because of carcinogenic or teratogenic properties, although it must be emphasized that hard evidence linking these to human health problems is scanty. Ochratoxin A is formed by *Aspergillus* and *Penicillium* species. It has been shown to be embryotoxic and teratogenic in several laboratory species of mammals (pig, dog, mouse, rat) and birds and is therefore viewed as a potential human teratogen. Acute effects in animals (including swine and poultry) include renal and hepatic destruction. Both humoral and cellular immune systems are adversely affected. Contamination of bread and cereals has been documented in parts of central Europe (Yugoslavia, Bulgaria, Poland, and Germany) and levels have been detected in human milk, urine, blood, and kidneys.

Patulin is potentially a carcinogenic toxin produced by several species of fungi, including some *Penicillium* spp. It is a highly potent inhibitor of RNA polymerase, having a strong affinity for sulfhydryl groups. It therefore inhibits many enzymes. Teratogenicity has not been demonstrated in mammals but embryolethality occurs at higher doses (2 mg/kg i.p.). A common source of patulin is *Penicillium expansum*, a common spoilage microorganism in apples. Apple juice can sometimes contain significant amounts of patulin. Acute toxicity in rodents is manifested primarily as gastrointestinal symptoms, including hemorrhage. Carcinogenicity studies in rats were negative, but clastogenic activity has been shown in some systems.

T-2 toxin is produced by various *Fusaria* and is both potent and common. It inhibits protein and DNA synthesis and is therefore potentially teratogenic and carcinogenic. Acute effects include loss of appetite and vomiting.

Fungi tend to produce mixtures of toxins so that exposure to a single agent is unlikely to occur. There is some experimental evidence that ochratoxin A potentiates the teratogenic effects of T-2. It must be emphasized that any mycotoxin that has been shown to be toxic in several mammalian species (including various farm livestock) must be regarded as potentially toxic for humans.

Economic impact of mycotoxins

In addition to their direct effects on human health, mycotoxins have a tremendous impact on agriculture. Through spoilage, field crops are rendered useless for animal or human consumption (the loss of 20% of the corn crop noted above is one example). Poor weight gain and outright illness occur in livestock that consume contaminated feeds. Losses are difficult to estimate but they undoubtedly run to many millions and possibly billions of dollars in North America. The presence of trace quantities in meat, dairy products, and eggs constitutes a further, if largely unconfirmed, health hazard to people. In the Great Lakes basin, various species of *Fusarium* are the most common offenders, especially in eastern Canada. They may produce a host of toxins with potent pharmacological actions.

Fusarium life cycle

Fusarium graminearum is the fungus responsible for maize ear rot in corn and head blight in wheat. Its life cycle is typical of all *Fusaria*. Spores survive in crop debris from the previous season (stubble, stalks and seeds) to re-infect the next year's crop. Intensive farming practices that involve planting susceptible species in the same fields year after year thus favor the spread of infection. Birds such as starlings and red-winged blackbirds, which puncture the corn kernels to eat the milk, may also spread spores. Insects such as the picnic or corn-sap beetle seek out damaged kernels and also may spread spores.

Certain weather conditions favor the spread of infection. Fungus growth is favored by warmth (15 to 35°C) and by surface wetness for more than 48 hr. After the infection is established, weather is not critical to the production of the toxin. Mold growth will continue throughout the season, and even afterward if not properly dried or if storage conditions are poor (too damp, too warm, poor air circulation). Mold growth can even occur during feed preparation and in poorly cleaned feed troughs. Late harvest may allow the growth of another fungus, *F. sporotrichiodes*, which produces the toxins T-2, HT-2, and diacetoxyscirpenol.

Contaminated grains thus will contain complex mixtures of toxins and metabolites. The following is a list of the other important ones and their effects.

Trichothecenes

Zearalonone

This estrogen-like toxin causes (in swine) swollen, red vulva, vaginal and rectal prolapse, vulval enlargement in piglets, and fertility problems. Developmental defects and lethality have been shown in some laboratory species.

Vomitoxin (deoxynivalenol or DON)

This trichothecene causes decreased feed intake and reduced weight gain in pigs at about 2 mg/kg of feed, vomiting, and refusal of feed at very high concentrations (>20 mg/kg feed). DON will be used as a "prototype" mycotoxin to illustrate agricultural problems associated with these agents (see below).

Species differences in DON toxicokinetics

Swine appear to be much more sensitive to the anorexic and weight loss effects of DON than ruminants (cattle, sheep) or poultry, which are very tolerant. In one study, laying hens actually preferred a diet containing 5 ppm DON in preference to clean feed. These differences are due, in part, to differences in absorption and, in part, to differences in biotransformation and elimination. Studies with radiolabeled DON indicated that sheep absorbed 9% or less of an orally administered dose, in turkeys 20% or less was absorbed, whereas pigs absorbed up to 85% of a single oral dose.

Intravenous administration of radiolabeled DON in sheep revealed an initial distribution phase ($t_{1/2}$ = 18 min) followed by an elimination phase ($t_{1/2}$ = 66 min). A glucuronide conjugate was formed and comprised 15 to 20% of plasma levels. In turkeys there was an extremely rapid distribution phase ($t_{1/2}$ = 3.6 min), a rapid elimination phase ($t_{1/2}$ = 46 min), and the formation of a conjugate (probably glucuronide) comprising up to 10% of the total dose. Again, swine showed a much different picture. There was a very rapid distribution phase ($t_{1/2}$ = 5.8 min), a secondary slower distribution phase ($t_{1/2}$ = 96.7 min), and a very prolonged terminal elimination phase (up to 510 min). There was no evidence of significant biotransformation in swine. Thus, it took 7 to 10 times longer for swine to clear the toxin than for the other two species.

Toxicity is a function of many factors: the concentration of toxin reaching the target organ (which in turn is affected by the rate of absorption at the portal of entry), the extent of distribution to non-target sites (i.e., where no toxic effects occur), the rate and extent of biotransformation to nontoxic metabolites (or to toxic ones as the case may be), and the rate of elimination in urine and feces. The effect of species differences in some of these factors was introduced in Chapter 2. Volume of distribution (Vd) and clearance data provide some information regarding the fate of the absorbed toxin. The apparent Vd is a mathematical calculation of the volume of diluent required to dilute an administered dose of a substance (usually intravenously) to the observed concentration. $Vd = M/C$ where M = mass (amount of substance) and C = concentration of substance. Calculations of Vd for DON yielded values of 0.167 L/kg for sheep vs. 1.3 L/kg for swine, suggesting that in the former DON was confined mainly to the extracellular fluid, whereas in the latter it was taken up by tissues. Initial systemic clearances were not all that different, being 1.37 mL/min/kg for sheep and 1.81 mL/min/kg for swine. An interpretation of this data suggests that DON is initially rapidly distributed to tissues and then slowly released back into the plasma, yielding the slow, terminal elimination phase. Turkeys also had a very large Vd (2.33 L/kg), but they also had an extremely rapid clearance (35.0 mL/min/kg), indicating that DON was rapidly distributed to tissue compartments but not held there.

Thus, the extreme sensitivity of swine to DON is the result of:

- High oral bioavailability
- Wide distribution to tissues
- Slow elimination from the body
- Minimal detoxification through biotransformation

Other trichothecenes

T-2 and HT-2 toxins and diacetoxyscirpenol are more toxic than DON and cause reduced feed intake, vomiting, irritation of the skin and gastrointestinal tract, neurotoxicity, teratogenicity, impaired immune function, and hemorrhage. Adverse effects seen in farm animals are generally caused by mixtures of these toxins rather than by a single toxin. Blending of several grains in the preparation of feed may further contribute to the toxic diversity of the mixture. Potentiation of effects may occur. Thus, DON at the subthreshold level of 1 mg/kg plus low (ppb) concentrations of T-2 and other unidentified toxins may cause severe toxic manifestations in a sensitive species such as swine. The chemical structures of some of these toxins are shown in Figure 38.

Zearalenone

Deoxynivalenol (DON, vomitoxin)

T - 2 : CH_3COO at R_1 and R_2, $(CH_3)_2CHCH_2COO$ at R_3

HT - 2 : OH at R_1, CH_3COO at R_2, $(CH_3)_2CHCH_2COO$ at R_3

Diacetoxyscirpenol (DAS) : CH_3COO at R_1 and R_2, H at R_3

Figure 38 Chemical structures of several mycotoxins.

Detoxification of grains

Because of the diverse chemical properties of the mycotoxins, physical and chemical procedures that are effective against one toxin may have little or no influence on the toxicity of others. Thus, there is no single process that can be used. The most important control factor must be the avoidance of conditions favoring fungal growth at all stages of food production.

Harvesting and milling

Infected kernels may represent less than 5% of all the grain. They may be broken or shriveled and in wheat may take on a "tombstone" appearance. In corn, the tips of cobs may have shriveled, highly infected kernels containing up to 3000 mg/kg of DON. Grain dust may be very contaminated. Screening and blowing will remove much of the dust, particles, and withered kernels.

Wet milling of corn has been shown to remove about two thirds of the T-2 toxin, but milling had little effect on the DON content of flour from hard wheat, nor did baking the flour into bread. Some milling procedures may actually increase the DON content of the finished product. In mild infections, washing and roasting may significantly reduce toxin levels.

Chemical treatments

Laboratory tests have shown that moist ozone, ammonia, microwaving, and convection heating reduce DON concentrations in moldy grain. Aqueous sodium bisulfite plus heat effected a complete detoxification. Studies have shown that this technique resulted in normal feed intake and weight gains when contaminated corn was treated and fed to swine.

Binding agents

The addition of binding agents, such as bentonite, anionic and cationic resins, and vermiculite-hydrobiotite were tested on the toxicity of T-2 in rats. Bentonite prevented T-2 toxicosis by blocking intestinal absorption. Polyvinylpyrrolidone or ammonium carbonate had no effect on DON toxicity in swine. Alfalfa fiber has been shown to partially overcome the growth-depressing effect of zearalenone in rats but not the estrogenic effects in swine.

Other techniques

Dilution of contaminated feed with clean feed will improve palatability and feed consumption. More concentrated diets with respect to calories, protein, etc. may overcome the effects of a moderate reduction in feed intake. Experimentally, antibodies against zearalenone have been raised in swine and shown promise in protecting against its toxic effects.

Regarding species differences in capacity to biotransform xenobiotics, evolutionary factors have played a role. Given the plethora of toxic substances in the plant kingdom, pure herbivores would have been exposed to the greatest risk of poisoning as compared to carnivores or omnivores and would therefore have needed to evolve nonspecific detoxifying enzymes. The toxin most likely encountered by pure carnivores, however, would have been the botulinum toxin, and their need would have been to evolve a method of eliminating it quickly before this potent neurotoxin could paralyze movement and respiration. Thus, canines and felines are very poor metabolizers of drugs (e.g., aspirin can be extremely toxic to cats and dogs) but they have exquisitely sensitive vomiting mechanisms to eliminate bad meat before the toxin is absorbed. Conversely, herbivores are efficient metabolizers but generally lack good vomiting reflexes. Emesis does not occur in equines, ruminants, or rodents. Omnivores, like humans and swine, fall in between. The importance of diet in the evolution of detoxifying systems applies even to primates. Despite their closeness to us on the phylogenetic tree, fruit- and plant-eating primates are generally more efficient metabolizers of xenobiotics than humans. In further support of this theory, it can be pointed out that the $t_{1/2}$ for amphetamine is about 86 min in both the rabbit and the horse, but is 390 min for the cat and 300 min for humans.

The question of mycotoxin residues in human food sources remains largely unanswered, but studies indicate that DON residues are not a problem. Feeding very high levels of DON to dairy cattle resulted in only trace quantities in milk; and when fed to poultry, no appreciable tissue residues were measured. Again, this is the consequence of the species pharmacokinetic characteristics.

Other toxins in unicellular members of the plant kingdom

Many soil organisms produce substances that are toxic to others and a continual state of chemical warfare exists in the microenvironment. These organisms provide the source of all of our antibacterial and antifungal antibiotics, many of which have significant mammalian toxicity. Many others were tested and discarded because of their high toxicity. Some antibiotics are teratogenic and are used in the treatment of cancer because of their effects on cell reproduction (e.g., actinomycin D, doxorubicin, adriamycin). Some organisms may be directly responsible for poisonings in humans.

The blue-green algae called *Cyanobacteria* produce pentapeptides that are hepatotoxic and have caused numerous deaths when they contaminate drinking water. Some strains of that ubiquitous organism *Staphylococcus aureus* are capable of producing a protein enterotoxin, 1 μg of which can induce vomiting, severe colic, and profuse diarrhea. The bacteria are usually introduced to foods from infected handlers, and they proliferate in warmth and especially in creamy foods (cream pies, salad dressings, etc.).

The most potent toxin known is the protein toxin from *Clostridium botulinum*, 1 μg of which can be fatal to a human. Lab animals may show symptoms at 10^{-6} μg/kg. Botulinum toxin blocks the release of acetylcholine from peripheral nerve endings.

Further reading

Eaton, L. and Gallagher, E.P., Mechanisms of aflatoxin carcinogenesis, *Annu. Rev. Pharmacol. Toxicol.*, 34, 135–172, 1994.

Marasas, W.F., Fumonisins: their implications for human and animal health, *Nat. Toxins*, 3, 193–198, 1995.

Nair, M.G., Fumonisins and human health, *Ann. Trop. Paediatr.*, 18 (Suppl.), S47–S52, 1998.

Pitt, J.I., Toxigenic fungi and mycotoxins, *Br. Med. Bull.*, 56, 184–192, 2000.

Rheeder, J.P., Marasas, W.F.O. et al., *Fusarium moniliforme* and fumonisins in corn in relation to human cancer in Transkei, *Phytopathology*, 82, 352–357, 1992.

Raney, K.D., Meyer, D.J., Ketterer, B., Harris, T.M., and Guengerich, F.P., Glutathione conjugation of aflatoxin B1 *exo-* and *endo-*epoxides by rat and human glutathione-S-transferases, *Chem. Res. Toxicol.*, 5, 470–478, 1992.

Raney, K.D., Shimada, T. et al., Oxidation of aflatoxins and sterigmatocystin by human liver microsomes: significance of aflatoxin Q1 as a detoxication product of aflatoxin B1, *Chem. Res. Toxicol.*, 5, 202–210, 1992.

Schroeder, J.J., Crane, H.M., Xia, J., Liotta, D.C., and Merrill, A.M., Disruption of sphingolipid metabolism and stimulation of DNA synthesis by fumonisin B_1. A molecular mechanism for carcinogenesis associated with *Fusarium moniliforme*, *J. Biol. Chem.*, 269, 3475–3481, 1994.

Shepphard, G.S., Theil, P.G., Stockenstrom, S., and Sydenham, E.W., World-wide survey of fumonisin contamination of corn and corn-based products, *J. AOAC Int.*, 79, 671–687, 1996.

Shier, W.T., Sphingosine analogs: an emerging new class of toxins that includes the fumonisins, *J. Toxicol.*, 11, 241–257, 1992.

Ueng, Y.F., Shimada, T., Yamazaki, H., and Guengerich, F.P., Oxidation of aflatoxin B1 by bacterial recombinant human cytochrome P450 enzymes, *Chem. Res. Toxicol.*, 8, 218–2235, 1995.

WHO, Ochratoxin A, in Evaluation of Certain Food Additives and Contaminants. 37th Report of the Joint FAO/WHO Expert Committee on Food Additives. WHO, Geneva, 1991, 29–31.

WHO, Patulin, in Evaluation of Certain Food Additives and Contaminants. 35th Report of the Joint FAO/WHO Expert Committee on Food Additives. WHO, Geneva, 1990, 29–30.

Review questions

For Questions 1 to 10, use the following code:

Answer A if statements a, b, and c are correct.
Answer B if statements a and c are correct.
Answer C if statements b and d are correct.
Answer D if statement d only is correct.
Answer E if all statements (a, b, c, d) are correct.

1. Regarding mycotoxins, which of the following statements is/are true?
   a. *Fusarium* species produce numerous mycotoxins that are hazardous for many species.
   b. Deoxynivalenol (DON) causes fertility problems in swine.
   c. DON causes vomiting and poor weight gain in swine.
   d. Other trichothecenes are less toxic than DON.
2. Which of the following statements is/are true?
   a. All varieties of wheat and corn are susceptible to fungal infections in varying degrees.
   b. Rapid drying of grain minimizes the risk of post harvest mould growth.
   c. Mycotoxin toxicity in livestock usually results from mixed fungal contamination.
   d. There is no practical way of detoxifying corn and wheat contaminated with mycotoxins.
3. Which of the following statements is/are true?
   a. There is no evidence that fumonisins are carcinogenic for humans.
   b. Zearalenone has estrogen-like activity.
   c. Aflatoxins are not toxic for poultry.
   d. Ochratoxin causes renal damage in swine and poultry.
4. The symptoms of ergot poisoning:
   a. Are similar for humans and animals.
   b. Include nausea, dizziness, burning pain in the extremities, and abortion.
   c. Arise primarily from consuming rye contaminated with ergot alkaloids.
   d. Are caused by ergot alkaloids produced by *Fusarium moniliforme*.
5. Which of the following statements is/are true regarding fumonisin B_2 (FB_2)?
   a. Is strongly suspected of being a carcinogen for humans.
   b. It produces brain damage in horses.
   c. Horses are very susceptible to FB_2 toxicity.
   d. In swine, the signs of toxicity are similar to those of DON.
6. Regarding ergot alkaloids:
   a. They are derivatives of lysergic acid.
   b. Chemically, they resemble catecholamine neurotransmitters.
   c. They are chemically related to lysergic acid diethylamide (LSD).
   d. They may produce hallucinations.
7. Which of the following statements is/are true regarding ergot alkaloids?
   a. Ergometrine (ergonovine) causes contractions of the gravid uterus.
   b. Ergometrine has never had any medical application.
   c. Ergotamine in low doses has been used to treat migraine headaches.
   d. Ergotamine is nontoxic except at very high doses.

8. Regarding aflatoxins:
   a. The mold that produces them is very hardy and survives in cold climates.
   b. They are oxygenated heterocyclic compounds.
   c. They cause hepatic damage but are not carcinogenic.
   d. Epidemiological studies have shown a high degree of correlation between liver cancer and the consumption of foods contaminated with aflatoxin B_1.

9. Regarding detoxification of grains contaminated with mycotoxins:
   a. A technique that works on one mycotoxin may not work on others.
   b. DON concentration may be reduced by heating.
   c. Binding agents may reduce the toxicity of T-2.
   d. Moist ozone and ammonia have reduced DON concentrations in lab tests.

10. Regarding DON toxicity:
   a. Swine are more susceptible than poultry or ruminants.
   b. Turkeys form a conjugate and rapidly eliminate DON.
   c. Sheep form a glucuronide conjugate.
   d. A high volume of distribution (Vd) is generally associated with higher toxicity in that species.

11. Match the statement with the correct toxin or effect.
   a. Cyanobacteria.
   b. Blocks acetylcholine release from peripheral nerve endings.
   c. Adriamycin.
   d. *Staphylococcus aureus*.
      i. Produces a protein enterotoxin.
      ii. Produce hepatotoxic pentapeptides.
      iii. A teratogen used in cancer chemotherapy.
      iv. Botulinum toxin.

12. Answer the following True or False.
   a. Aflatoxin B_1 8,9-epoxide is genotoxic.
   b. Aflatoxin B_1 epoxides are formed exclusively *after* ingestion of AFB_1.
   c. Conjugation with glutathione is an important means of detoxifying AFB_1 epoxide.
   d. The 8,9-epoxide is formed exclusively by Cyp 1A2.
   e. Cyp 3A4 is the main source of 8,9-*endo*-epoxide.
   f. Species differences in the rate and extent of glutathione conjugation of 8,9-epoxide may account for differences in toxicity.

Answers

1. B
2. A
3. C
4. A
5. A
6. E
7. B
8. C
9. E
10. A
11. i = d, ii = a, iii = c, iv = b
12. a. True
    b. False
    c. True
    d. False
    e. True
    f. True

chapter eleven

Animal and plant poisons

"He was a bold man that first ate an oyster"

— Jonathan Swift

Massasauga rattlesnake, eat brown bread.
Massasauga rattlesnake, fall down dead.
If you catch a caterpillar, give it apple juice.
But if you catch a rattlesnake, TURN IT LOOSE!

— Old Ontario skipping rhyme

Introduction

Chemical warfare is widely practiced in the animal and plant kingdoms. Just as microorganisms produce antibiotics that inhibit the growth and reproduction of competing organisms, more complex plants synthesize chemicals that render them unpalatable to potential predators or that are truly toxic, thus selecting for individuals that avoid them or producing aversive reactions that limit consumption to once or twice only. Occasionally, however, these plants accidentally enter the food chain of livestock or humans, directly or indirectly, leading to a toxic reaction.

Similarly, toxins and venoms are used in the animal kingdom for defense and prey capture. Toxins are consumed and operate much like those of plants. Toxins can also be secreted by special cells or glands in the skin, so that toxicity may occur simply by taking the intended victim into the mouth. Toxic reactions have occurred among students indulging in the fad of "toad-licking." These toxins tend to be low-molecular-weight peptides with neurotoxicity. Some fish actually swim in a "cloud" of toxin secreted by skin glands. Their toxins are usually steroid glycosides and choline esters. A neurotoxin secreted in this way by a species of flounder of the Red Sea is so powerful that a shark attempting to bite it is incapable of closing its jaws and instantly convulses. It has been proposed as a shark repellant for

241

downed fliers and divers. These skin toxins also serve as antibacterial agents to prevent infection, as the slime secreted by the skin of amphibians is an ideal culture medium for bacteria.

Venoms are injected in some way and can be employed both for defense and for prey capture. Humans and animals sometimes accidentally become the victims of envenomation or intoxication by poisons of animal origin. The term *toxin* is also used to refer to the individual components of venoms. Numerous texts have been written on these subjects, and this chapter can do no more than skim the field and discuss some of the more important, or more interesting, examples. Emphasis is placed on agents that have become important to the biological sciences, either as drugs or as research tools.

Toxic and venomous animals

Toxic and venomous marine animals

Venoms and toxins are distinguished by the manner in which they are inflicted upon the victim. A venom is a substance kept in a special poison gland and administered by a complex injection apparatus or by lacerating spines. This type of system can be employed in defense or in prey capture. A toxin is a poison usually ingested as an accidental component of tissues or organs. Toxins likely evolved as a species (rather than an individual) protection. Predators with a preference for that prey would be selected out if the toxin is fatal. Conditioned avoidance could occur in survivors. The term for poisoning by the muscle tissue of scalefish, as opposed to shellfish, is *ichthyosarcotoxism*. It includes toxins that are accumulated up the food chain from plankton, as well as toxins that are synthesized by the fish itself. In the former situation, poisoning is usually by a mixture of toxins.

Scalefish toxins

Ciguatoxin. The condition ciguatera is caused by this toxin, which probably concentrates up the food chain from a photosynthetic dinoflagellate to reach toxic levels in large marine fish such as grouper, snapper, amberjack, barracuda, and parrot-fish. These species are coral reef predators and browsers. Toxic symptoms are both gastrointestinal (nausea, vomiting, cramps, diarrhea) and neurological (numbness; tingling of lips, throat, and tongue; dizziness; headache). Ciguatoxin is thought to increase membrane permeability to sodium and be responsible for the neurological signs and symptoms. Ciguatoxin is very toxic to mice that are the test species for detecting its presence. It is a large, colorless, heat-stable lipid molecule, the structure of which has not been elucidated. Other unidentified agents are likely involved in ciguatera poisoning. Ciguatera is by far the most common cause of scale-fish poisoning. Any large, warm-water species is a potential source of the toxin. A component of ciguatoxin, maitotoxin, has recently been purified. Maitotoxin (MTX) is a water-soluble polyether that is a potent inducer of an increase in cytosolic free calcium ion $[Ca^{2+}]_i$ in a wide variety of organisms, including mammalian cells. MTX appears to activate a voltage-independent, nonselective cation channel that may require an extracellular source of Ca^{2+} for activation. MTX has become a useful research tool for studying ion channels (see also http://www.rehablink.com.ciguatera/poison.htm). Barracuda sometimes contain a related neurotoxin that is very toxic to cats. In some parts of the West Indies, it is customary to feed some of the meat from a large barracuda to a local cat before humans eat it.

Tetrodotoxin. The term comes from *Tetraodontidae*, meaning four teeth, the name of the genus that produces the toxin. This toxin is present in puffer fish, which are called "fugu" in Japan. It is a very potent blocker of fast sodium channels, and therefore inhibits nerve conduction in a manner similar to local anesthetics. It causes parethesias, paralysis, anesthesia, and loss of speech, but consciousness is retained. Prognosis is improved if vomiting occurs early after ingestion. In Japan there are about 150 cases of poisoning annually, with a 50% mortality. Fugu is considered a delicacy in Japan and special chefs are licensed to prepare it. The toxin is concentrated in the liver and gonads of the fish. Connoisseurs of fugu prefer that just enough toxin remains to cause a slight tingling of the lips when it is consumed. In Japan, this custom has spawned a somewhat macabre poetry form: "Last night he and I ate fugu; today I helped carry his coffin."

Tetrodotoxin is also found in some newts and frogs. The blue-ringed octopus, a small octopus that inhabits tidal pools and shallow waters around Australia and other central Indo-Pacific waters, produces a toxin, administered by a bite, that is believed to be identical to tetrodotoxin. The bite of a larger specimen can be fatal in minutes.

Tetrodotoxin is of interest as a research tool because of its potent sodium channel blocking activity. Its occurrence in such diverse species can be explained by the fact that it is not synthesized by the animal itself, but rather by certain species of *Vibrio* bacteria that exist in a symbiotic relationship with the host.

Scombroid poisoning. The name *scombroid* comes from *Scombridae*, referring to dark-muscled fish such as tuna and mackerel. This poisoning results from eating fish rich in histidine. Improper refrigeration results in the conversion of histidine to histamine by surface bacteria. The signs and symptoms resemble a histamine reaction. They may onset in minutes to hours and include dizziness, headache, diarrhea, facial flushing, tachycardia, pruritis, and wheezing. Fish contaminated in this way are often described as having a spicy or peppery taste. Levels of histamine in contaminated fish may exceed 100 mg/100 g. The FDA has set 50 mg/100 g as the hazard level. Histamine is not destroyed by cooking. There is some evidence that another product of decomposition, saurine, may contribute to the symptomatology.

Icthyotoxin. Many species of fish, including freshwater varieties, sometimes contain a toxin in the gonads that can cause severe gastrointestinal symptoms.

Shellfish toxins

There is some evidence that there has been a dramatic increase in toxic algal blooms, the so-called red tides. The first confirmed outbreak of diarrhoretic shellfish poisoning in North America occurred in 1990 and it was traced to dinoflagellates in Canadian waters. Brown pelicans eating anchovies off California were dying of domoic acid poisoning in 1991; saxitoxin has been found in Alaska crabs; and in 1987–1988, shellfishing off North Carolina was shut down because of a red tide of a dinoflagellate that produces a neurotoxin, brevitoxin.

Saxitoxin. The cause of paralytic shellfish poisoning, this toxin is produced by dinoflagellates (single-celled animals that are the cause of red tides) and it concentrates in bivalves such as oysters, clams, and mussels. The mechanism and symptomatology are the same as for tetrodotoxin. The minimum lethal dose is about 8 µg/kg for each. "Never eat oysters in a month without an R" is an old adage that remains good advice in the Northern Hemisphere as dinoflagellates bloom in the summer months.

Domoic acid. In 1987, there was an outbreak of mussel poisoning in Canada caused by mussels from the waters around Prince Edward Island. The toxin was domoic acid, which concentrates from algae called *chondria*. It is also found in a diatom and is rare in North Atlantic waters, being more common in Japan. The toxin was identified by the Canadian National Research Council Atlantic Laboratory. There were three fatalities in elderly nursing home residents in Quebec, and over 100 individuals were affected. Several have been left with short-term memory deficits, a condition called *amnesic shellfish poisoning*. It has been suggested that marine pollution might have changed the environment to favor the growth of the offending diatom. Domoic acid binds to a specific subset of glutamic acid receptors, known as kainate receptors, in the brain.

Okadaic acid. Also produced by a dinoflagellate, okadaic acid is responsible for a condition known as *diarrhoretic shellfish poisoning*. Okadaic acid is a selective inhibitor of phosphatases 1 and 2A with interference to protein activation/inactivation. It can induce contraction of smooth muscle and cardiac muscle. Experimentally, it is a tumor promoter but, unlike phorbol esters, it does not activate protein kinase C.

Direct toxicity from dinoflagellates

The dinoflagellates that cause red tides are pigmented, plant-like organisms. In addition to being toxic to a variety of marine life, they can be carried into the atmosphere in aerosols resulting from wave action. They can cause respiratory irritation in people and animals coming in contact with the aerosol; and in individuals with preexisting respiratory problems such as emphysema and asthma, the reaction can be quite serious.

Over the past decade, a genus of dinoflagellates called *Pfiesteria* has been responsible for numerous fish kills in North American estuaries, killing an estimated one billion fish in the Albemarle-Pamlico estuarine system and over 50,000 in Chesapeake Bay. *P. piscicida* and *P. shumwayae* have a strong attraction for live fish and produce complex toxins that affect neural function and can cause numerous other lesions. Fish behave erratically and have skin lesions described as "punched out." Neurotoxic aerosols can cause central nervous system damage in people, affecting memory, cognitive function, and both the autonomic and peripheral nervous systems. Contact with contaminated water can cause skin lesions and a burning sensation. The toxin is water soluble, affects calcium metabolism, and attaches to a purinergic receptor on immune cells of the brain. Commercial fishers, marine biologists, and lab personnel have been severely affected.

Stinging fishes

In venomous fish such as the stonefish, lionfish, scorpionfish, and stingray, spines are located ahead of the dorsal fin, on the tail, or around the mouth. A heat-labile protein causes intense pain and cardiac shock. Heat above 50°C may afford some relief. Stonefish and stingrays are most often stepped upon because they conceal themselves in the sand bottom. Fatalities have occurred from envenomation by these fish, generally as a result of cardiovascular shock, with AV block and bradycardia. Other symptoms include numbness, inflammation, and edema at the site of injury, severe pain in surrounding tissues, delirium, nausea, vomiting, and sweating. Stingray venom is a large, heat-labile protein (molecular weight > 100,000) with neurotoxic as well as cardiotoxic properties. There is an antivenin for stonefish venom.

Mollusk venoms

Conotoxins. These are found in marine cone snails. All are strongly basic peptides, highly cross-linked by disulfide bonds. Alpha-conotoxins are nondepolarizing, neuromuscular blocking agents like curare and, therefore, cause paralysis. They are 13 to 15 amino acid peptides. Mu-conotoxins are 22 amino acid peptides that block sodium channels, thus acting like tetrodotoxin and saxitoxin. These sodium channel blocking conopeptides are under investigation for use in treating chronic pain. Omega-conotoxins block presynaptic, voltage-dependent calcium channels. Recent research has identified a number of subtypes with specificity for various channel subtypes, making them useful research tools. Cone snails are univalve gastropods with a complex envenomation apparatus used for prey capture (see also http://grimwade.biochem.unimelb.edu.au/cone/newlog.html).

Coelenterate toxins

These are found in jellyfish and corals. Fire coral produces a protein venom that causes intense local burning when touched. It feels like a cigarette burn. *Physalis* (Portuguese man-o-war, bluebottle, mauve stinger) causes signs and symptoms similar to fire coral. Red streaks, called *straps*, occur where tentacles touch skin. Allergic reactions can occur, in addition to generalized symptoms such as fever, nausea, and cardiac and respiratory distress. Bluebottles washed up on beaches remain toxic until completely dried out. First aid consists of washing the affected area with vinegar, dilute acetic acid, or meat tenderizer (papain), all of which denature the protein. Even urine may help to relieve the pain. This species is common around the world.

The sea wasp (box jellyfish) is the most venomous marine animal known. It inhabits the Indo-Pacific region and accounts for many deaths annually. Contact with tentacles causes intense, agonizing pain, coma, and cardiac shock. Mortality is 25% and children, the elderly, and heart patients are most vulnerable. First aid is denaturation as above; removal of tentacles using forceps, gloves, a knife and fork, or any means to avoid touching the tentacles; and cardiopulmonary resuscitation (CPR). Local anaesthetic spray such as is used for sunburn may be helpful. An antivenin is available. The Irukandji is a tiny, four-tentacled, Indo-Pacific jellyfish that causes similar signs and symptoms, plus massive sympathetic discharge.

Echinoderm venoms

Sea urchins and sea anemones, such as the long-spined or black sea urchin and crown-of-thorns sea urchin, possess toxins. Injury usually occurs to the feet and lower limbs when a diver or swimmer steps on these bottom-dwellers. Spines are driven into the flesh and break off. The toxin produces local pain and burning similar to that of the bluebottle. It is heat labile so immersing the affected part in water as hot as the person can stand may help. Unlike the usual first aid for envenomations, movement and trauma may actually help by breaking up the spines so they can be absorbed more quickly. Vinegar or even urine may also help.

Sea anemones also produce very potent toxins that act when they are ingested. The best characterized of these are the equinatoxins (EqTs I, II, and III) from *Actinia equina*. In rats they cause coronary vasospasm, cardiac arrest, and other cardiorespiratory toxicity. Hemolysis also occurs, as does degranulation of blood platelets and white cells.

For more information on marine venoms and toxins, see http://www.pmeh.uiowa.edu/fuortes/63260/MARINEAN/index.htm and http://www.merck.com/pubs/mmanual/.

Toxic and venomous land animals

Venomous snakes

Many snakes are venomous, but most have rear fangs that are designed to paralyze prey after it has been taken into the oral cavity. These are incapable of inflicting a venomous bite. The four genuses of poisonous snakes that are a danger to humans include the *Viperidae, Crotalidae, Elapidae,* and *Hydrophidae.*

1. *Viperidae.* These old-world vipers have hollow, needle-like fangs that are set in short, movable maxilla which rotate to bring the fangs into the biting position as the mouth is opened. The head is large and triangular.
2. *Crotalidae.* These "pit vipers" are similar to the old-world vipers but they also have a deep, infrared-sensitive pit between the eye and the nostril that is used for tracking prey by body heat. This genus includes all rattlesnakes, the water moccasin (cottonmouth, etc.), and the copperhead. All are found in North America. Canada has the Western diamondback and, in Ontario, the Massasauga rattlesnake.
3. *Elapidae.* These are distributed worldwide and account for the most poisonous and feared snakes of the tropics and subtropics. This group includes the cobras, the boomslang, and many Australian species (taipan, tiger snake, brown snake, etc.). There are two species in the United States, the Eastern (Florida) and Arizona coral snakes. The rhyme "red-on-black, friend of Jack; red-on-yellow, kills a fellow" helps to distinguish the coral snake from a harmless look-alike.
4. *Hydrophidae.* These are the sea snakes. They are restricted to the warmer waters of the Pacific and Indian Oceans. Although they are highly venomous, they are not aggressive. Most bites occur to commercial fishermen because the snakes become trapped in the nets. For some years, sporadic discussions have occurred concerning the feasibility of digging a sea-level canal across the Isthmus of Panama to connect the Atlantic and Pacific Oceans. One environmental concern that has been raised is that such a canal could introduce the yellow-bellied sea snake to the Gulf of Mexico and the Caribbean Ocean where conditions are favorable for their proliferation.

Snake venoms.　　These are complex mixtures of proteins, many of which are proteolytic enzymes. In general, venoms of the *Elapidae* and the *Hydrophidae* tend to be neurotoxic with myonecrosis (breakdown of muscle tissue) occurring at the bite wound, whereas venoms of the vipers and crotalids are generalized coagulants with local anticoagulant activity to spread the venom, causing much local damage (pain, necrosis, bleeding). Neurotoxicity in less prominent.

Signs and symptoms of neurotoxic venoms include progressive paralysis, muscle spasm, respiratory distress or failure, muscle ache (myalgia), kidney failure with myoglobinuria (the appearance of myoglobin in the urine), and cardiac failure. These symptoms apply to coral snake bites in North America. The venom of the Banded Krait contains alpha-bungarotoxin, which is an irreversible blocker of acetylcholine receptors, and beta-bungarotoxin, which causes a massive release of neurotransmitter vesicles. Alpha-bungarotoxin is used as a research tool in biomedical research. The

Black Mamba (*Dendroaspis polylepis*) produces dendrotoxin-I, another K$^+$ channel blocker.

Signs and symptoms of viper and crotalid bites (including the Massasauga rattlesnake) include immediate, intense burning at the site of envenomation (like a bee sting), followed by numbness, swelling, shock, and hematuria. Necrosis and possibly gangrene at the bite wound may occur later. Subcutaneous hemorrhages may be present and severe cases may show signs of neurotoxicity. Table 10 lists some of the major components that have been identified in snake venoms. The venom of Russell's viper contains an activator of clotting Factor X and it is used in coagulation research.

Table 10 Some Components of Snake Venoms

Component	Elapids Hydrophids	Vipers Crotalids
Nicotinic blocker (neuromuscular blockade)	+	+/−
Cholinesterase (neurotoxic)	+	−
Coagulant protease (increases clotting)	−	+
Anticoagulant protease (decreases clotting locally)	+/−	+
ATP (shock, hemolysis)	+	+
Phosphatase (shock, hemolysis)	+	+
Phospholipase A (histamine and SRSA release[a])	+	−
Bradykininogen (forms bradykinin, causes pain at site)	−	+
Hyaluronidase (breaks down connective tissue, spreads venom)	+	+

[a] SRSA = Slow releasing substance of anaphylaxis (leukotrienes).

Phospholipase A$_2$ complexes in rattlesnake venoms are neurotoxic as well as inducing tissue damage. Myotoxic components have exhibited multiple mitogenic effects on growing cultured myocytes.

Elapidae and *Viperadae* both possess alpha-neurotoxins that act postsynaptically to prevent acetylcholine from binding to its receptor, as well as neurotoxins that affect transmitter release presynaptically. B-neurotoxins have phospholipase A$_2$ activity. The mambas and other African snakes have voltage-dependent K$^+$ channel blockers (dendrotoxins), noncompetitive inhibitors of acetylcholine (fasciculins), muscarinic toxins, and L-type Ca^{2+} channel blockers (caliseptins). All are small proteins containing about 60 amino acids and 3 or 4 sulfides.

There are a few venomous lizards, such as the Gila monster, that inhabit the American Southwest. They are not generally as dangerous as venomous snakes. The venom lacks neurotoxins but contains coagulants and enzymes as well as serotonin. The reaction tends to be more local. Lethal doses in animals lead to cardiorespiratory collapse.

First aid. Regardless of the type of snake bite, the most important first aid measure is a tension bandage applied to the entire affected limb with the same tension one would use to bandage a sprain. Rest and reassurance are important. Transportation to a medical center possessing antivenin should occur as soon as possible. Modern hospitals should have the antivenin (or polyvalent antivenin) appropriate to their area in stock. Forced exercise and alcoholic beverages are definitely contraindicated.

Venomous arthropods

Members of the order *Hymenoptera* (bees, wasps, ants) of the class *Insecta* and of the class *Arachnida* (spiders, scorpions) have venoms that contain substances commonly involved in the mammalian pain response. The Old World scorpion *Leiurus quinquestriatus* and the Israeli yellow scorpion (*Leiurus q. hebraeus*) produce charybdotoxin that affects calcium-activated K^+ channels, as does the bee venom apamin. The Mexican scorpion *Centroides noxius* produces noxiustoxin, which blocks voltage-dependent K^+ channels (see Table 11). Mast cell degranulating peptide in bee venom also blocks these channels.

Table 11 Components of Hymenoptera Venoms

	Histamine Releasers	Bradykinin	Serotonin	K^+ Channel Blockers
Bees				Apamin (Ca^{2+} activated), mast-cell degranulating (MCD) peptide (voltage activated)
Ants	+	+		
Wasps	+	+	+	
Scorpions	+	+	+	
Old World				Charybdotoxin Ca^{2+} activated R-agitoxin-2 (voltage activated)
Mexican				Noxiustoxin (voltage activated)

Allergic reactions to any of these insect venoms may occur and may be life-threatening if severe or multiple stings occur. The African honeybee is dangerous not because it is more venomous, but because it is extremely aggressive so that multiple stings are common.

Neurotoxins tend to predominate in spider venoms. Virtually all spiders are venomous, but only a few have an envenomation apparatus suitable for piercing human skin. So-called "widow" spiders (*Latrodectus* spp.) have a worldwide distribution in temperate and tropical climes. The Black Widow, also known as the death spider, as the hourglass spider, and by many other names, inhabits all of Ontario south of Sudbury but is rarely encountered since the demise of the outdoor privy. The venom is extremely toxic but very little is injected; thus, fatalities are rare. The toxin is complex. At least seven "latrotoxins" have been identified; five are specific for insects, one for crustaceans and one for vertebrates. The latter is alpha-latrotoxin (alpha-LTX). It is a medium-sized protein now available in pure form (see below) that induces the release of most if not all neurotransmitters. It bears similarity to beta-bungarotoxin (found in the Banded Krait). Both cause a massive release of cholinergic vesicles.

Fasciculations and board-like rigidity of the trunk muscles occur rapidly, followed by muscle cramps, pain in the muscles, and respiratory distress. The presence of a specific receptor on vertebrate neurones for alpha-LTX has been identified, and this is the likely explanation for the species specificity of the toxin. Recovery from Black Widow spider bite takes about 12 hours.

The Brown Recluse (*Loxosceles* spp.), known also as the violin spider, has been working its way north from Mexico and Florida and has reached New York state and probably southern Ontario. The bite is painless. Local necrosis develops and expands over the next week due to local blood clotting and microthrombosis. Occasional deaths have occurred from hemolytic anemia and kidney failure. The venom is complex and includes coagulants, enzymes, and a complement inhibitor.

The Australian funnel-web and red-backed spiders are claimed to be especially venomous. Two funnel-web toxins have been identified: a polyamine (FTX) and omega agitoxin. Both block high voltage-dependent P Ca^{2+} channels, and they are now used as research tools for this purpose. R-agitoxin-2, from the Israeli yellow scorpion, blocks some voltage-activated K^+ channels.

Toxic plants and mushrooms

Folk knowledge, passed orally from generation to generation, usually determines what humans can eat and what they cannot eat, and even what parts of a plant are edible. Thus, people make pies from rhubarb stalks but never make salad from the leaves, which are toxic. Nor do people eat the bulbs or stalks of the many ornamental flowers and shrubs in their gardens. Seldom does one consider the fact that these decisions are toxicologically based. The number of plants that are potentially harmful is too voluminous for extensive coverage. The subject has formed the basis of many texts. The following list contains examples of major groups of toxicants chosen because they are common or because of their pharmacological significance.

Vesicants

Many plants contain oxalates that can cause corrosive burns to the mouth, esophagus, and stomach. Symptoms also include vomiting and diarrhea. Because oxalates are anticoagulant, bleeding may also occur. *Dieffenbachia* contains calcium oxalate, which is a vesicant. Rhubarb leaves contain a variety of oxalates and may be fatal if consumed in quantity due to their renal toxicity (see also ethylene glycol in Chapter 7, "Glycols and glycol ethers"). May apple, buttercup, and philadendron contain vesicants. Philadendron sometimes causes poisoning in cats that chew the leaves. Poison ivy, poison oak, and poison sumac all contain urushiol, which is a phenolic vesicant.

Cardiac glycosides

White and purple foxgloves (*Digitalis lanata* and *D. purpura*) are the commercial source of medicinal digitalis. Many garden plants possess similar active components, including lily-of-the-valley, star-of-Bethlehem, and oleander. Symptoms are those of digitalis overdose: nausea, vomiting, visual disturbances (a green halo seen around objects), and cardiac arrhythmias.

Astringents and gastrointestinal irritants (pyrogallol tannins)

Acorns, geraniums, sumac berries, hemlock bark, rhubarb leaves, horse chestnut — all contain these irritants. North American natives learned to remove tannins from acorns by repeatedly steeping the crushed nuts in fresh water. This will not work for horse chestnuts, however.

Autonomic agents

Deadly and wooded nightshades contain atropine (hyoscyamine) and scopolamine (hyoscine). The signs and symptoms are those of blockade of muscarinic, cholinergic receptors plus CNS symptoms (disorientation, hallucinations). *Amanita muscaria* is the common poisonous mushroom. Signs of poisoning are those of massive cholinergic discharge due to the toxin muscarine, plus CNS effects like those of nightshade due the presence of anticholinergic agents. CNS depression and cardiac failure may follow. Symptoms may be delayed 12 to 24 hours.

Solanine and chaconine are glycogenic alkaloids that occur in potatoes and tomatoes (members of the *Solanaceae* family) when they are exposed to excessive sunlight, blight, sprouting, and prolonged storage. The compounds have cholinergic activity and have been shown to be teratogenic. Symptoms of poisoning include vomiting, diarrhea, cramps, dizziness, visual disturbances, and other CNS manifestations. In 1820, a man named Johnson defied conventional wisdom by publicly eating a tomato in Salem, New Jersey. To the surprise of onlookers, he did not die a horrible death. At the time, it was widely believed that tomatoes, because of their relationship to deadly night-shade, were highly toxic. Johnson single-handedly launched the North American tomato industry.

Another member of the *Solanaceae* is *Datura stromonium*, commonly known as jimsonweed or thornapple. Other names are devil's trumpet and angel's trumpet. This is a common weed throughout North and Central America, first introduced from Europe. It typically grows in rough ground such as vacant lots. It has a large white or mauve trumpet flower and a spiny seedpod. The seeds can contain significant levels of atropine and scopola-mine that can induce hallucinations. For this reason it is sometimes used as a recreational drug, usually by adolescents. Due to its anticholinergic activity, signs and symptoms of jimsonweed poisoning can include dilated pupils, dry mouth, rapid and irregular heartbeat, stasis of the bowel, agitation, and disorientation, as well as hallucinations. Several deaths have occurred.

Dissolvers of microtubules

Colchicine from the autumn crocus is used in the treatment of gouty arthritis and as an experimental tool. It dissolves microtubules to prevent mitosis and also phagocytosis. Overdose causes severe diarrhea. Vincristine and vinblastine from the periwinkle plant share this property and are used to treat childhood leukemias. Podophylotoxin from the May apple also dis-solves microtubules.

Phorbol esters

These are components of croton oil from spurge plants. These substances are cancer promoters. They directly activate protein kinase C (substituting for diacylglycerol) independently of extracellular calcium and are used as experimental tools for this reason. They act as drastic purgatives. The site of action of phorbol myristate acetate (PMA) is shown in Figure 39. It affects an important control mechanism for intracellular regulation.

Many other carcinogens, co-carcinogens, promoters, and anti-carcinogens exist in plants. Safrol is a liver carcinogen found in some spices (nutmeg, cinnamon) and in oil of anise (licorice flavoring). The use of anise oil and oil of sassafras has been banned. Some tannins are liver carcinogens, and the polyaromatic hydrocarbon (PAH) benzo[a]pyrene is a potent carcinogen that occurs in green vegetables, unrefined vegetable oils, coconut oil, and chicory. Benzanthracenes are other PAHs that occur in vegetables. Many others exist.

Thapsigargin is a plant-derived sequiterpene lactone capable of inhib-iting Ca^{2+}-ATPase and causing discharge of internal Ca^{2+} stores. It is widely used as a research tool because of this action. It also has tumor-promoting properties.

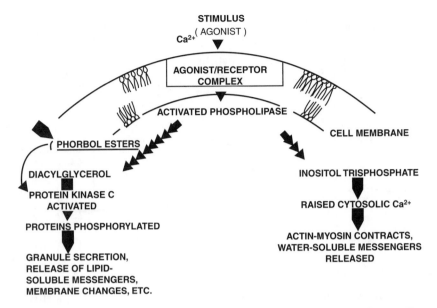

Figure 39 Site of action of phorbol esters as activators of protein kinase C.

Cyanogenic glycosides

Substances such as amygdalin in almonds, dhurrin in sorgham, linamarin and lotaustralin in cassava and lima beans, and prunasin in stone fruit (cherries, peaches, and chokecherries) are cyanogenic glycosides. They are capable of forming hydrogen cyanide (HCN) with the beta-glucuronidases from the plants when cells break down or from the microflora of the gastrointestinal tract. Cyanide poisoning can occur in ruminant animals from eating vegetation high in cyanogenic glycosides or in humans who have consumed improperly stored or prepared foods such as lima beans or cassava. In humans, HCN can be formed from organic nitriles by cytochrome P450-dependent monooxygenases, and from organic thiocyanates by glutathione-S-transferases.

Detoxification of hydrogen cyanide

Hydrogen cyanide (HCN) is detoxified by conversion to thiocyanate, which requires sulfur-containing amino acids and vitamin B12. Deficiencies of these increase the risk of toxicity. The metabolic detoxification system is overwhelmed and hydrogen cyanide interferes with electron transport in the cytochrome a-a$_3$ complex, resulting in tissue hypoxia. This leads to rapid failure of the CNS and death. Treatment is the administration of intravenous nitrites, which form methemoglobin. Methemoglobin has a high affinity for HCN and binds it to protect the cytochrome and allow time for biotransformation to occur. These events are summarized in Figure 40.

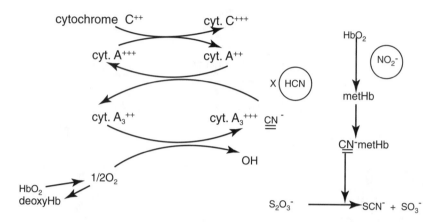

Figure 40 Site of action of HCN and detoxification by nitrates.

Convulsants

Water hemlock, the poison of Socrates, typifies this group. The toxin is cicutoxin (from the plant's Latin name *Cicuta maculata*). This plant resembles parsnips, smells like turnips, tastes sweet, and is the most toxic indigenous plant in North America. The toxin is present in all parts of the plant but is concentrated in the root. It is most toxic in springtime. Mild intoxication produces nausea, abdominal pain, epigastric distress, and vomiting in 15 to 90 minutes. Early vomiting may be protective. Severe poisoning produces profuse salivation, sweating, bronchial secretion, respiratory distress, and cyanosis. Convulsions occur and *status epilepticus* precedes death. Mortality rates are on the order of 30%. There is no known antidote. Fatal poisonings in children have occurred from using toy whistles made from the stem.

In the time frame 1978 to 1989, 58 people in the United States are known to have died from ingesting toxic plants mistaken for edible wild fruit or vegetables. Water hemlock was responsible for at least five of these.

Use in research and treatment

Many chemicals of animal and plant origin are useful as research tools in physiology and pharmacology. A partial list follows.

1. Tetrodotoxin from puffer fish and saxitoxin from shellfish block fast sodium channels and are used to study nerve conduction.
2. Alpha-conotoxin from cone snails is a non-depolarizng neuromuscular blocking agent.
3. Mu-conotoxin from cone snails acts like tetrodotoxin.
4. Omega-conotoxin from cone snails is a specific inhibitor of presynaptic, voltage-dependent Ca^{2+} channels.
5. Russell's viper venom activates Factor X in the clotting system and is used in certain clotting tests and coagulation research.

6. Alpha-bungarotoxin from the banded krait is an irreversible blocker of acetylcholine receptors.
7. Beta-bungarotoxin from the banded krait and alpha-latrotoxin from the Black Widow spider cause massive release of peripheral neurotransmitter vesicles.
8. Apamin from bees is a blocker of K^+ channels.
9. Charybdotoxin from the scorpion is also a potent K^+ channel blocker.
10. Digoxin from foxglove blocks Na^+/K^+ ATPase and is used to treat congestive heart failure.
11. Atropine from nightshade is a muscarinic blocker and has several uses.
12. Tannins from a variety of plants are used in astringent lotions.
13. Cytochalasins from fungi fix cell membranes and microtubules *in vivo*.
14. Phorbol myristate from croton oil activates protein kinase C and is used to study calcium intracellularly.
15. Colchicine from autumn crocus dissolves microtubules and arrests mitosis. It is also used to treat acute attacks of gouty arthritis.
16. Capsaicin from chili peppers is used as a counterirritant in lineaments and ointments to provide the heat by causing vasodilation.
17. Vincristine and vinblastine from periwinkle arrest mitosis and are used as anti-cancer drugs.

Recent research into the nature and chemical composition of polypeptide venoms has led to their availability in pure form as research tools (primarily from Alomone Labs in Jerusalem). Examples include:

1. From the eastern green mamba (*Dendroaspis angusticeps*), alpha-dendrotoxin blocks certain voltage-gated K^+ channels and beta-dendrotoxin blocks certain voltage-gated K^+ channels in synaptosomes and smooth muscle cells;
2. From the Australian taipan (*Oxyuranus scutellatus*), taicatoxin selectively blocks high-threshold voltage-gated Ca^{2+} channels in heart cells; and
3. From the Black Widow spider (*Latrodectus tradecemguttatus*), alpha-latrotoxin, a 130,000-Dalton protein, is the principal toxic component of the venom, causing massive exocytotic secretion of neurotransmitter vesicles both centrally and peripherally.

Numerous other examples exist. Some are discussed in the text.

Further reading

Anderson, D.M., Red tides, *Scientific American*, 271, 62–68, 1994.
Burkholder, J.M. and Glasgow, H.B., Jr., *Pfiesteria piscicida* and other toxic *Pfiesteria*-like dinoflagellates: behavior, impacts, and environmental controls, *Limnol. Oceanogr.*, 42, 1052–1075, 1997.

Burkholder, J.M. et al., Fish kills, bottom-water hypoxia, and the toxic *Pfiesteria* complex in the Neuse River and estuary, *Mar. Ecol. Prog. Ser.*, 179, 301–310, 1999.

Cataldi, M., Secondo, A., d'Alessio, A., Taglialatela, M., Hoffmann, F., Klugbauer, N., Di Renzo, D., and Annunziato, L., Studies on maitotoxin-induced intracullar Ca^{2+} elevation in Chinese hamster ovary cells stably transfected with cDNAs encoding for L-type Ca^{2+} channel sububits, *J. Pharmacol. Exp. Ther.*, 290, 725–730, 1999.

Culotta, E., Red menace in the world's oceans, *Science*, 257, 1476–1477, 1992.

Dickey, R.W., Fryxell, G.A. et al., Detection of the marine toxins okadaic and domoic acid in shellfish and phytoplankton in the Gulf of Mexico, *Toxicon*, 30, 355–359, 1992.

Edmonds, C., Lowry, C., and Pennefather, J., Dangerous marine creatures, in *Diving and Subaquatic Medicine*, 3rd ed., Butterworth-Heineman, Oxford, 1992, 318–347.

Escobar, L.I., Salvador, C., Martinez, M., and Vaca, L., Maitotoxin, a cationic channel activator, *Neurobiology (Budapest)*, 6, 59–74, 1998.

Grattan, L.M. et al., Learning and memory difficulties after environmental exposure to waterways containing toxin-producing *Pfiesteria* or *Pfiesteria*-like dino-flagellates, *Lancet*, 352, 532–539, 1998.

Hashimoto, Y., *Marine Toxins and Other Bioactive Marine Metabolites*, Japan Science Society Press, Tokyo, 1979.

Lampe, K.F., Toxic effects of plant toxins, in *Casarett and Doull's Toxicology: The Basic Science of Poisons*, Klassen, C.D., Amdur, M.O., and Doull, J., Eds., Macmillan, New York, 1986, 757–767.

Larm, J.A., Beart, P.M., and Cheung, N.S., Neurotoxin domoic acid produces cyto-toxicity via kainate- and AMPA-sensitive receptors in cultured cortical neu-rones, *Neurochem Int.*, 31, 677–682, 1997.

Malins, D.C. and Ostrander, G.K., Eds., *Aquatic Toxicology: Molecular, Biochemical and Cellular Perspectives*, Lewis, Boca Raton, FL, 1994.

Manners, G.D., Plant toxins: the essences of diversity and a challenge to research, Chapter 2 in *Natural Toxins 2: Structure, Mechanism of Action, and Detection*, Singh, B.R. and Tu, A.T., Eds., *Adv. Exp. Med. Biol.*, 391, 9–36, 1996.

Noble, R.C., Death on the half-shell: hazards of eating shellfish, *Pers. Biol. Med.*, 33, 313–322, 1990.

Olivera, B.M., Gray, W.R. et al., Peptide neurotoxins from fish-hunting cone snails, *Science*, 230, 1985, 1338–1343.

Ostrander, G.K., Ed., *Techniques in Aquatic Toxicology*, Lewis, Boca Raton, FL, 1996.

Scheuer, P.J., Marine natural products: diversity in molecular structure and biodiver-sity, Chapter 1 in *Natural Toxins 2: Structure, Mechanism of Action and Detection*, Singh, B.R. and Tu, A.T., Eds., *Adv. Exp. Med. Biol.*, 391, 1–8, 1996.

Scombroid Fish Poisoning, *Morbid. Mortal. Wk. Rep.*, 38, 140–147, 1989.

Suput, D. and Zorec, R., Eds., *Toxins and Exocytosis* (Proceedings of a symposium), Ann. NY Acad. Sci., 710, 1994 (numerous authors).

Water Hemlock Poisoning — Maine, 1992, *Morbid. Mortal. Weekly Rep.*, 43, 229–231, 1994.

Tu, A.T., Overview of snake venom chemistry, Chapter 3 in *Natural Toxins 2: Structure, Mechanism of Action and Detection*, Singh, B.R. and Tu, A.T., Eds., *Adv. Exp. Med. Biol.*, 391, 37–62, 1996.

WHO, Evaluation of Certain Food Additives and Naturally Occurring Toxicants, 39th Rep. Joint FAO/WHO Expert Comm. on Food Additives, 828, 30–33, 1992.

Review questions

Answer the following questions True or False.

1. Crotalid venoms are predominantly anticoagulant.
2. Hyaluronidase in snake venoms helps to disseminate the poison at the site of the bite.
3. Omega-conotoxin blocks presynaptic, voltage-gated calcium channels.
4. Ciguatoxin does not biomagnify up the food chain.
5. Beta-bungarotoxin is found in the venom of the banded krait.
6. Potassium channel blockers are found in the venom of the brown recluse spider.
7. Urushiol is an astringent found in the horse chestnut.
8. Hyoscyamine is the same as scopolamine.
9. Alpha-bungarotoxin is an irreversible blocker of acetylcholinesterase.
10. Cytochalasin fixes microtubles *in situ*.
11. The venom of vipers is predominantly neurotoxic.
12. Tetrodotoxin is a potassium channel blocker.
13. Saxitoxin is synthesized by dinoflagellates.
14. Lily-of-the-valley contains cardiac glycosides.
15. The bite of the brown recluse spider is extremely painful.

For Questions 16 to 23, match the statements with appropriate response from the list below.

a. Vinegar
b. Traumatizing the area
c. A tension bandage over the entire affected limb
d. Phorbol myristate acetate
e. Tetrodotoxin
f. Domoic acid
g. Ciguatoxin
h. Okadaic acid

16. Causes generalized paralysis due to fast sodium channel blockade.
17. The cause of amnesic shellfish poisoning.
18. Directly activates phosphokinase C.
19. General first aid for any snake bite.
20. First aid for the sting of a bluebottle (*Physallis*).
21. May reduce the pain of an embedded sea urchin spine.
22. The cause of diarrhetic shellfish poisoning.
23. May cause gastrointestinal and neurological symptoms when large marine scalefish are eaten.

Answers

1. True
2. True
3. True
4. False
5. True
6. False
7. False
8. False
9. False
10. True
11. False
12. False
13. True
14. True
15. False
16. e
17. f
18. d
19. c
20. a
21. b
-22. h
23. g

Case study 20

In 1988, several patrons of a restaurant experienced signs and symptoms of illness including nausea, headache, dizziness, facial flushing, and diarrhea. The symptoms began about 5 to 60 min after the meal (median 38 min) and persisted for about 9 hr. Only these six patrons (four males, two females) experienced problems although an estimated 50 to 60 people had partaken of the same buffet lunch.

Q. What questions would you wish to ask of the affected and the unaffected patrons?

Q. What possible causes of this reaction could there be?

   Several of the affected individuals noted upon questioning that a particular fish dish had a "Cajun" or peppery flavor.

Q. Does this help to identify the problem?

Case study 21

In June 1990, six fishermen aboard a private fishing boat off the Nantucket coast of Massachusetts developed symptoms that included numbness and tingling of the mouth, tongue, throat, and face; vomiting; loss of sensation in the extremities; periorbital edema; and 24 hr later, low back pain (in all six). The initial symptoms persisted for about 14 hr, the back pain for 2 to 3 days.

Q. What organ system is primarily affected?

Q. What information would you want to obtain from the victims?

It emerged that all six men had consumed blue mussels at the same meal. The blue mussels had been harvested in deep water about 115 miles offshore. The mussels had been boiled for about 90 min and were consumed with boiled rice, baked fish, and a salad. There appeared to be a correlation between the severity of the symptoms and the number of mussels consumed.

Q. What is the likely cause of the poisoning?

Q. What fish could have been responsible for the same array of signs and symptoms?

Case study 22

Over a period of 72 hr in August, eight seasonal tobacco workers were admitted to a regional hospital with a variety of signs and symptoms that included weakness, nausea, vomiting, dizziness, abdominal cramps, headache, and difficulty in breathing. They had all been working in the fields in the morning following an evening of steady rain. The average time of onset of the symptoms was 10 hr after commencing work. All patients were males, 18 to 32 years of age. All required hospitalization for 1 or 2 days.

Q. Is this likely an occupational disease, a food poisoning from something in the breakfast meal, or an infection?

Q. What occupational hazards might these workers encounter?

Q. What lab tests might help in the differential diagnosis?

Case study 23

On August 9th, eight people were admitted to the emergency department of a Florida hospital with one or more of the following symptoms: cramps, nausea, vomiting, diarrhea, chills, and/or sweats. All reported having eaten amberjack, a predatory scalefish, at a local restaurant within the preceding 9 hr (mean time to symptoms, 5 hr). Three of the victims required hospitalization. These symptoms persisted for 12 to 24 hr. Within 48 to 72 hr, most of these patients developed pruritis and parathesias of the hands and feet and muscle weakness.

Subsequent investigation uncovered 14 similar cases, all of which had eaten amberjack at one of several local restaurants. The restaurants had received the fish from the same supplier in Key West.

Q. What organ systems are involved in this intoxication?

Q. Does the evidence point to restaurant kitchens as the source of the toxin?

Q. What potential causes of this problem must be considered?

Q. Which is your choice?

Case study 24

In the fall of 1992, two young men were foraging in the Maine woods for wild ginseng. Several plants were collected. The younger man, aged 23, took three bites from the root and his 39-year-old brother took one bite from the same root. Within 30 min, the younger man vomited and began to convulse. They walked out of the woods and received emergency rescue within 45 min of the onset of symptoms. At this point, the younger man was unresponsive and cyanotic and had tachycardia, dilated pupils, and perfuse salivation. He had several clonic-tonic convulsions, developed ventricular fibrillation, and was dead on arrival at the local hospital despite resuscitative attempts. The older brother was not showing symptoms at this time and was given gastric lavage and activated charcoal. Some time later, he developed delirium and seizures. He recovered with symptomatic treatment.

Q. What is the likely source of this problem?

Q. What organ systems are involved?

Q. Which plant and toxin would cause this array of symptoms?

Case study 25

During the summer of 1999 in London, Ontario, several teenagers were admitted to the emergency department of a local hospital, one in critical condition. Signs and symptoms included stomach cramps, irregular heartbeat, hallucinations, and dilated pupils. One boy was found unconscious in the basement of his home. The teens admitted to eating the seeds from the seedpods of a wild plant with large, white trumpet flowers and spiny seedpods.

Q. What is the probable identity of this plant?

Q. What are the active ingredients that impart its toxicity?

Q. Where does this plant grow?

chapter twelve

Environmental hormone disrupters

> "Every man in this room is half the man his grandfather was."
>
> — Louis J. Guilette, Jr., zoologist, to Congressional Committee, on endocrine disrupters

Introduction

In 1950, Burlington and Lindeman made one of the earliest observations that synthetic chemicals could seriously impair normal reproductive function (see *Proc. Soc. Exp. Biol. Med.*, 74, 48–57, 1950). They showed that Leghorn cockerels exposed to DDT had impaired testicular growth and diminished secondary sexual characteristics. A decade or so later, reports began to emerge that women whose mothers had received diethylstilbestrol (DES) during pregnancy experienced difficulties in conceiving and an increased incidence of cervical deformaties and cancer (see also Chapter 8). The first reported case of clear-cell carcinoma in a DES daughter occurred in 1971. More recently, men exposed to DES *in utero* have also been shown to have an increased incidence of reproductive dysfunction. It is now well known but poorly publicized that the source of infertility problems can be traced to the male 50% of the time. About the same time that the DES problem was emerging, Glen Fox, a scientist with the Canadian Wildlife Service, discovered that gulls in Lake Ontario were showing numerous signs of disrupted reproductive function. Females were sitting on eggs that refused to hatch. Males were losing interest in sex, forcing females to pair up to brood over sterile eggs. The eggs themselves frequently were misshapen and fragile. Fox speculated that pollutants such as DDT and PCBs, persistent organic pollutants (POPs) that are now banned, could be responsible. In 1963, Rachel Carson, formerly an aquatic biologist with the U.S. Fish and Wildlife Service, published her book *Silent Spring*, in which she warned that the indiscriminate use of pesticides could have catastrophic environmental effects. By the mid-1970s, her words were appearing highly prophetic. Since then, numerous classes of chemicals have been shown to possess the ability to modulate or severely disrupt hormonal function, either experimentally or in wildlife.

The Lake Apopka incident

In 1980, a small chemical mixing company spilled massive amounts of sulfuric acid dicofol, a pesticide for mite control containing DDT into Lake Apopka, the fourth largest body of freshwater in Florida. Although the size of the spill was undetermined, 2 weeks after it, the pH of the lake water remained as low as 1.7 — about that of stomach acid. Six years later, a study by the University of Florida at Gainsville found high numbers of unhatched eggs in alligator nests and by 1988 only 4% of the eggs were hatching. In the hatchlings following the spill, both testes and ovaries showed anatomical abnormalities. In both sexes, estrogens had come to dominate. This species may be very sensitive to outside disturbances of hormonal balance. In females, the ratio of estradiol to testosterone was twice normal. In males, levels of testosterone were depressed and penises were abnormally small. This episode clearly indicated that exposure to high concentrations of synthetic chemicals could adversely affect reproduction. The question of the effects of long-term exposure to very low levels of such chemicals is not as clear.

A brief review of the physiology of estrogens and androgens

Estrogens and androgens are, respectively, the female and male steroid hormones that regulate reproductive function and behavior and the development of secondary sex characteristics. The most potent are 17β-estradiol (E2) (female) and testosterone (male). Estrogen synthesis, primarily by ovarian follicles, is controlled by follicle stimulating hormone (FSH) from the anterior pituitary gland (adenohypopheseal gland). Release of FSH is regulated by a negative feedback system. Androgen synthesis by interstitial Leydig cells of the testes is stimulated by luteinizing hormone (LH). This also is regulated by a negative feedback loop. FSH and LH, along with thyroid stimulating hormone (TSH), constitute the glycoprotein group of hormones.

Following synthesis, both sex hormones are transported in blood 99% bound to plasma proteins, primarily to sex hormone binding globulin (SHBG). In females of reproductive age, total E2 serum levels are 50 to 300 pg/mL, depending on the stage of the estrous cycle. Males also have circulating E2 levels (10 to 60 pg/mL). At the target cell, cytosolic receptors complex with their respective hormones and act as transcription factors, transporting the hormones to the nucleus where they bind to specific response elements and induce the transcription-regulated genes. When E2 binds to its receptor (ER), it induces a conformational change, forming a dimer with another ER complex. Testosterone is acted on intracellularly by

5-α-reductase to form 5-α-dihydrotestosterone. In some tissues (e.g., bone marrow and skeletal muscle), testosterone itself and other metabolites are the active forms.

Mechanisms of hormone disruption

Xenobiotic estrogens (xenoestrogens) belong to a group of chemicals known as persistent organic poisons (POPs) that are characterized by long biological half-lives and high lipid solubility. These characteristics contribute to their tendency to biomagnify in the environment and to accumulate and persist in lipid stores. There are several mechanisms by which xenoestrogens could, at least in theory, disrupt or modulate normal hormone function. With regard to estrogen (17β-estradiol, hereafter referred to as E2):

1. They could mimic estrogen by binding directly with its receptor.
2. By this same mechanism, they could also prevent E2 from interacting normally with its receptor.
3. They could react directly or indirectly with E2 carrier proteins.
4. They could react directly with free or bound E2 to alter plasma E2 levels.
5. They could interfere with E2 synthesis.
6. They could up- or down-regulate the number of E2 receptors available.
7. They could antagonize E2 by virtue of inherent androgenic activity.
8. Although seldom included in a discussion of environmental hormone disruption, any toxic agent that is present in sufficient concentration to cause overt toxicity, either acute or chronic, can have a negative effect on reproductive function.

The same mechanisms could also apply to interference with normal androgenic function. Most research, however, has focused on interference with E2 activity.

Methods of testing for hormone disruption

Two main types of laboratory studies have been used to assess the potential of xenoestrogens for hormone-modulating activity. The first involves receptor binding studies using E2 receptor proteins from a variety of species, including human. These may be in their natural cell membrane, as in cultured, E2-sensitive breast cancer cells such as the MCF-7 cell line, or they may be transfected into another cell type such as bacteria or cultures of human cell lines. The second involves *in vivo* studies in rodents, usually rats. Because much of the theoretical concern about environmental hormone disrupters centers on their potential to cause damage to offspring *in utero*, studies often involve administration of the test substance to pregnant rats, followed by examination of the pups for abnormalities. These include delayed vaginal opening, anatomical differences such as weights of the uterus or testes, and anatomical defects.

Field studies include the examination of wildlife species, notably marine and aquatic ones, that have been exposed to high concentrations of xenobiotics for evidence of impaired reproductive function. In humans, populations accidentally exposed to such agents (e.g., the women in Seveso exposed to high levels of TCDD) are monitored for reproductive problems. Epidemiological studies attempt to examine trends over a long time course during which it is assumed that exposure levels to hormone disrupters has increased. Sperm count studies are included in this category. Populations of workers or others known to have unusual exposures are also useful, especially if effects can be correlated with serum levels of the suspect agent.

Some examples of xenoestrogen interactions with E2 receptors or effects in vivo or in vitro

Bisphenol A (BPA), a chemical used in the manufacture of plastics, is an environmental estrogen. Its interaction with the human E2 receptor Er-alpha has been investigated using human hepG2 hepatoma cells with the transfected receptor. Compared to E2, it was 26 times less potent and acted as a partial agonist. By using ER-alpha mutants in which the AF1 or AF2 regions were inactivated, differing patterns of activity were demonstrated for BPA, weak agonists estrol and estriol, partial agonists, and antagonists. BPA had no effect on uterine weight when given to immature female rats but shared some effects with E2 on peroxidase activity.

Of the hundreds of PCBs examined, only a handful have shown estrogenic activity. When the effects on age of vaginal opening were compared in rats, their potency (as compared to diethylstilbestrol [DES]) was 1/80,000 to 1/1,000,000. Even the most potent required 4 to 5 times the dose of E2 to affect the pituitary secretion of LH and FSH. Estrogenic activity seems to require substitution in the *ortho*-position, and non-*ortho*-substituted PCBs may actually possess anti-estrogenic activity.

In studies using MCF-7 breast cancer cell cultures, technical-grade dichlorodiphenyltrichloroethane (DDT) and its metabolite *o,p'*-DDT had estrogenic activity about 1/1,000,000 that of E2. When 5 mg/kg/day were given for 27 days to female rat pups, vaginal opening was delayed and uterine and ovary weights were increased. Lower doses had no effect. Mature animals required a dose of 25 mg/kg/day for 7 days to develop an increase in uterine weight.

Several environmental chemicals were tested in a yeast-based, human E2 receptor assay. Most potent were the DDT metabolites *o,p'*-DDT, *o,p'*-DDD, and *o,p'*-DDE. Their potencies were, respectively, $1/8$, $1/15$, and $1/24 \times 10^{-6}$ less potent than E2. Other organochlorines such as dieldrin were of similar potencies. In the *in vivo* assay, dieldrin, aldrin, and mirex had no effect on time of vaginal opening, estrous activity, or ovulation. Kepone, however, increased uterine weight and had E2 receptor binding characteristics that were 0.01 to 0.04% that of E2.

Several compounds have been shown to reduce E2 binding to its receptor. Most potent of these were butyl benzyl phthalate (BPP) and di-*n*-butyl phthalate (DBP) from plastics and the anti-oxidant food additive, butylated hydroxyanisole (BHA). Their potencies were also about 1/1,000,000 that of E2. The plasticizer bisphenol-A was considerably more potent being 1/1000–5000 times less potent than E2 in the MCF-7 cell culture assay.

Some effects on the male reproductive system

Male mice exposed to DES *in utero* had, as adults, decreased sperm production and a high frequency of abnormal morphology. It is axiomatic that exposure to high levels of estrogenic substances will have a deleterious effect on the male reproductive system, whether such exposure occurs *in utero* or not. A variety of xenobiotics has been examined for their ability to cause DNA breaks in rat and human cultured testicular cells. Methoxychlor, benomyl, thiotepa, acrylonitrile, and Cd^{2+} had litle effect on either. Styrene oxide, 1,2-dibromoethane, and chlordecone induced a significant number of strand breaks in both types of cells, although not always at the same concentration. Several others, including aflatoxin B_1, induced strand breaks only in rat testicular cells. Octylphenol has been shown to impair spermatogenesis in male rats and to decrease the percentage of viable cultured spermatogenic and Sertoli cells.

One of the most potent environmental anti-androgens yet identified is the dicarboximide anti-fungal agent vinclozoline. Administration of 100 or 200 mg/kg/day to pregnant female rats did not produce any signs of toxicity — or affect the viability of pups. However, all male pups were misclassified as females at birth and had delayed puberty and impaired copulation. Numerous anatomical anomalies were seen at necropsy, including vaginal pouches and ectopic testes. These effects appear to be due to anti-androgenic activity and not to estrogenic activity. There was no evidence of estrogenic effects in female pups. No anatomical anomalies were noted and fecundity was normal. Primary metabolites of vinclozoline, but not vinclozoline itself, were shown to bind to the androgen receptor 2,2-bis(p-chlorophenyl)-1,1-dichloroethane. (*p,p*′-DDE) also has anti-androgenic activity and tributyltin is an environmental androgen.

Modulation of hormone activity through effects on the Ah receptor

Many compounds are capable of modulating both estrogenic and androgenic function, either by acting as ligands for the aryl hydrocarbon (Ah) receptor (AhR) or by acting as Ah inducers. In the former category are dibenzo-*p*-dioxins, dibenzofurans, biphenyls, diphenyl ethers, substituted PAHs, hydroxylated benzo[a]pyrenes, and many others. These may induce genes in a tissue- and species-dependent manner, including those responsible for the synthesis of estrogens and androgens. Thus, both estrogenic and androgenic effects can occur as well as effects relating to thyroxin.

A number of studies have shown hypothyroid effects of PAHs. Rat pups exposed perinatally to PCBs had T_4 levels that were depressed, whereas T_3 levels were not. Studies with TCDD required doses much higher than the no effect level to depress T_4 and, again, T_3 levels were not altered. Studies in adult rats have shown a depression of plasma T_4 following the administration of a number of PAHs and organohalogens. Both transport and metabolism of T_4 were felt to play a role. The results generally indicated that PCBs and related compounds affected thyroid function to a greater extent than did TCDD.

Other compounds bind poorly to the AhR but are capable of inducing it. These include imidazoles, pyridines, oxidized carotonoids, heterocyclic amines, fumonisin B_1, and others. TCDD has been shown to induce endometriosis in female monkeys and goiter, hypothyroidism, and hyperthyroidism in rats. See Chapter 5 for more on the reproductive toxicity of TCDD and other AhR ligands.

Phytoestrogens

Phytoestrogens are natural estrogenic substances found in cereal grains, vegetables, and fruit. Estrogenic activity in plant extracts was first identified in 1926 and by 1975, several hundred plants were shown to contain substances with estrogenic activity. Three principle classes of phytoestrogens occur in plants and their seeds: isoflavones, coumestans, and lignans. Different parts or stages of the same plant may contain different phytoestrogens. Thus, soy sprouts are rich in coumestrol, whereas the bean has high levels of isoflavones. Mycoestrogens such as zearalonone also exist and have caused reproductive problems in livestock (see Chapter 10).

Some relative potencies, taking that of E2 as 100 in the MCF-7 bioassay, are:

Coumestrol (the principle coumestan), found in sunflower, alfalfa, legumes, and soy products: 0.03–0.2

Genistein and daidzein, found in barley, oats, rye, rice wheat, and soy products: 0.001–0.01

By comparison, the mycoestrogens zearalonone and its metabolite zearalanol: 0.001–0.1

Coumestrol is thus the most potent of the phytoestrogens.

Results of human studies

Males

Golden et al. published an extensive review of the literature in 1998. Concern over the effects of environmental xenobiotic disruption of male reproductive function stems mainly from a report by Carlsen et al. in 1992 presenting evidence that sperm counts in men had declined steadily from 1938 to 1990.

It was assumed that, during this period, exposure to xenobiotics would have been high and that this could have been responsible for the decline. Recent findings that men exposed *in utero* to DES had impaired sperm quality lent weight to this supposition. The methodology of the Carlsen study was criticized for failure to control for such confounders as smoking, marijuana use, and environmental temperature, all known to affect sperm count. When their data were reexamined, it was found that all of the low sperm counts occurred prior to 1970. In a more recent study of men who banked sperm before vasectomy, there was a slight but significant increase in sperm count from 1970 to 1994 and no change in motility or volume. The Air Force Health Study of Vietnam Veterans exposed to Agent Orange during Operation Ranchhand (the application of herbicides to defoliate the jungle) found no correlation between serum dioxin levels and either sperm quality or the number of conceptions.

Experimental evidence that xenoestrogens increased the incidence of undescended testes (cryptorchidism) prompted epidemiological studies. A study of 6935 live births at Mount Sinai Hospital in New York did not indicate any change in the incidence of cryptorchidism between 1950 and 1993. The incidence at 3 months of age remained at about 1%. In Great Britain, there was a slight increase in the incidence but no conclusions could be drawn as to the cause.

The incidence of prostate cancer has been increasing for several years. Much of this increase is attributed to earlier and better diagnosis. Epidemiological studies, however, have shown that agricultural workers, golf course managers, and pesticide applicators have a slightly higher incidence of prostate cancer. The pesticide applicators had a relative risk factor of 2.38. The 95% confidence limit was 1.38 to 3.04.

There has been a dramatic increase in the incidence of testicular cancer over the past few decades. Studies in men exposed perinatally to DES do not indicate that it is a risk factor for testicular cancer. One study suggested an association between pesticide use and a slight increase in the incidence of both prostate and testicular cancer in Hispanic and Black agricultural workers in California. No firm conclusions can be drawn at this time regarding exposure to pesticides or other chemicals as a risk factor for testicular cancer.

Because organochlorines have been shown experimentally to produce a number of thyroid-related endocrine disturbances, there has naturally been great interest in the significance of this for humans. The Air Force Health Study, in 1987, examined TCDD serum levels in 866 Ranchhand veterans. For 319 officers, the median level was 7.8 parts per trillion (ppt, 0–42.6); for 148 enlisted air crew, it was 18.1 ppt (0–195.5); and for 399 enlisted ground crew, it was 24 ppt (0–617.8). Extrapolating back to the initial exposure levels, 130 had 52–93 ppt, 260 had 93–292 ppt, and 131 had greater than 292 ppt. A decade later, the highest level group had slightly elevated thyroid stimulating hormone (TSH) levels compared to normal background levels. This is suggestive of a slight decrease in thyroid function. T_3 uptake by the thyroid gland was 29.99% compared with 30.65% for normal subjects with background levels of TCDD.

Females

Xenoestrogens have been implicated as risk factors for breast cancer and endometriosis. To date, there has been no demonstrated association between risk for endometriosis and exposure to xenoestrogens other than the reported findings in DES daughters. Several studies have been conducted in which levels of PCBs and other xenoestrogens in breast fat, blood, or serum were measured in breast cancer patients. Cancer incidence was then correlated with levels of the xenobiotics. A study conducted at Mount Sinai Hospital in New York and published in 1992 found that women with the highest levels of DDE in their blood had a breast cancer incidence that was 4 times higher than that of women with the lowest DDE levels. By 1994, four studies had been published showing an association between levels of DDE and PCBs in fluids or tissues and an increased incidence of breast cancer. (It is a commentary on the ubiquitous nature of POPs that no population totally free of contamination can be found as a control group.) Some of these studies were criticized for their failure to control for known risk factors for breast cancer and for small numbers of participants. The New York study, for example, involved 58 women. In April 1994, a study led by Nancy Krieger of the Kaiser Foundation Research Institute examined blood that had been frozen and stored in the late 1960s from women who developed breast cancer, on average, 14 years later. There were 50 Caucasian, 50 African-American, and 50 Asian women. Blood levels of DDE in these samples were 4 to 5 times higher than those found in the New York samples (drawn 1984–1991), because the Kaiser samples were drawn prior to the 1972 ban on DDT. When the results were compared to matched controls, no association could be demonstrated between serum levels of DDE and risk of breast cancer. Similar findings were obtained for PCBs.

In 1999, the (U.S.) National Cancer Institute released results of a similar study. Women who donated blood over a 10-year period from 1977 were followed for up to 9.5 years. For each subject, two controls were selected and matched for age and date of blood collection. Of these, 105 donors who matched the selection criteria were diagnosed with breast cancer. Serum was analyzed for 5 DDT analogs, 13 other organochlorine pesticides, and 27 PCBs. For hexachlorobenzene, the women in the upper three quartiles of serum levels had twice the risk of breast cancer as those in the lowest quartile. There was no evidence of a dose response relationship, and the finding was limited to women whose blood was collected close to the time of diagnosis. Positive but weak associations were suggested for PCBs 118 and 138. Again, this association was noted only in women whose blood was collected close to the date of diagnosis. There would normally be a significant latency period between exposure to the carcinogenic agent and the development of malignancy. No other positive associations were found.

Two studies on organochlorines and risk of breast cancer in South America were published recently. The study from Brazil found no correlation between blood levels of DDE and breast cancer risk when 177 hospital patients with invasive breast cancer were compared with 350 controls (hospital visitors). The other study, from Columbia, compared 153 cases of breast cancer with 153 age-matched contols. Odds ratios were adjusted for a number of factors such as first-child breast feeding. An odds ratio of 1.95 (confidence interval 1.10–3.52) suggested increased risk for the higher category of DDE exposure, but the test for trend was not statistically significant ($p = 0.09$). Taken overall, there is no strong evidence for or against the premise that environmental exposure to xenoestrogens constitutes a significant risk factor for breast cancer. Further studies may clarify the situation. The massive exposures to PBBs that occurred in the Fukuoka region of Japan in 1968 and in Yucheng, Taiwan, in 1979 (see also Chapter 5) have revealed a number of developmental defects in children exposed in the womb. Total body burdens in the mothers were calculated to be 200 times higher than the average for North America. The Michigan study of the effects of pregnant women eating large amounts of lake fish on their offspring is discussed at the end of Chapter 3.

A 15-year follow-up study of the Seveso accident recorded the incidences of several cancers in three zones of differing rexposure to TCDD. The incidences of several cancers were increased (relative risk of multiple myeloma 6.6 in women, of rectal cancer 6.2 in men, 3.1 for leukemia in men), but the overall cancer mortality was not increased and there was no increase in breast cancer.

The situation with regard to phytoestrogens appears to be quite different. Epidemiological studies suggest that consumption of a diet rich in phytoestrogens, such as those eaten in Asian societies, is associated with lower incidences of breast, prostate, and colon cancers, as well as cardiovascular disease. Similar findings have been reported for people on strict vegetarian diets. As noted above, coumestrol is the most potent of the phytoestrogens and soy protein, a staple in Asian diets, is rich in coumestrol. People in Japan, Taiwan, and Korea are estimated to consume 20 to 150 mg isoflavones per day. Menopausal women in Japan are reported to have fewer hot flushes than their North American counterparts.

The phytoestrogen content of plants can vary widely with climate, time, and geography, making estimates of intake difficult. In one study, 12 oz. soy milk, three times daily decreased serum E2 and progesterone levels in 22- to 29-year-old women. Other effects of phytoestrogens on hormonal balance have been observed experimentally. While the clinical use of phytoestrogens has not yet occurred widely in Western medicine, their use in herbal medicine is becoming widespread. In his review of the subject, Sheehan points out the problems of lack of standardization of content and of the likelihood that herbal preparations can contain many other active ingredients, some of which could be toxic.

Effects in livestock and wildlife

In 1946, a report appeared in the literature of sheep in Western Australia developing infertility as a result of eating subterranian clover. This plant is rich in the equol precursor formonestein. Equol is a known phytoestrogen. Reproductive problems in swine consuming feed grains with high zearalonol (a mycotoxin) content have been discussed. There are numerous reports of livestock experiencing fertility problems as a result of grazing on poor pastures containing phytoestrogen-rich weeds. Wild quail in California also have been reported to have similar problems.

The reproductive problems in gulls and other fish-eating birds associated with DDT, as well as the Lake Apopka chemical spill and its effects on alligators, have been discussed. Vertebrate and invertebrate aquatic and marine animals are also vulnerable to the effects of POPs. Estrogenic activity has been detected in treated sewage outfalls and industrial effluents. Vertebrates such as fish downstream from such effluents have become feminized. Vitallogenin induction in males, morphological changes, and complete sex reversal have been reported. Xenoestrogens, natural sterols, organochlorines, and alkylphenols have been identified as suspected agents. Invertebrates have also been shown to be vulnerable to endocrine disruption by environmental xenobiotics. Intersex in crustaceans exposed to sewage discharges and endocrine disruptions in gastropod mollusks have been noted. Barnacle larvae displayed reduced resettlement activity following exposure to the xenoestrogen 4-nonyl phenol. Endocrine disrupters have been shown experimentally to have a negative effect on growth and reproduction of amphipods and marine worms, both of which inhabit and feed in bottom sediment.

Problems in interpreting and extrapolating results to humans

In both types of laboratory study, there is difficulty in extrapolating from the laboratory findings to the whole organism, whether it is mammal, fish, or reptile. Species differences abound with respect to all sites of potential interference, and differing sensitivities of E2 receptors have been recorded. One study compared the binding of 44 PCBs, 9 hydroxylated PCBs, and 8 arochlors to recombinant human, rainbow trout, and green anole (lizard) E2 receptors (fusion proteins) linked to the glutathione-S-transferase protein and expressed bacterially. Only PCBs 104, 184, and 188 competed effectively with E2 for the rainbow trout protein and displaced E2 by only 30% from the others. Other congeners showed varying affinities for the rainbow trout protein, but none for the others.

A further confounding factor is the fact that many agents have multiple effects that can all impact on reproduction and some of these may be mutually opposed. Thus, the isoflavone gentisein is a potent inhibitor of both E receptor positive and E receptor negative breast cancer cells of the MCF-7 strain. It is also a potent inhibitor of the tyrosine protein kinase activity of several growth factor receptors and oncogenes that might be involved in the growth of tumor cells. Other effects of genistein have been recorded.

Lack of potency is another factor that raises questions about the health impact of environmental hormone modulators. Much of the present concern stems from the DES experience where exposures *in utero* were much higher than what could be encountered environmentally. Coumestrol is the most potent of the phytoestrogens and the one most associated with mild estrogenic effects in women. Taking E2 potency as 100 in MCF-7 cell culture tests, coumestrol had a potency of 0.03 to 0.2. It is several orders of magnitude more potent than the xenoestrogens. In areas where the dietary intake of phytoestrogens is high, their metabolites are present in urine in easily measured amounts. This is seldom true of xenoestrogens, except in populations with abnormally high exposures.

Because humans are exposed to a veritable potpourri of environmental hormone modulators, many of them natural agents with potential benefit, the net effect is likely to be the result of interactions among many agents, some with mutually opposed activity. At present, one can only say that the evidence for environmental hormone modulators having a significant impact on human health is inconclusive and that more evidence is required. There is, however, a good case to be made for invoking the precautionary principle in calling for the eventual elimination of POPs. This is especially true with regard to the environmental damage POPs can cause.

Further reading

Bertazzi, P.A. et al., Dioxin exposure and cancer risk: a 15-year mortality study after the "Seveso" accident, *Epidemiology*, 8, 646–652, 1997.

Bjorge, C., Bruneborg, G. et al., A comparative study of chemically induced DNA damage in isolated human and rat testicular cells, *Reprod. Toxicol.*, 10, 509–591, 1996.

Carlsen, E., Giwercman, A., Keiding, N., and Skakkebaek, N.E., Evidence for decreasing quality of sperm during the last 50 years, *Br. Med. J.*, 305, 1228–1229, 1992.

Denison, M.S. and Helferich, W.G., Eds., *Toxicant–Receptor Interactions*, Taylor & Francis, Philadelphia, 1998.

Depledge, M.H. and Billinghurst, Z., Ecological significance of endocrine disruption in marine invertebrates, *Mar. Pollut. Bull.*, 39, 32–38, 1999.

Dorgan, J.F., Brock. J.W. et al., Serum organochlorine pesticides and PCBs and breast cancer risk: results from a prospective analysis (USA), *Cancer Causes Control*, 10, 1–11, 1999.

Fisch, H., Feldshuh, J. et al., Semen analysis in 1,283 men from the United States over a 25-year period: no decline in semen quality, *Fertil. Steril.*, 65, 909–911, 1996.

Fleming, L.E., Bean, J.A., Rudolph, M., and Hamilton, K., Mortality in a cohort of licensed pesticide applicators in Florida, *Occup. Environ. Med.*, 56, 14–21, 1999.

Golden, R.J., Noller, L. et al., Environmental endocrine modulators and human health: an assessment of the biological evidence, *CRC Crit. Rev. Toxicol.*, 28, 109–227, 1998.

Gould, J.C., Leonard, L.S. et al., Bisphenol A interacts with the estrogen receptor alpha in a distinct manner from estradiol, *Mol. Cell. Endocrinol.*, 142, 203–214, 1998.

Gray, L.E., Jr., Ostby, J.S., and Kelce, W.R., Developmental effects of an environmental antiandrogen: the fungicide vinclozolin alters sex differentiation of the male rat, *Toxicol. Appl. Pharmacol.*, 129, 46–52, 1994.

Krieger, N., Wolff, M. et al., Breast cancer and serum organochlorines: a prospective study among white, black and Asian women, *J. Natl. Cancer Inst.*, 86, 589–599, 1994.

Kurzer, M.S. and Xu, X., Dietary phytoestrogens, *Annu. Rev. Nutr.*, 17, 353–381, 1997.

Matthews, J. and Zacharewski, T., Differential binding affinities of PCBs, HO-PCBs, and arochlors with recombinant human, rainbow trout (*Onchorhynkiss mykiss*) and green anole (*Anolis carolinensis*) estrogen receptors, using a semi-high, throughput competitive binding assay, *Toxicol. Sci.*, 53, 326–339, 2000.

Matthiessen, P. and Sumpter, J.P., Effects of estrogenic substances in the aquatic environment, fish ecotoxicology, Braunbeck, T., Hinton, D.E., and Streil, B., Eds., Birkhauser Verlang, Basel, Series EXS 86, 319–335, 1998.

Mendonca, G.A., Eluf-Neto, J. et al., Organochlorines and breast cancer: a case-control study in Brazil, *Int. J. Cancer*, 83, 596–600, 1999.

Mills, P.K., Correlation analysis of pesticide use data and cancer incidence rates in California counties, *Arch. Environ. Health*, 53, 410–413, 1998.

Murkies, A.L., Wilcox, G., and Davis, S.R., Phytoestrogens, *J. Clin. Endocrinol. Metab.*, 83, 297–303, 1998.

Olaya-Contreras, P., Rodriguez-Villamil, J., Posso-Valencia, H.J., and Cortez, J.E., Organochlorine exposure and breast cancer risk in Columbian women, *Cad. Saude Publica*, 14 (Suppl. 2), 125–132, 1998.

Raychaudhury, S.S., Blake, C.A., and Millette, C.F., Toxic effects of octylphenol on cultured rat spermatogenic cells and Sertoli cells, *Toxicol. Appl. Pharmacol.*, 157, 192–202, 1999.

Safe, S.H. and Zacharewski, T., Organochlorine exposure and risk for breast cancer, *Prog. Clin. Biol. Res.*, 396, 133–145, 1997.

Sheehan, D.M., Herbal medicines, phytoestrogens and toxicity: benefit considerations, *Proc. Soc. Exper. Biol. Med.*, 217, 379–385, 1998.

Van Der Gulden, J.W. and Vogelzang, P.F., Farmers at risk for prostate cancer, *Br. J. Urol.*, 77, 6–14, 1996.

Review questions

Answer the following questions True or False.

1. Impairment of reproductive function and testicular growth was first demonstrated in chickens in 1950.
2. Some men exposed to DES *in utero* appear to have impaired reproductive function.
3. High levels of estrogens in male alligators leads to an increase in penis size.
4. Follicle stimulating hormone (FSH) primarily regulates the synthesis of testosterone.
5. Synthesis of sex hormones is regulated by negative feedback loops.

6. Males have circulating levels of estrogen (E2).
7. Testosterone is the active form of androgen in all tissues.
8. The estrogen (E2) receptor is located on the cell membrane.
9. Xenoestrogens have long biological half-lives.
10. The affinity of most xenoestrogens for the E2 receptor is about the same as that of E2 itself.
11. Only a handful of PCBs show estrogenic activity.
12. Non-*ortho*-substituted PCBs may possess anti-estrogenic activity.
13. DDT and its metabolite *o,p'*-DDT are equipotent with E2 in the breast cancer cell culture assay.
14. The insecticide kepone, one of the most potent xenoestrogens, has E2 receptor binding characteristics about 0.01 to 0.4% that of E2.
15. Aflatoxin B1 induced DNA strand breaks in rat testicular cells.
16. Vinclozoline is probably the most potent xenobiotic anti-androgen yet identified.
17. TCDD has no adverse effects on reproductive function.
18. Phytoestrogens tend to be less potent than xenoestrogens.
19. Pleiotropic effects modulated by the Ah receptor can include disruption of reproduction.
20. Fish and other wildlife are more vulnerable to xenoestrogen toxicity than mammals.

Answers

1. True
2. True
3. False
4. False
5. True
6. True
7. False
8. False
9. True
10. False
11. True
12. True
13. False
14. True
15. True
16. True
17. False
18. False
19. True
20. True

chapter thirteen

Radiation hazards

Introduction

The electromagnetic spectrum encompasses all forms of radiant energy. Table 12 lists the various components and their wavelengths in decreasing order.

Table 12 Components of the Electromagnetic Spectrum

Type of Radiation	Wavelength
Radio waves	30 km–3 cm
Microwaves	3 cm–10 mm
Thermal (heat)	0.078–0.001 mm
Infrared (includes thermal portion)	0.5 mm–10000 Å
Visible	7800–4000 Å
Ultraviolet	4000–1850 Å
Extreme ultraviolet	1850–150 Å
Soft X-rays	1000–5 Å
X-rays	5–0.06 Å
Gamma rays	1.4–0.01 Å
Cosmic rays (protons 85%, alpha particles 12%, electrons, gamma rays, etc.	

Ionizing radiation, that portion of the spectrum that can cause serious cell damage, includes all wavelengths of 1000 Å or less (see Table 12). Ionizing radiation, by stripping electrons from molecules as it passes through tissues, produces ionized species of everything from H_2O to macromolecules like DNA. These ionized species are unstable and reactive, and can produce dramatic disruptions in cell function, including mutation.

There is still controversy regarding the degree of risk. Radiophobia, an illogical fear of radiation hazards, has led to considerable controversy over the extent of the environmental risks of ionizing radiation. Consider the following conflicting statements from Cobb (1989).

- Dr. K.Z. Morgan, a pioneer in health physics: "It is incontestable that radiation risks are greater than published."
- C. Rasmussen, nuclear engineer at MIT: "There is a lot of evidence that low doses of radiation not only don't cause harm but may in fact do some good. After all, humankind evolved in a world of natural low-level radiation." (About 82% of our total radiation exposure come from natural sources.) This is another example of hormesis.
- R. Guimond, EPA: "We can't avoid living in a sea of radiation."

In some cultures, deliberate exposure to natural-source radiation is done in the belief that it has curative powers. In Japan, exposure to radon is courted in "radon spas" where natural hotsprings occur.

Sources and types of radiation

Sources

Natural sources of radiation

Natural sources of radiation include cosmic rays from space, solar rays that intensify during solar storms (sunspots), radiation emanating from rocks and groundwater, and radiation coming from within our own bodies, mainly from decay of radioactive potassium in muscle. Of considerable concern at present is the exposure to radon gas, a radioactive decay product of radium, a common radioactive element in soil and rock. This is discussed in more detail below.

Man-made sources of radiation

Man-made sources of radiation include medical X-rays and radioisotopes, ion sensors in smoke detectors, uranium used to provide the gleam in dentures, mantles in camping lanterns, radioactive wastes, nuclear accidents (e.g., Chernobyl), and careless handling of nuclear materials. Cesium-137 was discarded in a dump in Brazil and ultimately killed 4 people and contaminated 249. Until fairly recently, radium was used to hand-paint luminous watch dials. Workers used to "point" their brushes by running them between their lips, a practice that led to cases of radiation sickness.

The cause of radiation

Elements that exist in an unstable form are continually decaying to more stable ones. In the process, they give off energy in several ways. Ionizing radiation arises when an unstable nucleus gives off energy. An unstable nucleus is called a radionuclide. In contrast, X-irradiation is a form of cosmic ray and also occurs when a suitable target such as tungsten is bombarded with electrons. It does not arise from nuclear decay.

Types of radioactive energy resulting from nuclear decay

1. A nucleus can eject two protons and two neutrons to lose mass and convert itself into another element. The ejected components constitute an alpha-particle. An alpha-particle is slow moving and will not penetrate skin, but it can cause dangerous ionization if ingested. An alpha-particle is actually identical to the nucleus of helium.
2. The neutron of a nucleus can lose an electron to become a positron. The lost, negatively charged particle is called a beta-particle.
3. Even after emitting alpha- or beta-particles, a nucleus may remain in an agitated state. It can rid itself of excess energy by giving off gamma-rays. These are short, intense bursts of electromagnetic energy with no electrical charge. They can penetrate lead and concrete and can cause extensive tissue damage by ionization.
4. Neutrons are ejected from the nucleus during nuclear chain reactions. They collide and combine with the nuclei of other atoms and induce radioactivity of the above types. This is the primary source of radioactivity following a nuclear explosion.

Measurement of radiation

There are two different types of measurements of ionizing radiation. One is concerned with the level of energy actually emitted from the source, and the other is concerned with the amount of tissue damage that can be produced by a particular form of radiation. The field is unfortunately further confused by a more recent shift to the international (SI) system. The equivalent values are shown in Table 13. They are not easily interchangeable.

Table 13 Equivalent Values, Old and New Systems

Old System	New System (SI)
1 curie (Ci)	37 gigabecquerels (Gbq)
1 rem (rem)	10 millisieverts (mSv)
100 rem	1 Sievert (Sv)
1 rad (rad)	10 milligrays (mGy)
100 rad	1 gray (Gy)
1 roentgen (R)	258 millicoulombs/kilogram (mC/kg)

Measures of energy

1. As the nucleus of an atom decays, it gives off a burst of energy (the ionizing radiation) called a "disintegration." The number of disintegrations per unit time varies with the nature of the source. Various counting instruments (scintillation counters, etc.) measure disintegrations per minute (dpm). The basic unit of measuring radiation

energy is the curie (Ci) and it is the number of disintegrations (3.7×10^{10}) occurring in 1 second in 1 gram of radium-226. Radioisotopes used in science are usually provided in milliCi (mCi) amounts. The new unit is the Becquerel (Bq). A Bq represents one disintegration per second: 1 Ci = 37×10^9 Bq (37 gigabecquerels).

2. The first unit of radiation was the roentgen (R). It has a complicated definition based on the amount of X-rays or gamma-rays required to cause a standard degree of ionization in air.

Measures of damage

1. The earlier measure of radiation damage is the rem, which stands for roentgen-equivalent-man. It is that amount of ionizing radiation of any type that produces in humans the same biological effect as 1 R. The new international unit is the sievert (Sv): 1 Sievert = 100 rem.
2. The rad is the amount of radiation absorbed by 1 gram of tissue. The new international unit is the gray (Gy): 1 gray = 100 rad.

A rough scale of toxicity is as follows: 10,000 rem is rapidly fatal because of damage to the central nervous system (CNS). Whole-body exposure to 300 rem is approximately the LD_{50}. Between 100 and 300 rem, radiation damage occurs. The assumption is made that the risk associated with radiation is linear all the way to zero. When it comes to assessing carcinogenic potential, however, accuracy is extremely difficult. Below 10 rem, effects are unclear due to confounding factors such as smoking, pollution, and diet. Over 300 agents have been shown to be carcinogenic in animal tests. Residents of Denver have lower cancer death rates than those of New Orleans despite higher radiation exposures because of increased levels of cosmic radiation at their high altitude.

Major nuclear disasters of historic importance

Hiroshima

The people from whom the most reliable data have been gathered concerning radiation hazards are the survivors of the atom bomb dropped on Hiroshima at 8:15 a.m., August 6, 1945. In 1947, the Radiation Effects Research Foundation was established. Exhaustive studies have shown that the heavily exposed people, called the "hibakusha," had a 29% greater chance of dying from cancer than normal. Excess numbers of leukemia cases began appearing in the late 1940s and peaked in the early 1950s but by the early 1970s had dropped to levels near those of unexposed Japanese. Now the surviving hibakusha have longer life expectancies than the overall population, perhaps because of closer medical supervision.

One of the most feared hazards of radiation is that of congenitally deformed infants because of radiation-induced genetic defects in the mother.

While such defects have been demonstrated experimentally, the Hiroshima study compared 8000 children of hibakusha with 8000 of unexposed Japanese and found an incidence of chromosome damage in 5/1000 of the former and 6/1000 of the latter. Protective mechanisms may be functioning in humans (e.g., spontaneous early abortion). Fetuses exposed *in utero* are a different story. Dozens of mentally retarded infants were born in the areas around Hiroshima and Nagasaki (target of the second atomic bomb) in the months following the blasts. Spontaneous abortions were also numerous. The fetus appears to be most vulnerable between 8 and 15 weeks.

One recent development, however, is that a special panel formed by the U.S. National Research Council recently released a report following a reassessment of the Hiroshima and Nagasaki data and concluded that the levels of exposure were much lower than previously calculated. Original estimates were based on tests at the Nevada nuclear test site using much flimsier buildings than were actually present in those cities. As a consequence, the established safe limits have had to be revised downward. This is primarily a concern for people exposed to radiation on the job, but it has revived the controversy about whether there is any safe level of exposure. Ironically, evidence is now surfacing that scientists and technicians who worked on the atomic bomb project during the war are showing up with elevated incidences of cancer which, when adjusted for exposure level, may be even greater than those of the hibakusha. The new safe exposure limit is 20 mSv/yr averaged over 5 years with no more than 50 in any one year. The old level was 50 mSv/yr.

Chernobyl

The most recent and highly publicized nuclear disaster was Chernobyl in April of 1986 (a much worse but largely concealed disaster occurred in Russia in 1958). In this disaster, 20% of the plant's radioactive iodine escaped, along with 10 to 20% of its radioactive cesium and other isotopes. Approximately 135,000 people lived in a 30-km radius of the power plant. There were 30 deaths and 237 cases of severe radiation injury. Some 2000 children have been born to women who were living in the accident zone at the time of the disaster. No abnormalities have yet been detected in them. An examination of about 700,000 people over a wider area has thus far not revealed any physical problems. Russian scientists estimate an increase in the cancer rate of 0.04% over the next 20 years. In western Europe, exposed to the drift of radioactive dust, it is estimated that, over the next 50 years, 1000 additional cancer deaths will occur. Normally, there would be 30,000,000 cancer deaths in this period, so the increase is 0.003%. Aside from those people directly exposed to the effects of the explosion, the greatest risk of exposure appears to come from eating contaminated food. Thousands of reindeer had to be destroyed in northern Scandinavia because they had grazed on contaminated pasture.

Dr. Marvin Goldwin, chief of the Joint-U.S.-Soviet Medical Team, made the following points in a recent report.

1. Everyone in the Northern Hemisphere received a small dose of radiation. The degree of exposure of those people at highest risk cannot be accurately identified.
2. Radioactive iodine posed an early risk to the thyroid glands of exposed people.
3. As of 1991 their was no detectable increase in the incidence of cancer but leukemia may yet show up and solid tumors may not show up for 10 years.

Chernobyl provides another example of how psychological damage can often exceed physical damage in an environmental disaster. Thousands of people received a radiation exposure that exceeded the maximum recommended lifetime allowance. Because radiation levels in their locales have fallen to low levels similar to those of surrounding areas, it makes no medical or scientific sense to relocate them. Stress and fear, however, create an understandable desire in these people to be moved out of the area, and their wishes are presently being acted upon.

Three Mile Island

The partial core meltdown of the Unit 2 reactor at Three Mile Island in March 1979 was largely responsible for bringing the nuclear power program in the United States to a halt. Of the "defense in depth" safety features, all but the outer water shield failed. Some authorities claim that even a complete meltdown would not have breached this defense. Despite concerns of nearby residents, only 15 Ci of radioactivity were actually released. The news media exploited the event with sensational reports of "deadly clouds of radioactive gas" and made much of the potential for explosion of a large bubble of hydrogen in the reactor. In fact, there was none because no oxygen was present. The people at greatest risk from radiation were the workers who were involved in the clean-up. U.S. federal regulations limit the maximum exposure of workers in the nuclear industry to 12 rem per year. Workers in the "hot" areas receive about 1 mrem per hour, the equivalent of one chest X-ray. Totals for such workers in 1987 were about 710 millirem (see Chapter 2 for more on Three Mile Island).

The Hanford release

In contrast, massive amounts of iodine-131 were deliberately released (for purposes still classified as top secret) from the military nuclear facility at Hanford, Washington in the 1940s and 1950s. Some residents may have received as much as 2295 rem. Again, the greatest source of exposure may have been the consumption of contaminated meat and vegetables. Multimillion-dollar studies have recently been commissioned to seek answers to the degree of risk and to assign responsibility. Obviously, the potential for civil action is considerable.

The consistent elements present in all of these disasters, including the discarding of cesium-137 in Brazil, have been human error, misjudgment, and negligence. Deliberate callousness may have been involved at Hanford, where charges of unsafe practices and antiquated, dangerous equipment are still being made. Conversely, the real danger to the public has routinely been overblown, and events have been exploited by the media and by antinuclear groups. In the minds of many, nuclear reactors equate with nuclear bombs.

Radon gas: the natural radiation

As noted above, 82% of the ionizing radiation to which North Americans are exposed comes from natural sources. It has been estimated that there is enough natural radioactive material in the human body that, if it were a laboratory animal, it would have to be disposed of as hazardous waste. The average, annual, natural background exposure in Great Britain is about 1 mSv (0.1 rem or 100 mrem). In Canada, it is somewhat higher because the Canadian Shield (the band of granite rock that spans mid-northern Ontario, Quebec, and Manitoba) is rich in uranium deposits. Radon gas is by far the largest potential health hazard from natural radiation. It is the decay product of uranium and it seeps up through faults in the sub-strata of soil and may leak into houses through cracks in basement walls, drains, etc. The advent of airtight houses has increased the risk by trapping radon gas in the house. In the British study, it was calculated that 1,000,000 people were exposed to radon at levels of 5 to 15 mSv annually. In contrast, only 5100 workers in nuclear industry were exposed to levels as high or higher. This is the equivalent of 50 to 150 chest X-rays annually. Radon homes are distributed very randomly, with highly contaminated homes located right beside radon-free ones. The federal government conducted a survey of Canadian homes and found that Winnipeg had the highest radon levels of any Canadian city, with high levels also found in parts of Saskatchewan and northern Ontario and Quebec (see Table 14). Toronto (in Ontario) has low levels.

A map of radon risk areas in the United States, published by National Geographic, shows a band running slightly east of North through the middle of Ohio, and another running east-west through New York state. A U.S. federal study surveyed 20,000 homes in 17 states and found that 25% had potentially hazardous levels of radon. Radon was described as the largest environmental radiation health hazard in America. Debate over the degree of risk plagues this area, as it does others. The EPA study measured radon levels in basements where they would be highest. Calculations of risk have estimated the lifetime risk of dying from radon-induced cancer at 0.4% for exposed individuals, which is the same as one's chances of dying in a fire or a fall. If radon were a man-made carcinogen, it would unquestionably be banned, and most certainly would be the target of anti-nuclear activists. Nevertheless, there is still controversy regarding the degree of risk, or perhaps more correctly, the degree of exposure. A British study calculated that 6 to 12% of all myeloid leukemias might be attributed to radon exposure,

Table 14 Radon Concentrations in Canadian Homes in, pCi/L Air

City	No. Homes Tested	No. Homes >4.5 pCi/L	%
St. Lawrence, P.Q.	432	63	14.6
Sherbrooke, P.Q.	905	64	7.1
St. John, N.B.	866	51	5.9
Fredricton, N.B.	455	26	5.7
Thunder Bay, Ont.	627	29	4.6
Charlottetown, P.E.I.	813	35	4.3
Sudbury, Ont.	722	29	3.8
St. John's, NFL.	585	17	2.9
Montreal, P.Q.	600	9	1.5
Quebec, P.Q.	584	9	1.5
Toronto, Ont.	751	1	0.1
Vancouver, B.C.	823	0	0

Compiled from data reported by McGregor et al., *Health Physics*, 39, 285–299, 1980.

with levels rising to 23 to 43% in Cornwall, where the highest exposures occur. Worldwide, their calculations suggested 13 to 25% of all myeloid leukemias in all age groups could be due to radon.

In 1984, a construction worker at the Limerick nuclear generating station near Reading, Pennsylvania consistently set off radiation alarms despite the fact that he had never worked in a "hot" area. An examination of his home subsequently revealed radon levels of 2600 pCi/L, the highest ever recorded. North Dakota, with 63% of homes showing levels of 4 pCi/L or greater, leads the United States in radon exposure, followed by Minnesota (46%), Colorado (39%), Pennsylvania (37%), and Wyoming and Indiana (26%). After cigarette smoking, radon is probably the most common cause of lung cancer. Radon-222 (^{222}Rn) and ^{220}Rn are the only gaseous decay products of uranium. Their half-life is 3.8 days and they decay to particles (not gases), including radioactive poloniums, which actually are the alpha-emitting toxins that cause cell damage. Alpha-particles, unlike gamma-rays, can only cause cell damage for a radius of about 70 μ, hence the risk of lung cancer. The increased incidence of leukemias is difficult to explain on this basis.

Tissue sensitivity to radiation

In general, tissues with a high rate of turnover are more susceptible to the effects of ionizing radiation. Thus, thyroid, lung, breast, stomach, colon, and bone marrow have high sensitivity; brain, lymph tissue, esophagus, liver, pancreas, small intestine, and ovaries are intermediate; and skin, gall bladder, spleen, kidneys, and dense bone are low. This order of sensitivity roughly parallels the frequency of primary cancer in these tissues. If molecular disruption is sufficient, the cell will die. Because hair follicles and gastrointestinal mucosa have high turnovers, radiation sickness involves hair loss and severe diarrhea. Because bone marrow cells also have a high turnover, repair of

DNA may not be complete before replication occurs and the daughter cells may be malignant. This is why leukemia is the commonest cancer associated with radiation injury.

Questions regarding the safety of the nuclear industry continue to emerge. In August 1989, two workers at the Pickering Nuclear Plant in Ontario were mistakenly given unshielded practice equipment to change a new type of fuel rod just recently introduced. They received what were widely reported as the highest levels of radiation ever encountered by workers in the Canadian nuclear industry: 5.6 and 12.2 rem. The annual allowable limit set by the Atomic Energy Commission is 2 rem. The radiation exposure from an average chest X-ray is 15 mrem. This has been calculated (theoretically) to cause one additional cancer per 100,000,000 population. In other words, if the entire population of North America were to receive one chest X-ray, one could expect three additional cancer cases as a result. These workers received about 700 times this amount, which would cause one additional cancer per 143,000 people. This risk would be lessened if they were removed to areas where there is no possibility of additional exposure for at least 1 year. The safety maxim that should apply in all situations involving exposure to radiation is ALARA (as little as reasonably achievable). It should be noted that, once again, human error was responsible for this accident.

In Great Britain, a disturbing report was released in early 1990 to the effect that offspring of nuclear plant workers had an increased incidence of birth defects. Because these children are not directly exposed to radioactive material, the conclusion, if the data are correct, is that exposure of the parents (mostly men) caused genetic damage. These results contradict earlier studies, and it has been pointed out that clusters of birth defects occur geographically in the absence of nuclear generators (or other identifiable causes). The data will be of concern until they can be explained or disproved. Public pressure has resulted in the cancellation of 50 new nuclear power plants in the United States. As a result, there is greater reliance on coal-fired generators. Approximately 200 coal miners die each year in mine accidents and an equal number from black lung disease (pneumoconiosis). Recent (1992) mine accidents include 26 killed in Nova Scotia, 400 in Turkey, and 38 in Russia. Such events rarely cause a ripple of concern among opponents of nuclear energy. There is also the problem of acid rain resulting from sulfur pollution of the atmosphere by coal-fired generators. Has the public traded a potential but high-profile risk for a real and greater one that is less visible? Table 15 compares various sources of radiation encountered by Americans.

Microwaves

Microwaves are the shortest waves in the radio portion of the electromagnetic spectrum (1 mm–30 cm). They are at very high frequencies (1000–300,000 megacycles/s) and are used in radar, for long-distance transmission of phone and TV signals, and, of course, in microwave ovens.

Table 15   Average Annual U.S. Doses

Source	Dose
Natural sources	<100 mrem
Medical and dental X-rays	78 mrem
Radioisotopes	14 mrem
Weapons testing	4–5 mrem[a]
Nuclear industry	<1 mrem
Building materials (brick and masonry)	3–4 mrem
Total average annual exposure	~200 mrem
1 chest X-ray	10–15 mrem

[a] Much higher at downwind locations in Nevada, Utah, etc.

Because of its high dielectric constant, water dissipates energy as heat when exposed to microwaves. At high enough energies, thermal damage may occur in living tissues. Recent concern has been expressed over possible carcinogenic effects of microwaves given off by cellular phones. The energy level is so low (<5 watts), however, that no thermal effects can be detected. Little information is available concerning non-thermal effects of microwaves, but evidence for carcinogenicity is scanty and anecdotal.

Ultraviolet radiation

Ultraviolet radiation occupies the electromagnetic spectrum between 400 and 4 nm: UVA, 400–320 nm; UVB, 320–280 nm; UVC, <280 nm. An increase of 1 to 2% of UVB radiation is associated with an increase of 2 to 4% in skin cancer. UVB is in the ionizing radiation range and can therefore damage DNA, leading to mutations and cancer. The effect on melanocytes appears to be more complicated. Melanomas often appear first at sites not directly exposed to sunlight. Tropical and subtropical areas have much higher incidences of skin cancer than temperate zones. In North America, it has been claimed that the incidence of skin cancer has increased by 400% in recent years, presumably because of the destruction of the ozone layer.

Medical uses of UV radiation

1. Photophoresis, in which blood is removed from a patient and exposed to UV light, then returned to the patient, is useful for treating mycosis fungoides, a complication of skin cancer, and it is promising for some leukemias.
2. UVA is used in conjunction with a photosensitive drug called psorelen to treat the skin lesions of psoriasis. The technique is called PUVA (psorelen-UVA).

Extra-low-frequency (ELF) electromagnetic radiation

ELF waves are nonionizing radiation waves with extremely long wavelengths (several hundred kilometers) and very low frequencies (<300 Hz). Exposure to artificial ELF fields occurs near high-tension electrical lines and much lower levels emanate from household appliances. Electric blankets, in particular, are thought to be potent sources of ELF exposure because of the close proximity and prolonged contact they entail. Other types of low-frequency waves include microwaves, emissions from TVs, and radio-frequency fields. High-tension lines create both electrical and magnetic fields and, although insulation will shield the former, the latter have great penetrating powers and will penetrate the human body. The unit of measure of magnetic fields is the gauss, which has a very complicated definition. Electromagnetic fields are also measured in volts per meter (V/m). The field immediately below a 400,000-V transmission line is 10,000 V/m, dropping to 500–1000 V/m at a distance of 100 m.

Numerous studies have shown deleterious effects from low levels of ELF radiation, including deformities of chick embryos, behavioral alterations, and physiological changes. Mice and snails have shown increased sensitivity to heat following exposure to ELF. Experimental findings have not always been consistent, however. Calcium uptake by cells is inhibited in some systems. Most of these studies employed exposures ranging from 30 milligauss (the level emitted by some electric blankets) to 1 gauss, and exposure times ranging from 30 minutes to several days or weeks (in the case of the chick embryo studies). The field strength of the Earth is about 0.3 to 0.6 gauss.

The greatest controversy concerns the interpretation of epidemiological data. Several studies have been conducted that purport to show increased incidences of cancer, especially leukemia, recurrent headache and depression, congenital deformities, and other health problems in people living near high-voltage power lines or in high-exposure working environments. These studies have been reviewed by several epidemiologists and have been faulted on several counts, including failure to account for other risk factors, failure to balance control and test groups, and the use of inappropriate statistics. In particular, the use of the Proportional Mortality Rate (PMR) has been criticized. This is the practice of expressing mortality rates due to certain causes as a percentage of the total. With this method, a decrease in one cause, (e.g., traffic accidents) leads to an apparent increase in the other causes although none has actually occurred, or even when a decrease has occurred. Most reviewers concede that there is some evidence of a marginal increase in the risk of leukemia in electrical workers, but qualify this by stating that the question remains open. This has not discouraged Paul Brodeur, a staff writer for the *New Yorker*, from recently publishing a book on the subject entitled *Currents of Death*, to follow up on his previous success with *The Zapping of America* on the subject of the hazards of video display terminals.

Two studies in 1988 provided further conflicting evidence. One, conducted in Washington state, found no correlation between the incidence of acute lymphocytic leukemia and exposure to electromagnetic radiation from high-voltage power lines located within 140 feet of homes, whether compared to actual measures within the homes or to computed values based on the wiring configurations. The other, conducted in Colorado, looked at cancer incidences in children 14 years old and under and found no association between total cancer incidence and exposure levels, but did find a modest association for lymphomas and sarcomas. The situation was not clarified at a 1992 meeting in San Diego at which a Swedish group, using more accurate data regarding exposure levels than had been previously available, showed a correlation between exposure and cancer incidence. Conversely, a literature study commissioned by the U.S. Government found no convincing evidence of an increased cancer risk for exposure to ELF or electromagnetic fields (EMFs).

The results of a massive study of some 224,000 utility workers in Quebec, Ontario, and France were released in the spring of 1994. This was a case-control study in which 4151 new cancer cases identified during the study period were matched with 6106 cancer-free controls from the same population to eliminate confounding factors such as smoking, etc. Past exposure to electromagnetic fields (EMFs) was documented. There was no association observed between the overall cancer incidence and past EMF exposure. This was true for lymphoma, melanoma, lung, stomach, and colon cancer. There was a statistically significant association between cumulative exposure and a rare form of adult leukemia (acute non-lymphoid leukemia) and its subtype, acute myeloid leukemia. Risk was increased by a factor of 2.41 but a causal relationship was not demonstrated. The level of significance, moreover, was at the 95% confidence limit, which is usually considered to be the minimum acceptable. There was a nonsignificant association cumulative exposure to the highest levels of EMF and astrocytoma, a rare brain cancer. Given the size of this study, it seems unlikely that any more conclusive evidence will be forthcoming any time soon.

Conflicting findings continue to be reported. In 1999, a study in Toronto, conducted by the Hospital for Sick Children, reported that children exposed to high levels of electromagnetic radiation had an incidence of leukemia 2 to 4 times higher than did children not so exposed. In contrast, a similar study conducted in British Columbia did not find any association between exposure level and leukemia incidence.

One rather oddball theory attempts to associate geographic areas of high electromagnetic fields with deep underground tension between tectonic plates of the earth. One such area is in Canada, and the suggestion (largely unfounded) is that these areas are associated with luminous atmospheric phenomena, UFO reports, and increased incidence of cancer — notably brain tumors.

Irradiation of foodstuffs

The use of ionizing radiation to preserve foodstuffs by killing spoilage micro-organisms was proposed as early as 1905 when British Patent No. 1609 was issued to "J. Appleby, Miller, and A.J. Banks, analytical chemist." They noted that sterilization might be accomplished in the complete absence of foreign chemicals. It was not until the post-war arrival of the "atomic age," however, that sources of ionizing radiation were sufficiently abundant to make this procedure practical and economical. In the immediate post-war era, there was considerable public enthusiasm for the development of nuclear technology for peaceful purposes, and there was good acceptance of the notion of irradiated foods such as milk and vegetables. By 1965, the United States, Canada, and the Soviet Union had approved the marketing of certain foods treated with low-dose irradiation but no manufacturers had taken advantage of this situation. By this time, the health concern over strontium-90 levels in milk was widely known, the anti-nuclear movement was in full swing, and radiophobia was growing. Acceptance of irradiated foods was in rapid decline, and it became replaced with concerns over potential health hazards. These concerns tended to fall into three areas:

1. Foods would be made radioactive and those who consumed them might develop radiation sickness.
2. Foods might be altered in some way, rendering them toxic, even carcinogenic.
3. Microorganisms could mutate to new and horrific pathogenic forms.

The first objection is easily dismissed. Foods are not rendered radioactive by low-dose ionizing radiation. Extensive toxicological testing, including tests for carcinogenicity, over several decades led a United Nations Joint Expert Committee on Irradiated Foods to conclude in 1980 that "... the irradiation of any food commodity up to an overall average dose of 10 kGy presents no toxicological hazard; hence, toxicological testing of foods so treated is no longer required" (Diehl, 1990). The second objection is therefore readily disposed. The third concern stems from experimental evidence that pure, dilute solutions of glucose become mutagenic for *Salmonella typhimurium* after irradiation, and that polyunsaturated fats, in an oxygen atmosphere, form lipid peroxides when irradiated. Neither change has ever been noted in complex food materials. Nor is there any evidence of a direct mutuganic effect on microorganisms by ionizing radiation under the conditions in which it is employed. Nevertheless, resistance to irradiation varies greatly from bacterium to bacterium, with spore formers such as *Clostridium botulinum* being very resistant. Ionizing radiation is thus more likely to be more useful in preventing spoilage than in eliminating potential pathogenic organisms. Public acceptance of irradiated foods remains a problem.

The pendulum may be swinging back toward acceptance of irradiation of foods. Several large outbreaks of so-called "hamburger disease," resulting from contamination of ground beef by strain O157:H7 of *E. coli*, have led to the use of irradiation of ground beef. In June 2000, 77 metric tons of ground beef were recalled from a number of supermarket chains in Canada because of *E. coli* O157:H7 contamination. The public may be becoming aware that this is a far greater and very real health risk than any posed by irradiated food.

Further reading

Bailey, C.C., *The Aftermath of Chernobyl*, Kendall/Hunt, Dubuque, IA, 1989.

Bithell, J.F., Dutton, S.J., Draper, G.J., and Neary, N.M., Distribution of childhood leukemias and non-Hodgkin's lymphomas near nuclear installations in England and Wales, *Br. Med. J.*, 309, 501–505, 1994.

Bowie, C. and Bowie, S.H., Radon and health, *Lancet*, 1, 409–413, 1991.

Cobb, C.E., Living with radiation, *National Geographic*, 175 (Apr.), 403–437, 1989.

Cole, L., Much ado about radon, *The Sciences*, Jan/Feb, 19–23, 1990.

Diehl, J.F., *Safety of Irradiated Foods*, Marcel Dekker, New York, 1990.

Dowson, D.I. and Lewith, G.T., Overhead high voltage cables and recurrent headache and depressions, *Practitioner*, 232, 435–436, 1988.

Editorial. Living under pylons, *Br. Med. J.*, 297, 804–805, 1988.

Eijgenraam, F., Chernobyl's cloud: a lighter shade of gray, *Science*, 252, 1245–1246, 1991.

Henshaw, D.L., Eatough, J.P., and Richardson, R.B., Radon as a causative factor in induction of myeloid leukemia and other cancers, *Lancet*, 335, 1008–1012, 1990.

Jerrard, H.G. and McNeill, D.B., Eds., *Dictionary of Scientific Units*, 6th ed., Chapman & Hall, London, 1992.

McGregor, R.G. et al., Background concentrations of radon and radon daughters in Canadian homes, *Health Physics*, 39, 285–289, 1980.

Michaelson, S.M., Influence of power frequency electric and magnetic fields on human health, in *Environmental Sciences*, Sterrett, F.S., Ed., New York Academy of Sciences, Vol. 502, 55–75, 1987.

Pool, R., Electromagnetic fields: the biological evidence, *Science*, 249, 1378–1381, 1990.

Report of the Working Group on Electric and Magnetic ELF Fields. Electric and Magnetic Fields and Your Health. Publ. # H46-2/89-140E, Min. Supply and Services Canada, 1990.

Rutkowski, C.A. and Del Bigio, M.R., UFOs and cancer? *CMAJ*, 140, 1258–1259, 1989.

Savitz, D.A. et al., Case control study of childhood cancer and exposure to 60-Hz magnetic fields, *Am. J. Epidemiol.*, 128, 21–39, 1988.

Severson, R.K., Acute nonlymphocytic leukemia and residential exposure to power frequency magnetic fields, *Am. J. Epidemiol.*, 128, 10–20, 1988.

Shore, R.E., Electromagnetic radiations and cancer. Cause and prevention, *Cancer*, 62, 1747–1754, 1988.

Smith, H., ICRP publ. 50. Lung cancer risk from indoor exposures to radon daughters, *J. Can. Assoc. Radiol.*, 39, 144–147, 1988.

Stone, R., Can a father's exposure lead to illness in his children? *Science*, 258, 31, 1992.

Stone, R., Polarized debate: EMFs and cancer, *Science*, 258, 1724–1725, 1992.

Tar-Ching, A.W., Living under pylons, *Br. Med. J.*, 297, 1469–1470, 1988.

Theriault, G., Goldberg, M. et al., Cancer risks associated with occupational exposure to magnetic fields among electric utility workers in Ontario and Quebec, Canada and France: 1970–1989, *Am. J. Epidemiol.*, 139, 550–572, 1994.

Tronnes, D.H. and Seip, H.M., Health risks caused by indoor radon exposure, in *Risk Assessment of Chemicals in the Environment*, Richardson, M.L., Ed., Royal Society of Chemistry, London, 1988, chap. 16.

Wertheimer, N. and Leeper, E., Magnetic field exposure related to cancer subtypes, in *Environmental Sciences*, Sterrett, F.S., Ed., New York Academy of Sciences, Vol. 502, 43–54, 1987.

Review questions

Questions 1 to 8, use the following code:

A if statements a, b, and c are correct
B if statements a and c are correct
C if statements b and d are correct
D if only statement d is correct
E if all statements (a, b, c, d) are correct

1. a. Gamma-rays have a longer wavelength than microwaves.
    b. Ionizing radiation includes all wavelengths of 1000 Å or less.
    c. Only a small fraction of radiation comes from natural sources.
    d. Ionizing radiation strips electrons from molecules as it passes through tissues.
2. a. Alpha-particles consist of two neutrons and two protons.
    b. An alpha-particle is identical to the helium nucleus.
    c. Gamma-rays have no charge.
    d. Beta-particles have no charge.
3. a. A roentgen defines the amount of ionization in air caused by any radiation.
    b. rem stands for "reactive emission material."
    c. Predicting the carcinogenic potential for radiation is very difficult below 10 rem.
    d. ALARA stands for "always let active radiation alone."
4. a. The main source of radon in the environment is nuclear power plants.
    b. Alpha-emitting daughters of radon bind to dust particles that lodge in the lung.
    c. Strontium-90 poses no threat to human health.
    d. Radon daughters are likely the second-most common cause of lung cancer after smoking.
5. a. In some geographic areas, the public may get higher radiation exposures than most nuclear power plant workers.
    b. The thyroid gland is very sensitive to radiation damage.
    c. Neutrons given off during radioactive decay may collide with other atoms to induce alpha, beta, and gamma radiation.
    d. Smoke detectors contain a radioactive component in the ion detector.

6. a. Alpha-particles cause tissue damage only over a very short distance (70 μ).
   b. Tissues with a rapid turnover are most vulnerable to radiation damage.
   c. Tissues sensitive to radiation damage include hair follicles, bone marrow, and gastrointestinal mucosa.
   d. Safe limits for radiation exposure have been reduced recently because of new evidence that low exposures cause cancer in mice.
7. a. Extra-low-frequency (ELF) electromagnetic radiation is emitted by high-voltage power lines.
   b. Evidence that ELF radiation causes cancer is inconclusive.
   c. ELF radiation has been shown to adversely affect many biological systems in single-celled organisms and chick embryos.
   d. ELF radiation is given off by electric blankets.
8. a. Fruit are sometimes irradiated to prevent spoilage.
   b. Low-level irradiation has been shown conclusively to induce mutation in several species of bacteria.
   c. Lipid peroxides have been shown to form from polyunsaturated fats *in vitro* but not *in vivo*.
   d. Irradiated food becomes radioactive.

Answers

1. C
2. A
3. B
4. C
5. E
6. A
7. E
8. B

chapter fourteen

Gaia and chaos: how things are connected

"Life itself is a religious experience."

— **James Lovelock**

"When we try to pick out anything by itself, we find it hitched to everything else in the universe."

— **John Muir, 1911**

The Gaia hypothesis

Formulated in 1965 by the independent British biologist James E. Lovelock and elaborated by Lynn Margulis, distinguished biology professor at the University of Massachusetts, the Gaia hypothesis proposes that certain kinds of life on the planet grow, change, and die in ways that lead to the persistence of other life-forms. In some circles, this has been interpreted as meaning that life on Earth forms a single, complex continuum, one ecosystem throughout time and space. The Earth, according to this view, can thus be considered as a single organism and its various components as cells in that organism. The name is taken from the Greek earth goddess Gaea. Although Lovelock intended his first book to be taken as a scientific treatise, there was a considerable amount of mysticism and spiritual significance attached to it by segments of the public, and this tended to turn serious scientists away from the theory for a long time. As information accumulated about the role of rainforests in consuming CO_2 and producing O_2, of wetlands in purifying water, and of ocean phenomena such as El Niño in affecting climate, the idea of Earth as an integrated biosystem gained credibility. This was strengthened as it became evident that human disruption of its components, such as the ozone layer, could have serious consequences for life on Earth.

To a microbiologist like Margulis, the Gaia hypothesis made perfect sense as it seemed, when stripped of its earth goddess mystique, to be simply another, perhaps more complex, example of symbiosis, so commonly encountered in the world of microorganisms. As Margulis pointed out during an address to other microbioligists, "Without the few pounds of bacteria in each of our guts, no one would ever digest food; and without the nitrogen-fixing bacteria in the soil, no food would ever grow in the first place."

Both Lovelock and Margulis take considerable pains to point out that the Earth-as-single-organism view is not what Gaia is all about. To quote Margulis, "the surface temperature, chemistry of the reactive gaseous components, the oxidation-reduction state, and the acidity-alkalinity of the Earth's atmosphere and surface sediments are actively (homeorrhetically) maintained by the metabolism, behavior, growth, and reproduction of organisms (organized into communities) on its surface. Gaia is not an individual, it is an ecosystem."

One example of a self-regulating system involves dimethylsulfide (DMS) produced by algae in the ocean. Bubbles produced by wave action burst, injecting their DMS into the air. Solar heat causes rising convection currents to carry it into the stratosphere, where it oxidizes, providing nuclei for the formation of water droplets. The resulting cloud formation has a cooling effect on the ocean surface, reducing the transfer of DMS to the stratosphere by convection. Algae use DMS as a propellant gas. Its formation thus increases when the population of algae explodes, and the cooling effect can help to regulate algal growth. Algae inhabit coral reefs and other benthic invertebrates, often determining the color of the host organism. If too great an increase in temperature occurs, coral polyps may expel the exploding algae in the phenomenon known as coral bleaching. If the problem is short-lived, the algae will reestablish in the coral. If the situation persists, however, the coral will die. Because coral reefs cover only 10% of the ocean floor but account for 25% of its fish life, the consequences for the entire reef ecosystem are serious. Global warming, regardless of the cause, can thus have catastrophic effects on coral reef systems.

Another example of how one organism can affect others is 2 billion years old. At that time, a family of organisms called cyanobacteria were dominant on the Earth. They were photosynthetic; and in the process of consuming CO_2, they produced large quantities of O_2, which was poisonous to most other species. Those with resistance to O_2 survived and gradually evolved into the aerobic organisms that came to dominate the planet's species. There has never been as radical a change in the Earth's population since then.

The resurgence of interest in the Gaia hypothesis as a result of environmental concerns has had both good and bad consequences. On the plus side, there is increased awareness of the interconnectedness of life on Earth and the possibility that a disruption in one part of an ecosystem can have far-reaching consequences. Less desirable is the proliferation of fuzzy-minded philosophies ("we are all one with the universe," etc.) that has led to such New-Age phenomena as crystals and prior lives. (Why was everyone a princess or a warrior in a previous life, but never a toilet cleaner?) Beliefs such as these tend to erode interest in, and trust of, science and this may be reflected in declining enrollments in science programs at a time when society needs to be improving and increasing its science and technology.

Of course, the science of ecology is based on knowledge of the interconnectedness of the components of an ecosystem. An example of the practical importance of such knowledge was recently noted in the May/June 1999 environmental issue of *Canadian Geographic*. A decline in pollock stocks in the Bering Sea, attributed to over-fishing, has led to a decline in the population of Steller's sea lions, which are now considered a threatened species. These sea lions depend on pollock as a food source. Their decline has forced orcas to prey on sea otters instead of sea lion pups. As a result, sea urchins have proliferated and have decimated the kelp beds that constitute a nursery for many other species.

Chaos theory

In the introduction to his book, *Chaos: From Theory to Applications*, Tsonis points out that simplicity and regularity are associated with predictability, whereas complexity and irregularity are virtually synonymous with unpredictability. *Chaos*, in the language of mathematics, is defined as "randomness" generated by simple deterministic systems. The word *randomness* is presented in quotation marks to suggest that the random nature may be apparent and that the determinism may persist, albeit in forms difficult to recognize.

Chaos theory has its roots in bifurcation theory first formulated by Poincare, but it was developed by Edward Lorenz who showed that nonlinear differential equations exhibited final states that were nonperiodic (i.e., apparently random). During investigations using computer simulations of computer networks, it was discovered that, under some conditions, routings became random and chaotic instead of following the orderly sequence that the system had been designed to use. For example, if a request was placed to use a specialized computer in location A (e.g., designed for theoretical mathematics) during a slack period (lunch hour), the request would be honored and several minutes of use might be available. If, however, the request clashed with several other simultaneous ones, it might be rerouted to location B along with other requests and disrupt use of this facility with resulting further rerouting to tertiary locations and the subsequent production of ripples throughout the system. Graphic plots of usage reveal oscillating patterns that are neither organized nor completely random. They have been described as "organized complexity."

An important feature of chaos theory is the existence of so-called low-level attractors. Using the example of a free-swinging pendulum, where x_1 = position and x_2 = velocity, the tendency to return to the equilibrium state, where both x_1 and x_2 = 0, is defined as the trajectory, and the equilibrium state is the attractor. In a system in which the effect of friction is offset by a mainspring, a disturbance in the motion of the pendulum will eventually be overcome and it will return to its periodic state. In this case, the *cycle* is the attractor. When chaotic processes are plotted on phase-space (three-dimensional) graphs, patterns are produced that are not truly random, as they would be if the process were completely disorganized. These patterns are called "strange attractors."

Chaos theory is now being applied to such diverse fields as the physics of fluid mixing and weather forecasting. In regard to the latter, computer models literally suggest that a butterfly beating its wings in China can influence weather patterns in North America. While this may seem far-fetched in the real world, it once again illustrates the interconnectedness of nature, and it helps explain why accurate weather forecasting is a difficult goal to achieve over a time span of more than a few days. There are a number of models used for weather forecasting. They differ in the physics and other parameters. Chaos theory may provide a means of exploring which model is most appropriate under given conditions.

Other examples of interconnected systems

Students of ecology should be highly familiar with the concept of symbiosis and how important it is in the maintenance of an ecosystem. But for those whose education has been centered largely on human health, a few examples of how biological events can be interconnected might be illustrative.

A vicious circle

In Chapter 11, the subject of fungal infections of cereal grains was discussed, as well as how these caused economic losses measured in billions of dollars in Canada alone. Such economic loss to farmers means fewer dollars to spend on consumer goods. This leads to a slowdown in the manufacturing sector, leading to higher unemployment, leading to further declines in the purchase of consumer goods (a vicious circle) and, eventually, contributes to a recession. This is an economic vicious circle, but a biological one can also occur. Figure 41 illustrates how this might work. Past practice was to plough under corn stubble in the fall so as to expose the roots and allow frost to kill fungal spores. If the following spring was dry, wind would erode the topsoil, reducing its fertility. This would require the increased use of chemical fertilizers that could, in turn, increase water pollution and possibly cause eutrophication. Other possible consequences are also illustrated.

Domino effects of global warming

Even a small increment in mean annual temperature, including a cyclical one, could have profound effects on the biosphere. The current "hotspot" for fungal infestations of grain in Canada is southwestern Ontario. Warming would move the demarcation line northward, so that *Aspirgillus flavus*, the

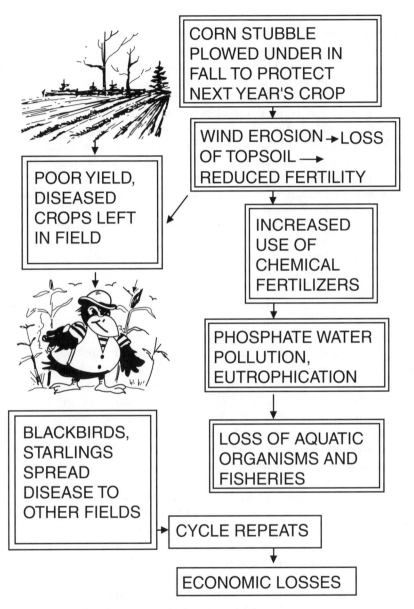

Figure 41 A vicious circle of fungal infection.

source of the carcinogenic mycotoxin aflatoxin, which cannot survive cold winters, could change its distribution.

Certain insects could also move north. The Africanized honey-bee is already present in the southwestern United States, and it would follow a warming trend northward. Mediterranean fruitflies, screwworm flies, and even malarial mosquitoes could follow. Malaria-carrying mosquitoes have been detected as far north as New York state. Insects are vectors of many diseases of animals and humans. These would spread along with their hosts. Venomous insects also could move northward. The brown recluse spider has moved from Florida to Pennsylvania and will undoubtedly cross the Great Lakes at some time, if it has not already done so.

There is some evidence that ecological disturbances caused by anthropogenic activity may result in viral infections jumping species barriers, especially from animals to humans. Recently, the Marburg virus and the Ebola virus have emerged as life-threatening infections of humans and are believed to have originated in monkeys. Every variety of influenza antigen has been identified in ducks and other waterfowl, and a recent outbreak of flu in Hong Kong was traced to chickens and resulted in the slaughter of thousands of birds in an effort to contain the outbreak.

Swine have long been known to harbor flu viruses and are thought to have been the reservoir of the Spanish flu epidemic during World War I. The plague bacillus *Yersinia pestis* periodically jumps from rats to humans, carried by fleas. Slash-and-burn agriculture, practiced in Africa for eons, creates an environment favorable to the Anopheles mosquito that carries malaria. These mosquitoes are present in the Great Lakes Basin, where malaria was endemic in the 1800s, and the possibility that there may be a resurgence of malaria associated with a warming trend in the climate cannot be dismissed. Lyme disease has already reached southern Ontario from its original identification site in New England. An outbreak of a viral infection in 1993 killed 12 people in New Mexico. The virus (Hantavirus) was identified as belonging to the Hantaan group, which is spread in the feces and urine of rodents. It was first identified during the Korean War as the cause of hemorrhagic fever in soldiers. The condition observed in the American Southwest is called Hantavirus pulmonary syndrome (HPS). An exceptionally high yield of pinon nuts, a staple diet for rodents, led to an explosion in the rodent population and a resulting increase in the exposure of humans to their droppings. It has since been identified as a cause of human infection in many states, including Florida. The deer mouse appears to be the most common carrier, but other rodents such as the cotton rat have also been identified as carriers.

Many of these problems are exacerbated by the ease with which people can now move from one locale to another, and some medical experts are warning that their profession must be on the alert for diseases not normally seen in northern climes. Global warming may facilitate easier movement of birds. Recent cases (1999) of West Nile encephalitis in New York have been traced to contaminated wildfowl that apparently brought the disease from North Africa. Also attributed to the movement of birds was an outbreak of Salmonella type DT104 in Vermont. This strain of bacteria, prevalent in cattle in Great Britain, has been responsible for an epidemic there. The organism was isolated from the milk tank on a dairy farm. One family member became critically ill but recovered.

On a more cheerful note, a longer growing season could mean increased crop yields and more arable land.

A feedback loop

An elegant example of how an ecosystem can self-regulate is the manner in which water temperature is regulated in small bodies of freshwater. It has been shown recently that bioregulation of water temperature occurs in a manner reminiscent of negative feedback control systems in mammals and, similar to the DMS feedback system, in oceans. The system is shown in Figure 42. Again, algae play an important role. In an algal bloom, water turbidity is increased and penetration of sunlight lessens. The cooling effect inhibits algal growth. The same effect can occur if the pond is stocked with predatory fish. These will eat smaller fish that prey on algae eaters (small crustaceans, etc.), allowing uninhibited growth of the algae.

Food production and the environment

This text is primarily about the relationship between the environment and human health. It is undeniable that starvation and malnutrition are the greatest killers of humankind and that they relate to the ability of the Earth to feed its population. Some consideration of this question is thus not inappropriate.

Meat vs. grain

It is often stated that a vegetarian diet is environmentally friendlier than a diet containing meat because one can produce more food by growing plants than by grazing animals. In addition, animals produce methane, which is a greenhouse gas. This argument is frequently put forth by animal rights activists to support their philosophical position, which also draws heavily on the mystical side of Gaia. This conventional "wisdom" has even appeared in the popular press in articles written by nutritionists. But does it stand up to scrutiny? Consider the following:

1. Livestock can be, and are, grazed on grasslands unsuitable for cultivation and in colder climates with a very short growing season. This occurs in several locations in North America and northern Europe. In Australia and New Zealand, sheep are grazed extensively on land that is harsh and inhospitable to cultivation.
2. Before the North American prairies were plowed to plant grain, they supported an estimated 50,000,000 bison and millions of pronghorn antelope, elk, and caribou (at least seasonally) without damage to the soil. Cultivation coupled with drought brought the dustbowl of the 1930s. In southern Ontario, a dry spring and high winds may result in thick clouds of brown dust coating autos, houses, and grassland as topsoil is blown from surrounding farms that were fall-ploughed to eliminate the spores of the *Fusarium* mold.

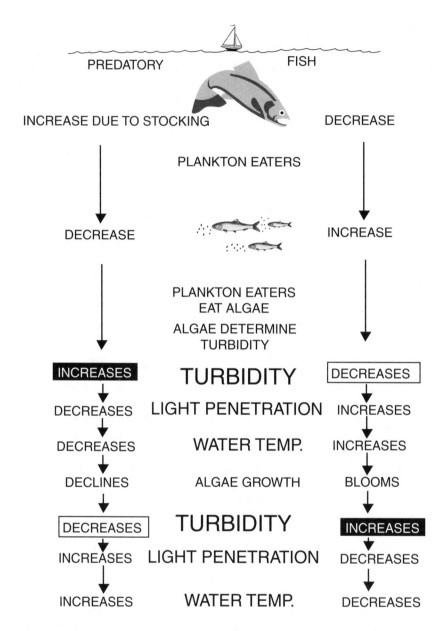

Figure 42 Thermoregulation in a small body of water. If fish-eating predators (piscivores) proliferate, they deplete the zooplankton that consume algae. The algae bloom, increasing turbidity and hence lowering temperature which, in turn, inhibits algal growth allowing more light to penetrate and the temperature to return to a warmer state. A decline in piscivores has the opposite effect.

3. Even in semi-arid areas such as the American Southwest, studies have shown that the footprints of wild ungulates and cattle form little traps for seeds and water. Grass tufts develop in these that help stabilize the soil. It must be kept in mind that there is a vast difference between intelligent grazing and over-grazing. Much of the evidence against livestock grazing is taken from underdeveloped countries where over-grazing occurs extensively and agricultural technology lags. Another bit of evidence is taken from the Amazon Basin, where deforestation for cattle ranching destroys the soil and the rainforest. This is quite true, but deforestation also occurs for paper production (some years ago, Japanese entrepreneurs floated an entire pulp and paper mill up the Amazon) and for crops such as sugarcane. All are equally destructive because the culprit is the deforestation.

4. What about that methane? Best estimates are that the rice paddies of the world produce as much methane as all of the animals combined, both domestic and wild (about 100 megatons per year). Because rice is a staple grain for much of the world, shifting reliance to it from meat would not prevent as much methane pollution as one might think. Besides, if one were to increase the wildlife population by protective measures (a laudable goal), one would also be increasing methane production. Is there good methane and bad methane? How much methane can an elephant make?

5. The average daily caloric intake of 40 countries from the poorest (Bangladesh) to the richest (United States) is 2571 calories of which meat provides 205 and vegetables only 41. Even in Guatemala, meat provides 49 of 2020 calories per day, vegetables only 18. The bulk is from maize (corn), which probably is high in carcinogenic mycotoxins.

6. Water pollution by animal wastes is often cited as evidence that livestock are less environmentally desirable than crops, but we have already seen how the latter can pollute through chemical fertilizers.

There is mounting evidence that world food supplies are declining at a time when the population is exploding. These two factors appear to be on a collision course. It is estimated that by the year 2015, population growth will outstrip food production unless production can be increased. The production of cereal grains seems to be declining faster than the production of meat, but fish stocks are also being depleted. A prime example of this latter point is the decline in North Atlantic cod stocks. Over-fishing has been singled out as the major culprit, but competition from an expanding seal herd for caplin, a herring-like fish that is the principal food source for both species, has also been incriminated. Surveys indicate that *all* ground species, including noncommercial ones such as monkfish and eelpouts, are in decline from waters off northern Newfoundland all the way up the coast of Labrador, suggesting that natural phenomena may be contributing to the decline. Levels of pollutants such as dioxins and heavy metals are negligible in Atlantic cod, making it unlikely that pollution is the culprit. A possible explanation may be the 0.5 to 1.0°C cooling of northern spawning waters, which may dramatically reduce the survival of cod fingerlings. El Niño, which involves a massive pooling of warm water in the mid-Pacific, has caused marked reductions in the fish catches off South America.

Genetically modified plant foods

With regard to cultivated foodstuffs, there is now great concern that the development of special strains of food grains such as rice and wheat, with increased resistance to specific diseases, may render them more susceptible to other diseases, some of which may not have emerged as yet. There is now an effort to collect and preserve the wild strains of important food sources such as rice, corn, potatoes, and fruit to constitute a genetic library that can be called upon in the future when it is needed.

A more recent concern over the environmental impact of human-created species relates to the development of genetically engineered species of macro- and microorganisms. Herbicide resistance has been conferred genetically on cultivated plants, resulting in concern that their survival advantage might lead to their invasion of inappropriate habitats. Insect resistance is usually imparted by incorporating a gene from the bacteria *Bacillus thuringiensis*. This gene controls the production of insecticidal proteins. The technique employed to transfect the target plant involves the use of an *Agrobacterium* species as a "gene taxi." This bacterium contains a plasmid that has a gene for a tumor gall that is replaced by the *B. thuringiensis* gene. There is concern that rDNA-modified species might interpollinate with wild ones (in the case of plants), attack benign insects (in the case of predatory or parasitic pest control species), or protect undesirable species (as in the case of the anti-frost bacteria *Pseudomonas syringae*). None of these scenarios has yet been identified as a practical problem, but there have been calls for improved methods of risk assessment of genetically engineered species before they are turned loose in the world.

A further concern is that allergens might be introduced to another species, resulting in an allergic reaction in an unsuspecting individual who consumes the altered food item. This has been demonstrated to be a real possibility when a gene from the Brazil nut was introduced into a strain of potato. The presence of the allergen was detected during safety testing and the product was not introduced to the market. Nevertheless, the public is becoming increasingly vociferous in its demand that genetically altered foods should be labeled as such. While there may be little reason for the public to be concerned about what is on the shelves of the supermarket, there is good reason to demand a high level of vigilance on the part of regulatory agencies. In the past, major corporations have submitted proposals to market products containing bacterial and insect toxins for pest controls without due regard for their allergenic potential. Fortunately, they were not approved; but given the ease with which transgenic plants and animals can be produced (there are over 10,000 already), the possibility of a hazardous transgenic species falling through the cracks must be guarded against. For more on genetically modified foods, visit www.epa.gov/scipoly/sap/2000/february/12-9.pbf; www.epa.gov/scipoly/sap/2000/june/mammaltox.pdf; and http://biosci.umn.edu/~pregal/biosafety.html.

Not all authorities are in agreement regarding the degree of crisis in food production. J. Ausubel, writing in *The Sciences*, predicts that the Earth will hold 8 billion people by the year 2020. While others believe that the current 5.5 billion is already overstressing our food-producing ability, his premise is that advanced technology will cope with the larger numbers. Ausubel believes that we are currently in a period of "creative destruction" brought on by the flagrant abuse of credit, both public and private, in the 1980s. The cycle of bankruptcies, fiscal crises, and unemployment that followed will in turn be followed by a period of sustained growth that will foster the emergence of new technologies to solve the food crisis. His recommendations, however, focus primarily on the need to foster cooperation among nations to promote sustainable development and reduce the competitive nature that dominates most current international affairs. Apart from stressing the need to reduce our dependence on polluting energy sources, as by substituting natural gas for coal and oil, no specific innovative measures are offered. Not surprisingly, Ausubel's article generated a flurry of letters in support of the conventional wisdom that zero population growth must be achieved within the span of the present generation if there is to be any chance of reversing the pollution and starvation generated by overpopulation. There was no discussion, however, of the fact that much of the world's food-producing capacity is under-utilized because of market forces that prevent its redistribution to areas of need. In 1992, farmers in Prince Edward Island buried millions of kilos of potatoes because there was no market for them and no agency could be found to convert them to potato flakes for shipment to war-torn areas such as Bosnia, even if they were donated at no cost. American–Canadian trade disputes over wheat exports further illustrate the fact that cost is often the limiting factor in getting food to those who need it.

In his reply to his critics, Ausubel uses a quotation that bears repeating. It goes as follows:

> "The most convincing examinations of the phenomenon of overpopulation hold that we humans have by this time become a weight on the Earth, that the fruits of nature are hardly sufficient for our needs, and that a general scarcity of provisions exists, which carries with it dissatisfaction and protests, given that the Earth is no more able to guarantee the sustenance of all. We thus ought not to be astonished that plagues and famines, wars and earthquakes come to be considered as remedies, with the task, held necessary, of reordering and limiting the excess population."

These words were written by Tertullian, a priest, around 200 A.D. Were they prophetic or merely alarmist?

The environment and cancer

The public fear that anthropogenic chemicals in the environment may be contributing to cancer incidence has already been noted, as has the existence of natural carcinogens such as certain mycotoxins and radon gas. Statements in the press that a high percentage of cancers are "environmentally produced" are usually taken to mean "produced as a result of anthropogenic activity," with no reference to the existence of natural carcinogens. Indeed, anthropogenic carcinogens may be of greater significance to aquatic organisms than they are to human beings. Cancer statistics for North America indicate that, in fact, the incidence of most cancers is declining. There are some noteworthy exceptions, such as smoking-related lung cancer. In contrast, the apparent "epidemic" of breast cancer in North America has been discounted by both the American Cancer Society and the National Cancer Institute as being a statistical aberration due to better detection methods and a population bulge in the over-40-year-old age group of women. Despite the generally encouraging statistics regarding cancer incidence, the (U.S.) National Academy of Sciences recently issued a report "Pesticides in the Diets of Infants and Children," in which it is stated that allowable levels of pesticides may be several hundred times too high for these age groups because of their age-related susceptibility (they consume more food per unit weight, and may not be efficient detoxifiers) and because their eating habits may lead them to consume many times more of a particular food than the amounts used to calculate allowable levels. The economic need for pesticides in agriculture may not be as great as previously thought, and the pressure is increasing to limit their use.

Evidence has recently emerged in Utah of an epidemic of bone cancer, including chondrosarcoma (cartilaginous tumors), which is believed to be associated with an environmental carcinogen. The victims were dinosaurs. They lived 135 million years ago. The problem of cancer obviously is not a new one.

Further reading

Ausubel, J.H., 20/20 vision, *The Sciences*, 33 (6), 14–19, 1993.

Comments on above and response, *The Sciences*, 34 (3), 7 and 51–52, 1994.

Fisher, D.E., *Fire and Ice: The Greenhouse Effect, Ozone Depletion and Nuclear Winter*. Harper and Row, New York, 1990.

Hantavirus infection: southwestern United States: interim recommendations for risk reduction, *Morbid. Mortal. Wk. Rep.*, 42, RR-11, 1993.

Hantavirus pulmonary syndrome: United States, *Morbid. Mortal. Wk. Rep.*, 42, 816–820, 1993.

Huberman, B.A., An ecology of machines: how chaos arises in computer networks, *The Sciences*, July/Aug, 39–44, 1989.

Joseph, L.E., *Gaia: The Growth of an Idea*, St. Martin's Press, New York, 1990.

Lloyd, S., The calculus of intricacy: can the complexity of a forest be compared with that of Finnegan's wake? *The Sciences*, Sept/Oct, 38–44, 1990.

Lovelock, J., *Gaia: A New Look at Life*, Oxford University Press, Oxford, 1979.

Lovelock J., *The Ages of Gaia*, W.W. Norton & Co., London, 1988.

Margulis, L., Gaia in science (letter), *Science*, 259, 745, 1993.

Margulis, L. and Dobb, E., Untimely requiem, *The Sciences*, Jan./Feb., 44–49, 1990.

Marshall, E., Hantavirus outbreak yields to PCR, *Science*, 262, 832–836, 1993.

Mazumder, A., Ripple effects: how lake dwellers control the temperature and clarity of their habitat, *The Sciences*, Nov./Dec., 39–42, 1990.

Miller, H.I. and Gunary, D., Serious flaws in the horizontal approach to biotechnology risk, *Science*, 261, 1500–1501, 1993.

Morse, S.S., Stirring up trouble: environmental disruption can divert animal viruses into people, *The Sciences*, Sept./Oct., 16–21, 1990.

News and comment: Experts clash over cancer data, *Science*, 250, 900–902, 1990.

Regush, N., Border crossing pathogens: microbes on the march, *Can. Geograph.*, 120 (6), 62–66, 2000.

Tsonis, A.A., *Chaos: From Theory to Applications*, Plenum Press, New York, 1992.

What We Eat. *Geo: The Earth Diary*, 3 (July), 139, 1981.

Yanko, D., Are animal disease patterns changing because of global warming? *Vet. Mag.*, 2(3), 18–21, 1990.

chapter fifteen

Case study reviews

All case studies are based on actual occurrences.

Case study 1

This population of laboring miners has an incidence of a rare cancer many times that of the general population, and those who live near the mine, as well as family members of miners, have a lower, but still elevated, incidence of the same cancer. The incidence of lung cancer in smoking miners is 60 times that in nonsmoking miners and several times higher than in smokers who are not miners.

This is actually the situation in asbestos miners who were first employed 30 years ago or more. The cancer is mesothelioma. The miners carried home the asbestos fibers on their clothing. Family members, especially the wives who did the family laundry, inhaled these fibers. Living in close proximity to the mine was also a risk factor, as particles were airborne over short distances. Cigarette smoke and asbestos fibers were acting as co-carcinogens and/or promoters to greatly increase the risk of lung cancer.

Socio-economic factors that might increase cancer incidence, in addition to smoking, would include diet (high saturated fats and low fiber), high alcohol consumption, lack of exercise, and other workplace hazards (e.g., benzene). Miners would likely be 18 to 65 years of age and predominantly male, so that the elderly and the very young, as well as women, would be excluded from the group at risk.

Case study 2

Working in a confined space always constitutes a risk whenever volatile solvents or gases, explosive or otherwise, are in use. In this particular case, explosion-proof electrical devices (lights, ventilating fans) should have been available. Failing that, the ventilation fan could have been placed at a distance from the site and flexible, large-bore conduit used to conduct air into the tank. The tank should have been pumped dry before attempting to work in it. A safety person should have been left at the surface to summon help (not to enter the tank alone), and safety lines should have been attached to the workers in the tank so that they could be recovered if they lost consciousness. Breathing apparatus could have been used if available. Ideally, air quality should have been tested before any workers were allowed into the tank (O_2 > 19.5%; flammable substances < 10%; toxic chemicals to meet published standards).

This accident took place in the Philippines, in a remote area. Epoxy paint solvents may contain a variety of volatile substances that are sedating and heavier than air. In this case, volatile glyceridyl ether was identified.

Case study 3

This 4-year-old boy had acrodynia, a rare form of childhood mercury poisoning. His 24-hr urine mercury level was 65 µg/L, as were those of his mother and siblings. The house was air-conditioned and therefore sealed. The interior latex paint used on the walls of the house was identified as containing about 950 ppm mercury. Seventeen gallons had been applied to the interior walls. The U.S. EPA allows the addition of 300 ppm mercury as phenylmercuric acetate to interior latex paint as a fungicide and bactericide to prolong shelf life. This limit was exceeded by more than threefold.

Case study 4

The manure pit is another example of the dangers of working in a confined space. In this case, methane gas from the decomposition of the manure displaced O_2 so that the workers died of asphyxiation. All of the safety measures noted for Case study 2 apply here as well. It is almost unbelievable that four men would consecutively enter the confined space, but it is a testimony to how panic can overcome training and common sense. Over 25 people in the United States have died in this type of accident in the last 5 years.

Case study 5

Yet a further example of a "confined space" problem, these workers were using a chlorofluorocarbon (Freon-113) as a degreasing agent in a pit beneath a large piece of machinery when they were overcome and one died on route to hospital. The portal of entry was obviously by inhalation, and the target organs were the brain and the heart. CFCs are CNS depressants, causing narcosis, stupor, and loss of consciousness. Asphyxiation may occur from respiratory depression as well as from displacement of O_2. These agents, like the related anesthetics halothane and chloroform, sensitize the heart to adrenaline, with resulting arrhythmias and ventricular fibrillation. A number of years ago, sniffing CFCs in aerosol propellants became a cheap way of getting "high" in some adolescent cultures. A number of deaths resulted from cardiac arrest. From 1983 to 1990, 12 deaths were recorded in the United States from the industrial use of CFCs.

Case study 6

These Inuit soapstone carvers are exposed to the dust of their carving stones. This can cause pneumoconiosis, similar to silicosis. Like the asbestos miners, they could carry the dust home on their clothing and thus put their families at risk as well. The use of respirators capable of filtering out the dust, and of special workclothes left at the site, is a necessary safeguard.

Case study 7

The variety of signs and symptoms, not all of which are infectious, in these agricultural workers do not suggest that the primary cause is a microorganism, but rather an inhaled pollutant. Potential airborne causative agents in this environment would include animal dandruff, dried feces, grain dust, dust from feed additives (antibiotics, sulfa drugs, minerals, etc.), parasites, bacteria, bacterial endotoxins, fungi, ammonia (adsorbed onto dust particles), and methane. Many of these agents are allergens and an allergic component may be present in some individuals.

Corrective measures could include improved ventilation, the use of filter masks or respirators, damping down floors to control dust, and removal from this work environment of anyone suffering from allergies or respiratory disease. In many jurisdictions, agricultural workers are not protected by workplace legislation.

Case study 8

The array of respiratory symptoms is strongly suggestive of an inhaled toxicant. Because ice-surfacing machines use internal combustion engines, CO poisoning should be suspected. Carboxyhemoglobin levels in blood should be measured. This was done and values ranging from 10 to 20% were found (normal levels are <2% for nonsmokers, 5 to 9% for smokers). CO would account for the headache, nausea, and dizziness, but not for the difficulty in breathing nor the coughing up of blood. This is a strong indication of nitrogen dioxide (NO_2) poisoning.

The following day, tests were conducted with an ice resurfacing machine equipped with an internal combustion engine. The machine had not been serviced for some time. Levels of NO_2 reached 1.5 ppm. The standard set by the U.S. Office of Safety and Health Administration (OSHA) is 1 ppm for the Short Term Exposure Level (STEL), which has a 15-min limit. The CO reached 150 ppm and may have been higher the previous evening. The maximum recommended for ice arenas is 30 ppm.

Case study 9

These family members became dizzy and nauseated after eating a snack food purchased at a local convenience store. Two of them suffered convulsive seizures. The brain is obviously the primary organ of toxicity. The rapid onset (< 1 hr) indicates a preexisting toxin rather than an infectious agent. Because common bacterial toxins do not produce this type of reaction, and because the preparation did not involve seafood or mushrooms, the likely offender is a chemical contaminant.

It is difficult to identify the source of the contaminant because all cases involved the same retail outlet and the same manufacturer. The fact that the snacks were prepackaged would suggest the manufacturer as the source, but the nature of the package was not identified and it is not impossible that some chemicals could penetrate paper or polyethylene bags.

Endrin is an organochlorine cyclodiene insecticide. No traces were found in the store or the manufacturing plant. Another possibility is that the flour from which the taquito snacks were made was contaminated. There have been previous incidents of such contamination.

Signs of acute organochlorine intoxication include headache, nausea, vomiting, dizziness, clonic jerking, and epileptiform seizures.

Case study 10

This is another case of a worker dying from exposure to fumes from a volatile chlorinated hydrocarbon solvent being used as a degreaser. Trichloroethane acts like CFCs to sensitize the heart and depress the CNS. Inadequate ventilation is the critical factor in all such cases (refer to Case Study 2). Chronic intoxication would have probably involved central necrosis of the hepatic lobules.

Case study 11

These demolition workers were cutting up an old iron bridge. The organ systems involved in the toxic reaction were the musculoskeletal system (joint and muscle pain), the CNS (headache), and the gastrointestinal system (nausea). This last symptom could also be central in origin, through stimulation of the chemoreceptor trigger zone. The most likely portal of entry is the lungs. These torch cutters would be wearing coveralls and gauntlets to protect against sparks. The source of the toxicant is thus likely to be vapors from the cutting process.

An old bridge of this nature is unquestionably going to be coated in many layers of lead-based paint. Blood lead levels were performed and ranged from 60 to 160 µg/dL. U.S. regulations require that workers having levels >60 µg/dL be removed from the work site. The highest level was detected in the barge worker who did not benefit from the nearly constant breeze encountered on the superstructure of the bridge. The paint from the bridge was found to contain 30% lead by weight.

Treatment was initiated with chelation therapy (EDTA) and substantial amounts of lead were excreted in the urine with an accompanying reduction in symptoms.

The employer was fined for not providing appropriate respirators, clean work clothing, and facilities for washing up before lunch and at the end of the day.

Case study 12

This is another case of lead poisoning. The blood lead level was 70 µg/dL in the primary patient and elevated in other family members. In chronic gastric pain that cannot be attributed to an ulcer or cancer, a blood lead determination is useful, as gastric distress is a common symptom of lead poisoning.

The portal of entry in this case is almost certain to be oral because no activities were in place that could have caused lead vapors. The ceramic jug was suspect because it was imported from Mexico where lead glazing still is used. The jug had been used to store a fermenting beverage, so that considerable time was available for leaching to occur. The entire family partook of this beverage over the course of the summer.

Case study 13

An investigation of the home environment revealed that the house had been built after the banning of lead paints. Water was obtained from a well by means of a galvanized pipe system, eliminating solder joints as a potential source of the lead. No suspicious ceramic or pewter utensils were used for food storage or preparation. All of this pointed to the work environment as the source of contamination.

A 19-day course of treatment was begun with dimercaptosuccinic acid (DMSA), an orally administered lead chelating agent. The patient was instructed to remain off work during the treatment period. At the end of this period, his blood lead level (BLL) fell to 13 µg/dL. One month after returning to work it was back up to 53 µg/dL, confirming this environment as the source of the lead.

Further careful questioning elicited the information that he habitually chewed on bits of insulation cut from the ends of electrical wires. Analysis of this colored plastic (white, blue, and yellow) indicated that it contained 10,000 to 39,000 µg of lead per gram of plastic. He was instructed to desist from this habit and within 4 months his BLL fell to 24 µg/dL and he reported a subjective improvement in his symptoms.

Lead, usually as lead chromate, is used in pigments employed in the manufacture of colored plastics. Lead salts are used in the manufacture of polyvinylchloride (PVC) plastics. Thus, any colored plastic may contain lead and should not be chewed or ingested. Cadmium is also present in a number of pigments.

Lead sometimes appears as an unexpected contaminant in unusual circumstances. Ethnic health remedies are such examples. Lead, in amounts up to 90% by weight, has been detected in "Azarcon" from Mexico (used as a digestive aid), "Greta," also from Mexico (same use), "Paylooah" from Southeast Asia (applied to inner lower eyelid to improve vision), and a substance from Tibet given to improve development. All of these have resulted in raised blood lead levels in children (20 to 80 µg/dL) and symptoms of lead poisoning (*Morbidity and Mortality Weekly Rep.*, 47, 522, 1993).

Case study 14

These five steam press operators became ill with a variety of signs and symptoms suggesting involvement of the CNS, the blood, and the heart. The use of a solvent-borne adhesive raises questions about ventilation of the workplace; the nature of the solvent and its toxicity; whether respirators were in use and, if so, were they approved for the task; and whether protective gloves and clothing were in use. Because the steam presses had been in use for many years without incident, it is important to carefully probe to identify any changes in work habits or materials that might have taken place recently.

A wide variety of aromatic nitro compounds are capable of causing methemoglobinemia. Remember that nitrites are administered deliberately in cases of cyanide poisoning to form methemoglobin, which has a higher affinity for CN than does reduced hemoglobin. The top ten aromatic nitro compounds implicated in methemoglobin formation include, in order of decreasing potency, *ortho*-chloraniline, dinitrobenzene, *meta*-nitroaniline, *para*-toluidine, nitrobenzene, *meta*-toluidine, *ortho*-nitrochlorobenzene, aniline, *para*-dinitrosobenzene, and *ortho*-toluidine.

Because analysis of air samples was not helpful, analysis of the actual solvent should be undertaken if possible. Analysis of the adhesive in use at the time of the toxic event was undertaken and the results compared with a newly delivered lot. Samples were extracted with carbon disulfide/methanol and analyzed by gas chromatography with flame ionization detection. The "old" sample was found to contain 1% by weight *para*-dinitrobenzene (pDNB) vs. 0.03% in the new sample. Tracing the product back to the manufacturer, it was found that a proprietary solvent used in the preparation of the adhesive was contaminated with pDNB. MetHb levels were monitored after the removal of the contaminated lot and found to be normal. Periodic monitoring was instituted, and workers were required to wear butyl rubber gloves because it was felt that significant skin absorption may have been occurring.

Case studies 15 and 16

The rapid onset of symptoms in both cases — the hemodialysis patient and the people who attended the picnic — makes it unlikely that an infectious agent is involved. There are no water-borne bacterial toxins that could account for the symptoms. The CNS appears to be the organ system to which most symptoms are related (sleepiness, dizziness, etc.).

In both cases, it emerged that the potable water system was cross-connected with a chilled water system used for air conditioning. Ethylene glycol was used as an antifreeze. The ethylene glycol diffused into the tap water and was consumed by those at the picnic and diffused across the dialysis membrane into the bloodstream of the patient with renal disease. Ethylene glycol acts first as an intoxicant similar to alcohol. Subsequently, the formation of oxalate occurs. This chelates calcium to form calcium oxalate, which precipitates in the kidneys and other tissues to cause renal failure and other tissue damage. Blood calcium levels are low also because of the binding of calcium to oxalate. Ethanol is administered because alcohol dehydrogenase is the metabolic enzyme for ethylene glycol but ethanol is the preferred substrate.

Case study 17

These patrons of a Chinese restaurant most likely took in something orally. Again, the rapid onset of symptoms suggests a preexisting substance rather than an infectious agent. The signs and symptoms relate to the CNS. The symptoms are not compatible with known pesticide toxicity.

The physician inquired about fish and mushroom consumption because a type of scalefish poisoning produces similar symptoms. Toxic mushrooms (toadstools) also might have been responsible.

The fact that only these three patrons consumed a particular dish is helpful. Inquiry revealed that this particular dish, pork chow yuk, had been prepared by three different chefs. Each had added monosodium glutamate to the full amount, resulting in concentrations three times normal. This is a classical case of "Chinese restaurant syndrome."

Case study 18

This crop-duster pilot is displaying characteristic signs of excessive cholinergic activity. The signs and symptoms are opposite to those of atropine overdose and include pupillary constriction, visual disturbances, mental confusion and other mental disturbances, profuse sweating, dizziness, weakness, and diarrhea. Parathion is an organophosphorus insecticide. Questioning revealed that the pilot had been careless in handling the concentrated stock solution, had not worn protective clothing, including gloves, and had not worn a respirator. The signs and symptoms of organophosphorus poisoning result from its irreversible inhibition of cholinesterase and are frequently delayed. Treatment would consist of atropine to block the effects of acetylcholine overload and pralidoxime to reactivate the acetylcholinesterase enzyme by removing the phosphate group.

Pralidoxime would be contraindicated if a non-organophosphorus inhibitor of acetylcholinesterase had been used, such as a carbamate like Sevin. Because no phosphate is involved, no antidotal response occurs, and the situation could be worsened because pralidoxime has some anticholinesterase activity in its own right.

Plasma cholinesterase levels and red-cell acetylcholinesterase would be depressed in this individual.

Case study 19

The signs and symptoms of this outbreak are characteristic of poisoning with a cholinesterase inhibitor. Questioning revealed that 23 of the men had been involved in the mixing, loading, or application of mevinphos, an organophosphorus insecticide being used for aphid control. It has high toxicity (EPA Class 1). The remaining men had entered and worked in orchards within 24 hr of spraying.

Questioning of the orchard operators revealed that protective clothing or equipment either was not available or had not been used. Respirators, gloves, goggles, coveralls, and rubber footwear are recommended and, in some jurisdictions, required by law.

Seven individuals required hospitalization. Their plasma and red blood cell cholinesterase levels were depressed 75 to 95%. Other workers were treated as outpatients and had levels depressed by 15 to 25%. Atropine was given to a total of 11 patients.

Case study 20

This restaurant-related syndrome involved a small number of patrons (6) out of 48 who had eaten lunch there. Careful inquiry should be undertaken to determine whether these six people had eaten anything different from the others. It emerged, after questioning these and other available patrons, that the six had consumed yellow-fin tuna. On the surface, this resembles a monosodium glutamate (MSG) reaction, but the chefs adamantly denied using this flavor enhancer in the fish dish. The persistence of the symptoms, up to 9 hr, also argues against MSG as the cause of the reaction.

Three of the affected individuals reported that the tuna had a distinctly peppery or "Cajun" taste; but again, the chefs claimed not to have used any such spicing.

This is a typical case of scombroid fish poisoning. Analysis of the yellow-fin tuna revealed histamine levels of 50 to 160 mg/100 g (normal <1 mg/100 g). A telephone survey of hospital emergency departments in the city uncovered nine more cases over a period of two days. All had eaten yellow-fin tuna, either in restaurants or in the home. Investigation did not elicit any evidence of a serious lapse in refrigeration or handling. The fish were cleaned and packed in ice onboard the fishing boat, and delivered by refrigerated vehicle to a distributor where they were repacked in ice in smaller lots for delivery to retailers and restaurants. It was extremely hot during the period of the outbreak, however.

All cases were treated successfully with oral antihistamines. In asthmatics and cardiac patients, the condition can be life-threatening, requiring more intensive emergency treatment.

Case study 21

One would want to inquire carefully as to recent ingestion of food, especially seafood, given the environment. These six sport fishermen consumed blue mussels that had been collected in deep water far offshore by commercial fishing boats. This makes bacterial contamination from sewage effluent or other human sources unlikely. The toxin is obviously a neurotoxin. Shellfish toxins that must be considered include domoic acid (amnesic shellfish poisoning), okadaic acid (diarrhetic shellfish poisoning), and saxitoxin (paralytic shellfish poisoning). Tingling and loss of sensation are strongly indicative of saxitoxin poisoning. Significantly, the episode occurred in June, when red tides would be more common in the Northern Hemisphere.

The symptomatology and mechanism of action of saxitoxin are identical to those of tetrodotoxin from fugu (puffer fish).

Analysis of the remaining blue mussels revealed saxitoxin levels of 25,000 µg/100 g in uncooked mussels and 4300 µg/100 g in cooked mussels. The boiling destroyed much of the saxitoxin; otherwise, the attacks would have been fatal, as they were in a similar incident in which clams were steamed but not boiled. The raw clams contained up to 13,000 µg/100 g of saxitoxin, most of which survived the steaming.

Case study 22

In this case, the differential diagnosis must include food poisoning (some of the symptomatology could be due to *Salmonella* infection or *Staphylococcus* toxin), a chemical contaminant such as a pesticide, or some other cause. The time to onset is not suggestive of poisoning by a preexisting toxin as in staph food poisoning. Careful questioning might reveal whether a possible source of *Salmonella* existed in the diet (e.g., undercooked eggs, chicken, or beef). The symptomotology is reminiscent of poisoning by an organophosphorus insecticide (see Case Study 16) but it is unlikely that spraying to control insects would occur so close to harvesting the tobacco.

This is a condition known as green tobacco disease. It is due to the absorption across the skin of nicotine and it characteristically occurs in periods of wet weather. The signs and symptoms are those of stimulation of nicotinic receptors in the ganglia of the autonomic nervous system and of the neuromuscular junction. Some of the symptoms will relate to para-sympathetic stimulation, as do those of organophosphorus poisoning.

Case study 23

This poisoning obviously involves the gastrointestinal tract and the peripheral nervous system with different times-to-onset of signs and symptoms. Without knowing the common denominator (amberjack), one would initially have to consider an infectious agent, a contamination occurring in the restaurant, and a preexisting toxin. Stool and vomitus cultures were negative for all common bacterial causes of gastroenteritis (*Salmonella*, *Shigella*, *Campylobacter*, and *Yersinia*). Monosodium glutamate (MSG) has a different array of symptoms and a more rapid time-to-onset. Seafoods that could be responsible include shellfish (diarrhetic shellfish poisoning, paralytic shellfish poisoning) and scalefish (scombroid and ciguatera poisoning). Shellfish poisoning is ruled out by the commonality of amberjack consumption. The rapid onset of histamine-related symptoms of scombroid poisoning are lacking. The combination of gastrointestinal and neurological symptoms is characteristic of ciguatera poisoning. Analysis of samples of the amberjack by mouse bioassay was positive for ciguatera-type biotoxins.

Case study 24

The rapidity of symptom onset after ingesting the root points clearly to this as the source of the toxin. There is also a clear dose dependency because the man who consumed three bites died, whereas his brother, who ate only one bite, survived. This is a classical case of water hemlock poisoning, with cicutoxin as the causative agent. The involvement of the gastrointestinal tract and the central nervous system is typical.

Case study 25

These teenagers had collected some jimsonweed from an empty lot taken over by weeds. They had consumed some of the seeds from the seed pod. These are rich in scopolamine and atropine, muscarinic blocking agents with central effects. This accounts for the hallucinations (typically involving insects), confusion, combative behavior, disorientation, seizures, and coma. Peripheral effects include dry mouth, blurred vision and photophobia, and urinary retention.

Treatment is supportive and may include induced emesis and activated charcoal to decontaminate the gastrointestinal tract, plus physostygmine (a cholinesterase inhibitor) to increase the availability of acetylcholine.

Index

A

Abiotic modifiers, 76–77
Absorption, 4–5, 310
Acceptable daily intake (ADI), 58–59
Acclimation, 78
Acclimatization, 78
Acetone, 170
Acidifiers, 180
Acidity and toxic metals, 84–86
Acidulants, 180
Active transport, 5
Acute toxicity, 26–28
Additive interactions between drugs,
    25–26
Adjuvants, 180
Aerating agents, 180
Aflatoxins, 227–228
Age
  as a biotic modifier, 77
  xenobiotics, 20–21, 52
Agency for Toxic Substances and
    Disease Registry, U.S., 2
Agricultural runoff, 80
AIDS, 61–62
Air
  asbestos, 108–109
  atmospheric distribution of
      pollutants, 105–106
  barn dust, 110
  chlorine, 116
  Gaia hypothesis, 123
  gaseous pollutants, 103–104, 106–107,
      114–119
  global cooling, 119–121
  health hazards of polluted, 107–108
  methane, 118–119
  motor vehicle exhaust, 120–121
  ozone, 115
  particulates, 104, 107
  pollution
    case studies, 307

  multiple chemical sensitivity
      (MCS), 111–113
  types of, 103–104
  pyrolysis of plastics, 109–110
  silicosis, 109
  smog, 104
  stratosphere, 106
  waste disposal, 105
Alcohol, 36
Aleukia, 227
Algae, 235
Aliphatic alcohols, 169–170
Aliphatic hydrocarbons, 168
Alkalies, 180
Allergies, 193
  genetically modified foods, 300–301
Aluminum silicates, 85–86
Ames, Bruce, 55, 56, 65
Amitrole, 215
Androgenic hormones, 36
Aneuploidization, 29
Animals, *see* Livestock; Wildlife; *see also*
    Marine and aquatic species
Antagonistic interactions between
    drugs, 25–26
Antibacterial disinfectants, 130–131
Antibiotics, 181, 188–192
Anti-browning agents, 181
Anti-caking agents, 181
Antifoaming agents, 181
Anti-mold agents, 181
Antimony, 159
Antimyotic agents, 181
Anti-oncogenes, 32
Antioxidants, 181, 184–185
Antirope agents, 181
Antistaling agents, 181
Aquatic environments, *see also* Marine
    and aquatic species
  DDT, 209–210
  feedback loops, 297
  Great Lakes, 90–93

format   apply design
template